GCSE 9-1 geography OCR B

Series editor
John Widdowson

Andy Crampton
Catherine Owen

OXFORD
UNIVERSITY PRESS

OXFORD
UNIVERSITY PRESS

Great Clarendon Street, Oxford, OX2 6DP, United Kingdom

Oxford University Press is a department of the University of Oxford. It furthers the University's objective of excellence in research, scholarship, and education by publishing worldwide. Oxford is a registered trade mark of Oxford University Press in the UK and in certain other countries

Series editor: John Widdowson

Authors: Andy Crampton, Catherine Owen

The moral rights of the authors have been asserted

Database right of Oxford University Press (maker) 2016

First published in 2016

British Library Cataloguing in Publication Data
Data available

ISBN 978-0-19-836665-2

10 9 8 7 6 5

Paper used in the production of this book is a natural, recyclable product made from wood grown in sustainable forests. The manufacturing process conforms to the environmental regulations of the country of origin.

Printed in India by Multivista Global Pvt. Ltd

Acknowledgements

The publisher would like to thank the following for permission to use photographs and other copyright material:

Cover: Rob Hyrons/Shutterstock; Art Painter/Shutterstock: **p4:** Rob Hyrons/Shutterstock; Art Painter/Shutterstock; **p6:** FrankvandenBergh/ iStockphoto; **p8:** cristianl/iStockphoto; **p11:** Universal Images Group/DeAgostini/Alamy; **p11:** BRITISH ANTARCTIC SURVEY/ SCIENCE PHOTO LIBRARY; **p12:** Antonio Gravante/ Shutterstock; **p12:** Gary Weathers/Tetra Images/Corbis; **p12:** Bartosz Hadyniak/iStockphoto; **p12:** Newspix/Rex/Shutterstock; **p12:** mdekeyser (mdekeyser@weather4expeditions); **p12:** OAR/ERL/National Severe Storms Laboratory (NSSL)/NOAA; **p13:** Courtesy of the Arklay S. Richards Co., Inc.; **p16:** Jose Goitia/ dpa/Corbis; **p16:** STRDEL/AFP/GettyImages; **p18:** Image courtesy Jacques Descloitres, MODIS Land Rapid Response Team at NASA GSFC; **p19:** CLAUDIA DAUT/Reuters/Corbis; **p22:** Nick Gibson/age fotostock/Alamy; **p23:** Ezra Acayan/NurPhoto/NurPhoto/Corbis; **p24:** ERIK DE CASTRO/Reuters/Corbis; **p25:** Sherbien Dacalanio/Pacific Press/LightRocket via Getty Images; **p26:** Vincent Lowe/Alamy; **p29:** David Holmes; **p35:** Tom Bean/Alamy; **p38:** Jim Sugar/Getty; **p39:** STRINGER/MEXICO/Reuters/Corbis; **p40:** NASA images by Jeff Schmaltz, LANCE/EOSDIS Rapid Response.; **p40:** Suvra Kanti Das/ZUMA Press/Corbis; **p41:** Athit Perawongmetha/ Reuters; **p42:** PRAKASH MATHEMA/AFP/Getty Images; **p42:** Suvra Kanti Das/ZUMA Press/Corbis; **p43:** PRAKASH MATHEMA/AFP/Getty Images; **p44:** PRAKASH MATHEMA/AFP/Getty Images); **p45:** YOSHIKAZU TSUNO/AFP/Getty Images; **p46:** Thelma Sanders/Eye Ubiquitous/Alamy Stock Photo; **p48:** Frederique Olivier/ Hedgehog House/Minden Pictures/Corbis; **p50:** THIERRY BERROD, MONA LISA PRODUCTION/ SCIENCE PHOTO LIBRARY; **p50:** Ben Kennedy/U.S. Geological Survey; **p51:** NASA/Goddard Scientific Visualization Studio; **p53:** Heritage Images/Getty; **p55:** NASA/ SDO; **p56:** David Harlow/US Geological Survey/The LIFE Picture Collection/Getty Images; **p57:** PhotoEdit/Alamy; **p57:** Stephen McCluskey/Shutterstock; **p57:** Robert Brook/Alamy; **p57:** Leadinglights/iStockphoto; **p58:** Scott Willis (scott@scottwillismurals.com); **p60:** Prabhat Kumar Verma/ZUMA Press/Corbis; **p61:** A9999 DB Rep. of Maldives/Dol/Ho/dpa/ Corbis; **p61:** Sipa Press/REX Shutterstock; **p62:** Guenter Guni/ iStockphoto; **p63:** NASA/ Science Photo Library; **p64:** Ashley Cooper/Corbis; **p64:** Esther Havens (ester@ estherhavens.com 214.762.8968; **p67:** PA Archive/Press Association Images; **p68:** PA Archive/Press Association Images; **p69:** Jeff Gilbert/Alamy; **p70:** Jeff J Mitchell/Getty Images); **p70:** Environment Agency; **p71:** author; **p72:** Travel Pictures/Alamy; **p74:** Mel Thompson/Shutterstock; **p74:** The Hay Wain, 1821 (oil on canvas) by John Constable/ The National Gallery, London, UK/Getty Images; **p75:** Julie Meech/Ecoscene/Corbis; **p75:** Paul White Aerial Views/Alamy; **p75:** David Peter Robinson/ Shutterstock; **p75:** Skyscan/Corbis; **p76:** Derek Croucher/Alamy; **p76:** NigelHigson/iStock/Thinkstock; **p79:** NATURAL HISTORY MUSEUM, LONDON/SCIENCE PHOTO LIBRARY; **p79:** Anthony Hatley/Alamy; **p80:** Ashley Cooper/Alamy; **p80:** Mark Bulmer/ Shutterstock; **p83:** FLPA/Alamy; **p84:** Derek Stone/Alamy; **p84:** MSB Photography/Alamy; **p84:** mrtom_uk/iStockphoto; **p86:** kodachrome25/iStockphoto; **p86:** Liquid Light/Alamy; **p87:** Janet Baxter Photography; **p88:** kodachrome25/iStockphoto; **p89:** David Bagnall/Alamy; **p90:** peachroomtom/iStockphoto; **p90:** Chris Cooper-Smith/Alamy; **p91:** Jack Sullivan/Alamy; **p91:** Jo Chambers/ Shutterstock; **p92:** Getty Images/Flickr Open; **p92:** Julian Simmonds/The Telegraph; **p93:** claffra/Shutterstock; **p94:** author; **p95:** WWPhotography/Alamy; **p96:** Tim Wright/Alamy; **p98:** Martin Shields/Alamy; **p99:** Ordinance Survey; **p100:** Patrick Ward/Alamy; **p100:** Rob Read/ Nature Photographers Ltd/Alamy; **p101:** Getty Images/Moment Open; **p101:** Commission Air/Alamy; **p102:** Skyscan Photolibrary/Alamy; **p104:** Jeff Gilbert/Alamy; **p106:** Hoberman Collection/Alamy; **p107:** Ordinance Survey; **p108:** Frans Lanting/Corbis; **p110:** jeffdalt/iStockphoto; **p112:** Footage.Pro/Shutterstock; **p112:** sittitap/ Shutterstock; **p113:** spider.de/Shutterstock; **p113:** Vlad61/Shutterstock; **p113:** erichon/Shutterstock; **p113:** Muznabutt/Shutterstock; **p115:** fazeful/iStockphoto; **p115:** Dr. Morley Read/ Shutterstock; **p117:** Ethan Daniels/Shutterstock; **p118:** UniversalImagesGroup/Getty; **p119:** combomanbo/iStockphoto; **p119:** Shaiith/ iStockphoto; **p119:** jlanfel/ iStockphoto; **p119:** HANDBAOLUAN/iStockphoto; **p120:** 68/Ocean/Corbis; **p120:** Robert Harding World Imagery/Alamy; **p121:** Asahi Shimbun via Getty Images; **p121:** Kevin Schafer/Alamy; **p122:** Frans Lanting Studio/Alamy; **p124:** Courtesy of Samasati Retreat and Rainforest Sanctuary; **p124:** Courtesy of Samasati Retreat and Rainforest Sanctuary; **p125:** Radius Images/Getty; **p126:** Armin Rose/ Shutterstock; **p127:** Arndt Sven-Erik/Arterra Picture Library/Alamy; **p129:** PavelSvoboda/Shutterstock; **p129:** Ashley Cooper/Alamy; **p130:** Steven Kazlowski/Science Faction/Corbis; **p131:** Accent Alaska.com/Alamy; **p132:** Mark Votier/Hulton Archive/Getty Images; **p134:** Geoff Renner/Robert Harding World Imagery/Corbis; **p135:** Frank Whitney/Getty; **p136:** Josef Friedhuber/iStockphoto; **p136:** NG Images/ Alamy Stock Photo; **p138:** Danny Lehman/ Corbis; **p141:** Eric2x/iStockphoto; **p141:** fuyu liu/Shutterstock; **p143:** Chris Harrison; **p144:** afitz/Shutterstock; **p144:** Quagga Media/Alamy; **p145:** Mark Steward/Alamy; **p146:** Sean Sprague/Alamy; **p147:** AKINTUNDE AKINLEYE/X02000/Reuters/Corbis; **p148:** Malcolm Linton/Getty Images; **p150:** David Hughes/ Shutterstock; **p151:** DavidCallan/ iStockphoto; **p152:** fuyu liu/Shutterstock; **p152:** Malcolm Linton/Getty Images; **p156:** Quagga Media/Alamy; **p157:** Andrew Fox/Alamy; **p158:** Popperfoto/Getty Images; **p158:** David Bagnall/Alamy; **p163:** dynasoar/iStockphoto; **p161:** Jeff Morgan 11/ Alamy; **p161:** Ewelina Wachala/Shutterstock; **p159:** Concept design (by BCC) for the Birmingham City Centre HS2 Terminus Station. © Copyright GeoPerspectives, supplied by Bluesky International Ltd 2012. © Crown Copyright and database right Birmingham City Council Licence No.100021326, 2015. © Zmapping Ltd 2015 Context Model.; **p164:** author; **p165:** David Bagnall/Rex Shutterstock; **p166:** Chris Schmidt/Getty; **p167:** Dominic Lipinski/PA Archive/PA Images; **p170:** Picade LLC/Alamy; **p172:** Angus McComiskey/Alamy; p173: author; **p174:** Murad Sezer/Reuters; **p175:** author; **p175:** askihuseyin/Fotolia; **p176:** author; **p177:** author; **p178:** MCT/Getty Images; **p180:** Ashley Cooper Pics/Alamy; **p180:** Hung Chung Chih/Shutterstock; **p180:** Goran Tomasevic/Reuters/Corbis; **p180:** African Way/iStockphoto; **p184:** Massimo Borchi/ Atlantide Phototravel/Corbis; **p186:** Tanya Bindra/AP/Press Association Images; **p187:** Sam Tarling/Corbis; **p188:** Trevor Smithers ARPS/Alamy; **p192:** Erik De Castro/Reuters/ Corbis; **p190:** John Panella/Shutterstock; **p194:** epa european pressphoto agency b.v./ Alamy; **p195:** Goran Tomasevic/Reuters/Corbis; **p196:** Christopher Scott/Alamy; **p197:** Naashon Zalk/Bloomberg/Getty; **p198:** GavinD/iStockphoto; **p200:** Jake Lyell/Alamy; **p200:** Blaize Pascall/Alamy Stock Photo; **p202:** Stefanie van der Vinden/Shutterstock; **p203:** Paul Springett B/Alamy; **p204:** HO/Reuters/Corbis; **p206:** Eitan Simanor/Alamy; **p208:** imageBROKER/Rex Shutterstock; **p209:** Jerome Favre/Bloomberg/Getty Images; **p209:** Richard Gardner/REX Shutterstock; **p210:** Lynn Y/Shutterstock; **p212:** Room to Read; **p213:** Precious Habeenzu; **p214:** dynasoar/iStockphoto; **p214:** stephenmeese/ iStockphoto; **p214:** Thinglass/iStockphoto; **p214:** Pell Studio/Shutterstock; **p216:** NASA; **p217:** alice-photo/Shutterstock; **p217:** kodachrome25/iStockphoto; **p217:** Chris_Elwell/ Shutterstock; **p217:** Daniel_Kay/iStockphoto; **p217:** Cezary Zarebski/iStockphoto; p217: kodachrome25/iStockphoto; **p222:** Joanna Pecha/Getty Images; **p222:** G Baden/Corbis; **p222:** Alexander Raths/Shutterstock; **p224:** Diego Cervo/Shutterstock; **p226:** Miles Davies/Alamy; **p229:** The Royal Commission on the Ancient and Historical Monuments of Wales; **p230:** wavebreakmendia/Shutterstock; **p231:** Brian A Jackson/Shutterstock; **p232:** Justin Kase zsixz/Alamy; **p235:** Chris Ratcliffe/Bloomberg/Getty Images; **p235:** epa european pressphoto agency b.v./Alamy; **p236:** Visit Britain/Eric Nathan/Getty Images; **p237:** author; **p238:** Courtesy of Here East; **p241:** author; **p241:** author; **p241:** author; **p241:** Ordinace Survey; **p244:** Ramzi Haidar/AFP/Getty Images; **p245:** Marketa Jirouskova/Oxford Scientific/Getty Images; **p246:** A F Archive/Alamy; **p247:** Mark Bourdillon./Love Productions; **p248:** Stu/Alamy; **p249:** travellight/Shutterstock; **p250:** Sean Sprague/Alamy; **p252:** OMAR TORRES/AFP/Getty Images; **p254:** Friedrich Stark / Alamy; **p254:** Tyler Olson/Shutterstock; **p255:** Greenpeace; **p256:** Collage Rettet den Regenwald/IAR; **p256:** ROMEO GACAD/AFP/Getty Images; **p257:** Roger Hutchings/ Alamy; **p257:** Melanie Stetson Freeman/The Christian Science Monitor via Getty Images; **p258:** NASA/Goddard Space Flight Center Scientific Visualization Studio Unite; p259: Xinhua/Alamy; **p260:** Mike Goldwater/Alamy; **p261:** JEAN-CHRISTOPHE VERHAEGEN/AFP/Getty Images; **p261:** ABDURASHID ABDULLE/AFP/Getty Images; **p263:** WILL OLIVER/AFP/Getty Images; **p265:** Lurin/Shutterstock; **p266:** David Keith Jones/Images of Africa Photobank/Alamy; **p266:** Grigorev_Vladimir/ iStockphoto; **p268:** Herb Swanson/Portland Press Herald via Getty Images; **p269:** Peter Groenendijk/ robertharding /Alamy; **p270:** Courtesy of Halvdan Jakobsen; **p271:** blickwinkel/ Alamy; **p274:** Courtesy of Fairtrade; **p275:** Photographer: Patrice De Villiers Agency: Marcel Worldwide; **p276:** B Brown/ Shutterstock; **p276:** Greg Taylor/Alamy; **p277:** Mark William Richardson/Shutterstock; **p278:** JUAN MABROMATA/AFP/Getty Images; **p278:** Sion Touhig/Getty Images; **p279:** ERIK DE CASTRO/Reuters/Corbis; **p280:** TONY KARUMBA/AFP/GettyImages; **p280:** ChinaFotoPress via Getty Images; **p280:** keith burdett/Alamy; **p282:** Minerva Studio/Shutterstock; **p283:** Gary Braasch/CORBIS; **p284:** KEITH BEDFORD/Reuters/Corbis; **p286:** Scott Olson/Getty Images; **p287:** ImagineGolf/ iStockphoto; **p288:** Reuters/CORBIS; **p289:** David McNew/Getty Images; **p290:** The Rocky Mountain News, George Kochaniec, Jr./AP/Press Association Images; **p292:** www. webbaviation.co.uk; **p293:** ordinance Survey; **p295:** incamerastock/Alamy; **p296:** Paul White/Alamy Stock; **p296:** REX Shutterstock; **p296:** ANDREW HAYES/Alamy; **p296:** Alan Payton/Alamy; **p298:** Ashley Cooper/Global Warming Images/Alamy; **p299:** Bjanka Kadic/Alamy; **p300:** TerraColor/Apollo Mapping; **p302:** Londonstills.com/Alamy; **p305:** Xavier TESTELIN/Gamma-Rapho via Getty Images; **p306:** Matthew Taylor/Alamy; **p307:** Courtesy of Transport for London; **p309:** Jeffrey Blackler/Alamy; **p310:** Marcin Rogozinski/Alamy; **p311:** Anthony Palmer/Alamy; **p313:** Sean Pavone/Shutterstock; **p313:** Reuters/Corbis; **p315:** Colin Garratt/Milepost 92 1/2/Corbis; **p316:** Hung Chung Chih/ Shutterstock; **p317:** STRINGER/CHINA/Reuters/Corbis; **p318:** Imaginechina/REX Shutterstock; **p321:** Gianni Ferrari/Cover/Getty Images; **p321:** TonyV3112/ Shutterstock.

Design by Ian Foulis. Artworks are by: Lovell Johns, Mike Parsons, Giorgio Bacchin, Ian West, Ian Foulis, Q2A Media

Contents

Component 1 Our Natural World

Contents

Component 2 People and Society

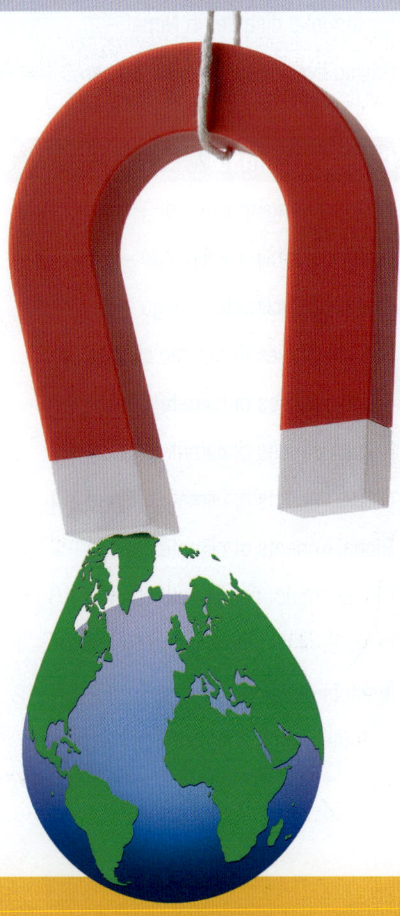

OCR endorsement statement

This resource is endorsed by OCR for use with specification *J384 GCSE (9-1) Geography B (Geography for Enquiring Minds)*. In order to gain OCR endorsement, this resource has undergone an independent quality check. Any references to assessment and/or assessment preparation are the publisher's interpretation of the specification requirements and are not endorsed by OCR. OCR recommends that a range of teaching and learning resources are used in preparing learners for assessment. OCR has not paid for the production of this resource, nor does OCR receive any royalties from its sale. For more information about the endorsement process, please visit the OCR website, www.ocr.org.uk.

Component 3 Geographical Exploration

1 Global Hazards

What is this topic about?

When a large earthquake or tropical storm strikes, images of suffering hit the news and we may dip our hands into our pockets to help. Geographers play an important role in understanding, analysing and planning for these **hazards**. We can't stop the hazards happening, but can reduce the damage they cause.

Hazards are extreme events that pose a threat to people's lives and property. Natural hazards are caused by physical factors such as the weather or movement within Earth's structure. But human factors, including poverty and how well prepared we are, can have a significant impact on how much damage natural hazards cause. You are going to find out about the causes and impacts of weather and tectonic hazards, and our responses to them.

Your enquiry

In this chapter you will investigate the specification's two key questions.

How can weather be hazardous? You will:

- Explore the role of the global circulation system in causing extreme weather conditions around the world.
- Find out about extreme temperatures, winds and precipitation in different countries.
- Study the distribution and frequency of tropical storms and drought, and whether these have changed over time.
- Outline the causes of the extreme weather conditions related to tropical storms.
- Outline the causes of the extreme weather conditions of El Niño and La Niña that lead to drought.
- Investigate the causes, consequences and responses to Typhoon Haiyan, which hit the Philippines in 2013.
- Investigate the causes, consequences and responses to the UK drought in 2012.

How do plate tectonics shape our world? You will:

- Describe the structure of Earth and explain how it is linked to the processes of plate tectonics.
- Understand the processes that take place at different types of plate boundaries and hotspots.
- Explain how the movement of tectonic plates causes earthquakes and volcanoes.
- Investigate the causes, consequences and responses to the earthquake that struck Nepal in 2015.
- Explore how technological developments can mitigate the impact of earthquakes.

Look at the photo opposite.

- Why do people live here?
- What hazards could they face?
- How could these hazards affect their lives?
- What could they do to reduce the impact of these hazards?

1.1 Global circulation system

On this spread you will start to explore how weather can be hazardous.

Living with hazards is risky

The people in the squatter settlement on page 6 are living with a constant hazard. Look at the steep slope. If there's a landslide, it could destroy buildings and bring injury and death. An **earthquake** (see spread 1.15) or even heavy rain could trigger that sort of disaster.

Are the people living in the apartments in photo A at similar risk? The hazards there are **tropical storms** (see spreads 1.5–1.6) and rising sea levels (see spread 2.7). Storms with high winds would be just as destructive as a landslide and there is a high risk of flood damage too.

Some hazards happen very rarely, but many people live with the risk that disaster will strike.

A Very desirable homes along the beach in Florida

Weather and climate

Many natural hazards are caused by the **weather** – the day-to-day condition of the atmosphere. **Climate** is the average weather recorded over at least a 30-year period. It varies around the world, but even in one place it can change over long periods of time (see Chapter 2).

The weather and climate are driven by the way the Sun's energy is transferred around the world. When the Sun's rays hit Earth, the **insolation** is stronger in some places than others because of the angle of the rays (diagram B). Insolation is the radiation (energy) from the Sun that reaches Earth.

Too hot to handle?

The Equator receives more heat than it can radiate away, whilst the poles lose more heat than they receive. So why doesn't the Equator get hotter and hotter, and the poles get colder and colder? That doesn't happen because the heat is transferred from the Equator towards the poles by ocean currents and winds. This keeps Earth's temperature in balance. The movement of air is called the **global circulation system** (diagram B).

B The global circulation system and why heat from the Sun varies around the world

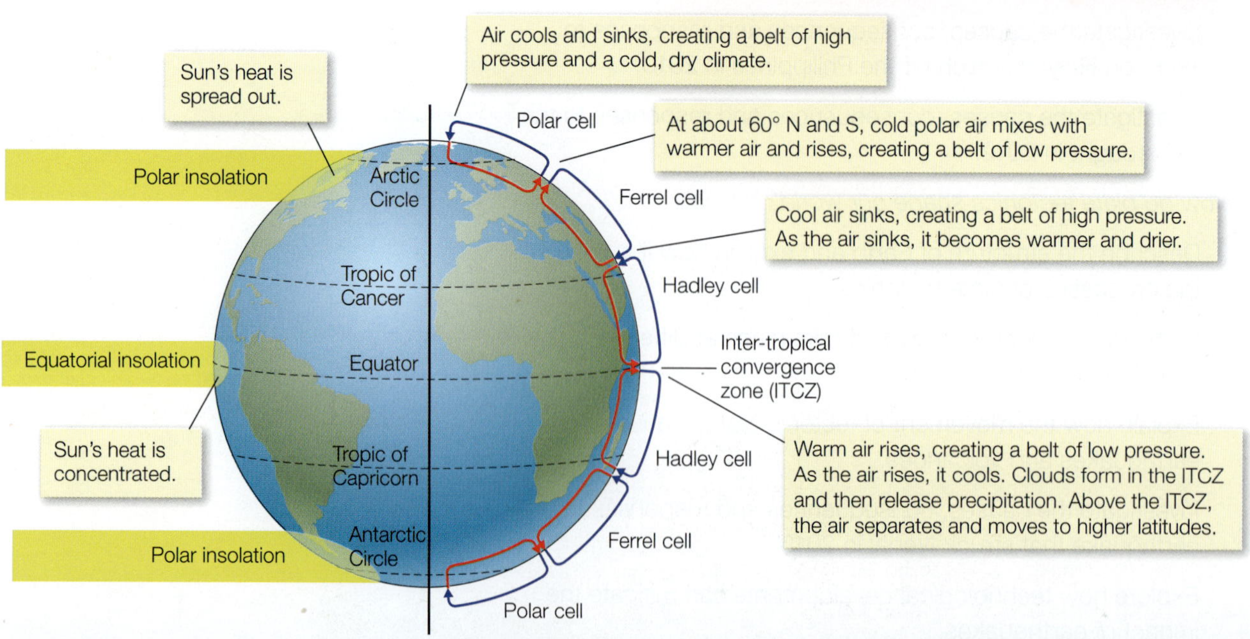

The global circulation system and pressure belts

The global circulation system is driven by the fact that warm air rises and cool air sinks. When an air mass descends, it presses down on Earth, resulting in **high pressure**. When an air mass rises, there is **low pressure**. This movement creates three interconnected cells of air in each hemisphere. These cells form belts of high and low pressure (map C).

The three cells shift north and south of the Equator during the year as the position of Earth changes in relation to the Sun. This results in seasonal weather and is why the band of rain caused by low pressure moves.

Climatic zones

Around the world, the global circulation system controls temperatures, and influences precipitation and the **prevailing winds**. This creates distinctive **climate zones**, including:

- a hot, wet climate around the Equator as warm air rises where the **Hadley cells** meet
- a hot, dry climate as air sinks where the Hadley and Ferrel cells meet. This explains why there are deserts around the tropics
- unsettled weather at about 60° N and S as cool air mixes with warm air where the Polar and Ferrel cells meet. This includes the wet, mild climate in the UK.

C *High and low pressure belts help to explain these world climatic zones*

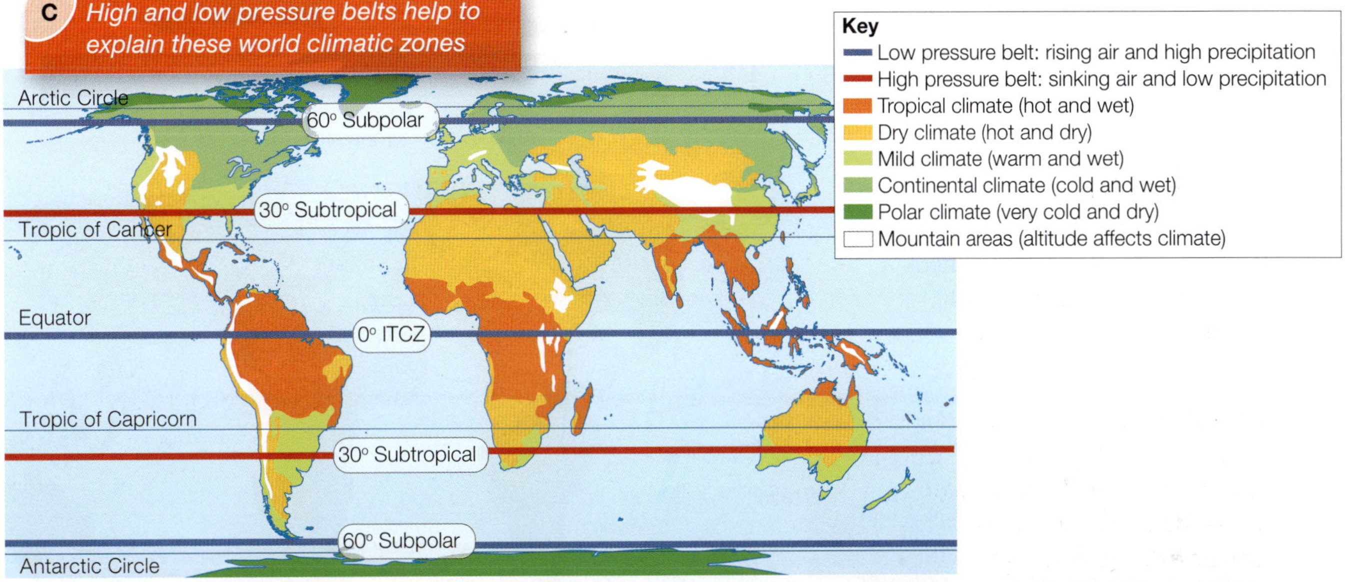

Activities

1. Look at photo A. Why do you think people choose to live in such risky places?
2. Study the left-hand side of diagram B.
   - **a** How could you use a torch and a football to show how the Sun's rays are stronger in some places than others?
   - **b** Explain why the polar regions are not warmed as much by the Sun as the Equator.
3. Study the right-hand side of diagram B.
   - **a** Draw a diagram to show how the Hadley cell works.
   - **b** Label your diagram to show the high and low pressure zones. Make sure it shows why areas of high pressure are usually dry and areas of low pressure are usually wet.
4. Study map C. Imagine you are taking a trip from pole to pole.
   - **a** What climatic zones would you pass through?
   - **b** Write one or two sentences each to explain why:
     - there is a hot, wet climate around the Equator
     - there are deserts around the tropics
     - the UK has a wet, mild climate.

Your enquiry

How can weather be hazardous?

As you work through this section, make notes about anything that will help you to answer the question. You will bring your ideas together to do an activity at the end of spread 1.10.

Extreme temperatures

On this spread you will find out about extreme temperatures in contrasting countries.

Temperature patterns

There is a clear pattern of temperature increasing as you move away from the poles, as shown on map A. What is the average annual temperature in the UK?

The Sun's energy

The Sun radiates energy in heat and light. Only some of this **radiation** reaches Earth (see Chapter 2). This is called **insolation** and it heats Earth's surface.

The Equator has the potential to be the hottest part of the world because the Sun's rays are strongest there (remember diagram B on spread 1.1). At the poles, the Sun's rays hit Earth at an angle, resulting in lower temperatures. However, other factors also affect temperatures around the world.

A *Annual temperatures around the world*

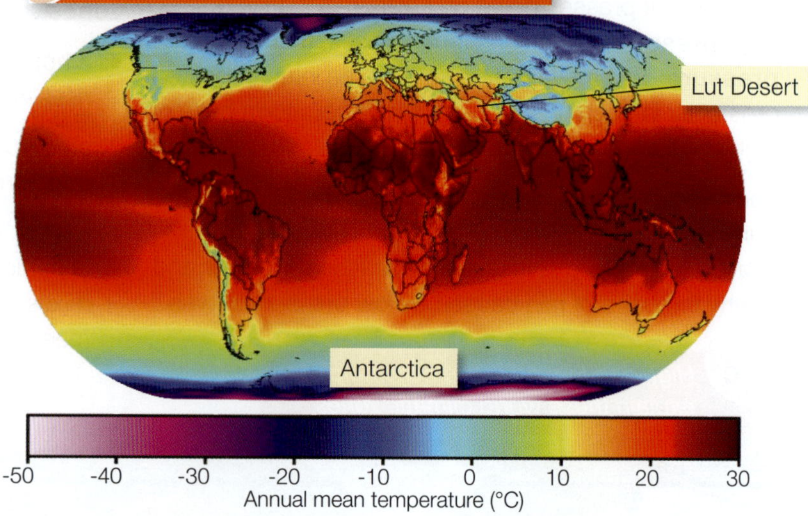

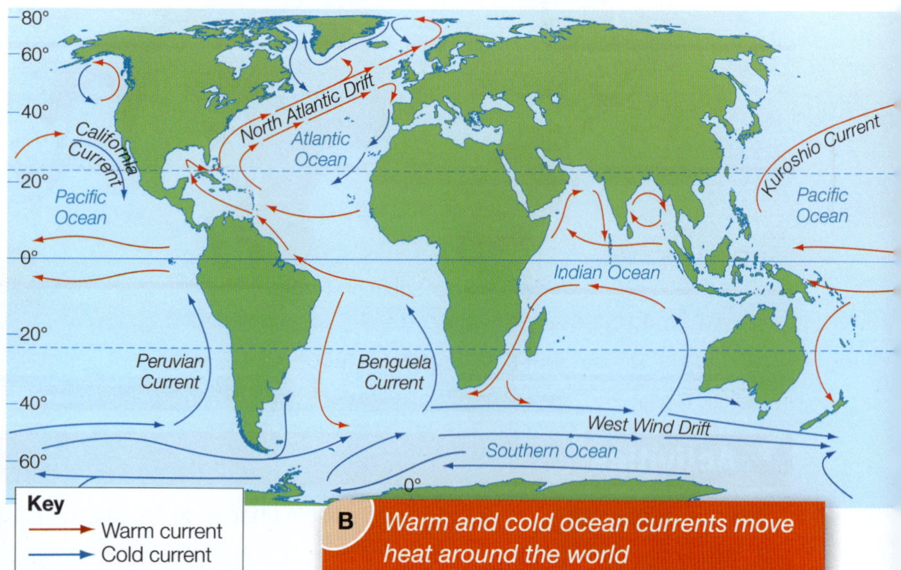

B *Warm and cold ocean currents move heat around the world*

Other factors affecting temperatures

- **Albedo effect:** This is how much a surface reflects or absorbs the Sun's rays. For example, polar ice has a high albedo, reflecting more heat and making the poles colder. The oceans and rainforests have a low albedo and absorb more heat.
- **Cloud cover:** Clouds also reflect the Sun's rays. Although potential insolation is stronger at the Equator, thick clouds reduce the temperature. Insolation is actually stronger close to the tropics, which are largely free from cloud cover.
- **Winds:** Surface winds move heat around the world (see spread 1.3).
- **Ocean currents:** Because water is fluid, ocean currents move heat easily (map B). The UK is much warmer than other countries at the same latitude because it is warmed by the North Atlantic Drift, a warm ocean current that originates in the Caribbean.
- **Land and sea:** On land, insolation is concentrated on the surface. At sea, it reaches deeper because water is transparent. Although water is slower to warm up than land, it stays warm for longer.
- **Altitude:** Higher up, air pressure is lower, which makes temperatures colder. They drop by about 1°C for every 100 metres in altitude.

The hottest place on Earth

The highest temperature recorded on Earth is 70.7°C. It was recorded in the Lut Desert in Iran (photo C). The area where it was recorded is covered in dark lava. Temperatures are high because there is intense sunshine but also because the surface absorbs the heat.

No one lives permanently in the Lut Desert because it is too hot and dry, but nomads and tourists pass through. The climate is hazardous, so visitors must be careful to carry enough water and to protect their skin from the Sun's harsh rays.

C *Strong sunshine on dark lava makes the Lut Desert the hottest place on Earth*

The coldest place on Earth

Antarctica is the coldest place, with surface temperatures dropping as low as –89.2°C. Temperatures are low because the Sun's rays are less intense in polar regions. Also, 98% of Antarctica is covered with ice, which reflects the heat rather than absorbing it.

These extreme conditions mean that no one makes a permanent home in Antarctica. Tourists may visit Antarctica to enjoy its beauty and scientists spend several months at a time at research stations (photo D), investigating the unique environment.

D *British Antarctic Survey research station*

Activities

1 Look at map B. Identify the areas where:
   - **a** warm ocean currents start and end up
   - **b** cold ocean currents start and end up.

2 Look at map A.
   - **a** Describe the location of the Lut Desert and Antarctica. Include where they are in relation to the Equator/poles, landmasses and oceans.
   - **b** Suggest reasons for the extreme temperatures found in these two locations.

3 On map A, find the area of blue to the north of India. This is the very high mountains of the Himalayas, which is colder than the surrounding areas. This an anomaly, which doesn't fit the usual pattern.
   - **a** Identify at least one other anomaly on map A.
   - **b** Suggest reasons for any anomalies. You could use an atlas to help you.

4 Look at photo C. Imagine you are visiting the Lut Desert. Write a postcard home including:
   - a description of the climate
   - an explanation of why it is hot
   - an account of how the heat is affecting you.

5 Look at photo D. The British Antarctic Survey employs people to spend six months at a time working in research stations in Antarctica. Write a job advertisement for a science post, describing and explaining the temperatures.

Go further

Find out more about weather hazards in Antarctica by visiting the discoveringantarctica.org.uk website. The 'Living there today' section gives information about what it is like to work in Antarctica and includes activities for you to try.

Practice question

Describe the pattern of global temperatures shown on map A. Use data to support your answer. (3 marks)

1.3 Extreme winds

On this spread you will learn about extreme winds in contrasting countries.

What is wind?

If you blow up a balloon and let it go, what happens to the air inside the balloon? The air inside is at higher pressure than the air outside, so when you let it go the air rushes out.

Wind is the movement of air from an area of high pressure to one of low pressure. The greater the difference in pressure, the stronger the wind. Usually winds at ground level are slowed down by friction.

A *What happens to the air inside the balloon?*

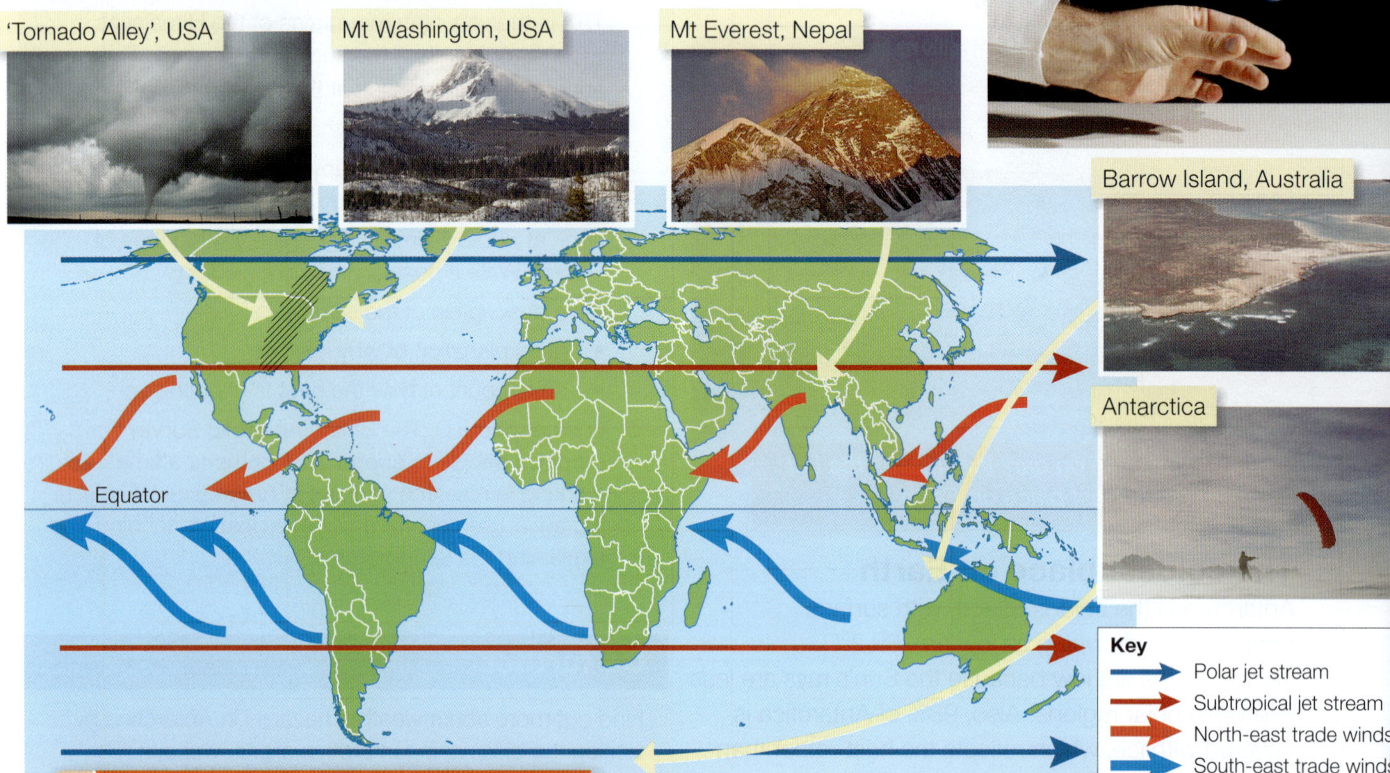

B *Some of the windiest places in the world*

Types of wind

Trade winds

The winds that blow from high pressure belts to low pressure belts are known as **trade winds**. Historically, trading ships used them to sail around the world. Mt Washington is one of the windiest places in the world, with winds up to 372 km/h, because it forms a barrier to the westerly trade winds.

Katabatic winds

These are caused by air flowing downhill. In Antarctica large ice sheets sloping gently into the sea contribute to wind speeds of over 320 km/h.

Jet streams

Winds high in the atmosphere, called **jet streams**, are very strong as there is little to slow them down. Wind speeds of 225 km/h have been recorded on Mt Everest because it reaches up into the jet stream. Most people who want to climb Everest make their attempt in May because the jet stream usually moves to the south then, making the winds less fierce and the climb safer.

More wild winds

Tornadoes and tropical storms

Tropical storms (see spreads 1.5–1.6) and **tornadoes** are very strong rotating winds. They start as smaller storms, but can develop into the strongest winds ever recorded.

Tornadoes are much smaller than tropical storms and occur over both land and sea. They can be extremely hazardous and can reach 400 km/h. The Moore tornado, with winds up to 340 km/h, killed 24 people in Oklahoma in May 2013. Many people living in 'Tornado Alley' build shelters in their gardens or strong rooms in their houses in case a tornado hits.

Tropical storms are bigger and die out when they reach land. The highest wind speed ever recorded – 408 km/h – was on Barrow Island during Hurricane Olivia in 1996.

The UK's wildest winds

The strongest winds in the UK have been recorded on mountain tops, usually in the west. But some strong gusts have been recorded in coastal locations such as Cornwall. There, the wind is not slowed down because there is less friction when it passes over the sea.

Activities

1 Write a sentence explaining the link between air pressure and wind speed.

2 Make or design your own anemometer. Use paper cups, straws, a drawing pin and a pencil. Then test it in a wind!

3 Look at map B here and diagram B on spread 1.1. How can the direction of the trade winds be explained by the global circulation system?

4 Choose two of the locations shown around map B. Compare and contrast the extreme wind conditions in these locations, considering:
- the nature of the wind
- the causes of the strong winds
- the impact the winds have on people.

5 Study chart D.
   a Which direction is the prevailing (most common) wind coming from?
   b Which direction are the strongest winds coming from?

Measuring wind speed and direction

Anemometers are used to measure wind speed by counting the number of times they rotate in a minute (photo C). Wind speed and direction in a location can be shown using a rose chart (chart D). Wind speed is measured using the **Beaufort scale**, which ranges from 0 (calm) to 12 (hurricane).

C *This anemometer measures the wind speed at Mt Washington*

D *Wind rose for Mt Washington, July–August 2015*

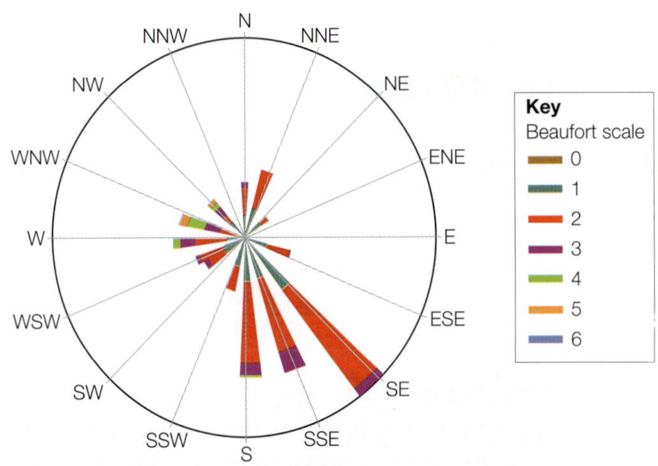

Practice question

Explain why mountainous areas are often very windy. (3 marks)

1.4 Extreme precipitation

On this spread you will find out about extreme precipitation in contrasting countries.

What is precipitation?

When warm air rises, it carries water vapour. As it rises, the air cools and the water vapour **condenses** (turns from a gas back into a liquid) to form clouds. Eventually the clouds release **precipitation**, which falls to Earth as rain, sleet, snow or hail.

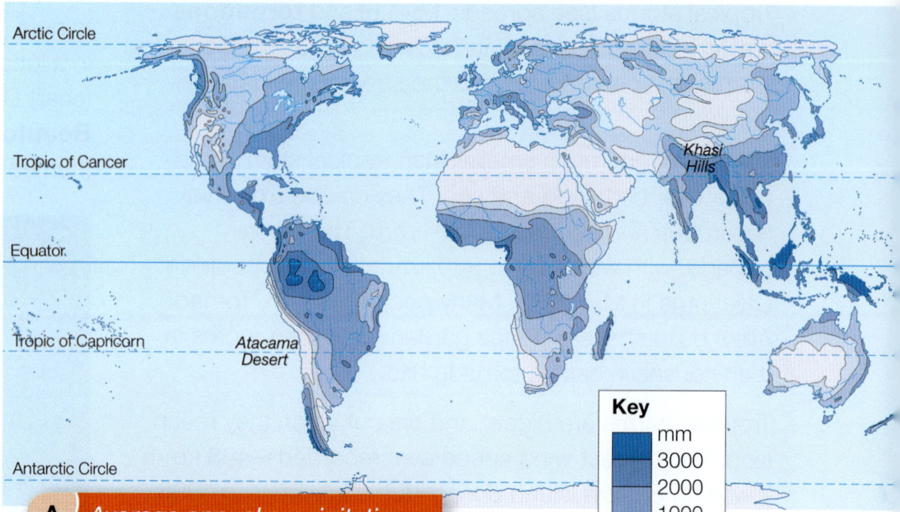

A *Average annual precipitation*

Rain around the world

The amount of rainfall varies around the world (map A). The global circulation system plays an important role in producing this pattern (see spread 1.1). Areas of low pressure have rising air and high levels of precipitation. Areas of high pressure have descending air and low precipitation.

On a global scale, rain falls heavily in a band around the Equator, but the tropics are much drier. On a regional scale, the coasts of continents can be particularly wet. On a local scale, precipitation is influenced by altitude.

Convectional rainfall

Strong sunshine warms up the ground and air above it. Pockets of warm air rise as **convection currents**. At high altitude, the air cools and its water vapour condenses to form clouds. When rain falls, it can be heavy, sometimes thunderstorms. This type of rain is significant at the Equator where temperatures are high.

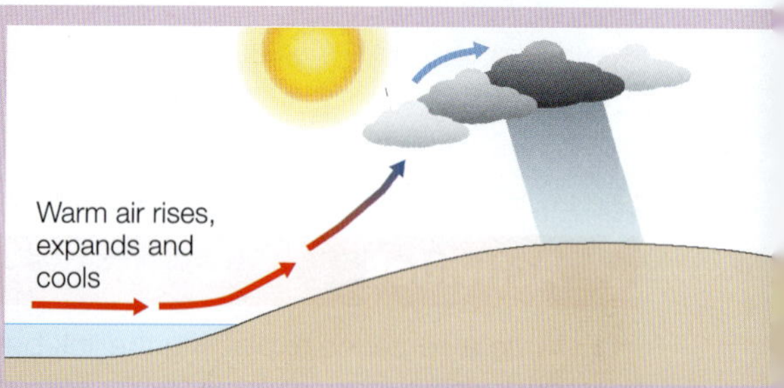

Frontal rainfall

When warm air meets cool air, it is called a **front**. The warm air rises over the cool air and clouds are formed. Eventually there is steady rain. Frontal rainfall is common in the UK because it is situated where warm air from the tropics meets cool air from the North Pole.

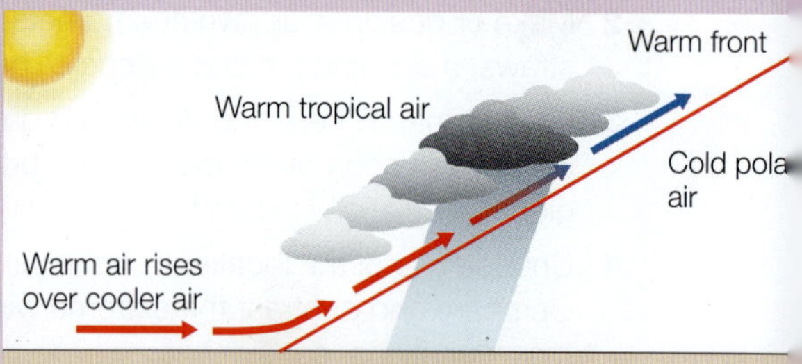

Relief rainfall

When wind gets to mountains, the warm air is forced to rise. This leads to cooling, condensation and precipitation on the windward side of the mountain. When the air descends on the leeward side it will be dry, creating a **rain shadow**. In the UK, mountains like the Pennines tend to be wetter on their western sides and drier on the east.

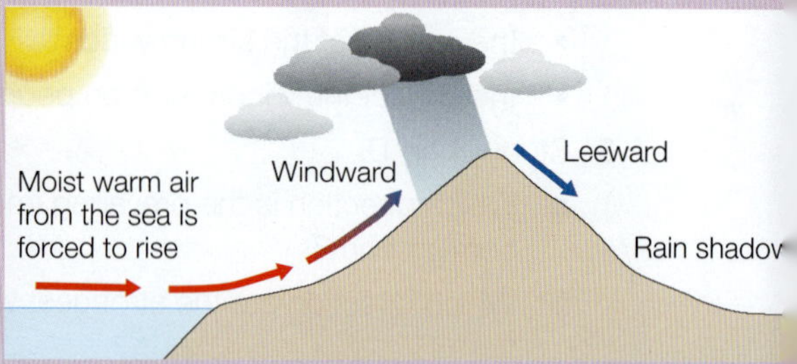

The Khasi Hills – the wettest place on Earth

The town of Cherrapunji in the Khasi Hills of northern India is a contender for the title of the world's wettest place. It holds the record for the most rainfall in a month – it received 9300 mm of rain in July 1861. Clouds are blown in from the Bay of Bengal and travel over the flat land of Bangladesh. When the clouds reach the Khasi Hills, they are forced to rise rapidly, resulting in relief rainfall.

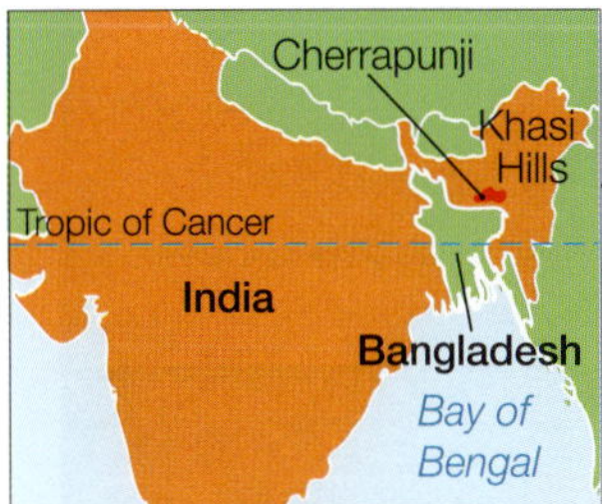

B *Average rainfall in the Khasi Hills*

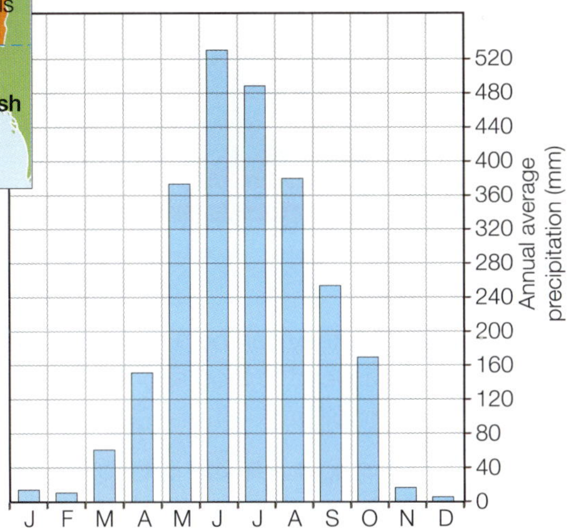

The Atacama – the driest hot desert on Earth

All deserts are dry, but the Atacama in Chile is the driest hot desert. It has an average annual rainfall of about 15 mm. Some weather stations there have never received any rain. The Atacama lies on the Tropic of Capricorn in Chile. It is in the rain shadow of the Andes mountains. But it also has a cold ocean current flowing alongside it, which means there is no moist warm air to form clouds. Few people choose to live in the Atacama, but there are some mining towns.

Antarctica is a cold desert and is even drier than the Atacama because the air is so cold that it holds very little water vapour.

C *Average rainfall in the Atacama Desert*

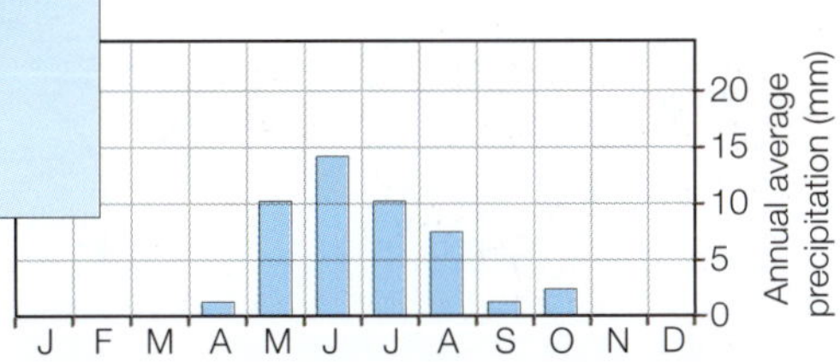

Activities

1 Use map A and an atlas.
   - **a** Identify three regions with annual rainfall over 2000 mm.
   - **b** Identify three regions with annual rainfall below 250 mm.
   - **c** Suggest reasons for the amount of rainfall in these regions.

2 The Atacama Desert is on the Tropic of Capricorn, but that isn't the only reason it is so dry. Draw a labelled diagram to explain other factors influencing its rainfall, including:
   - **a** the Andes rain shadow
   - **b** the influence of the cold ocean current (see map B on spread 1.2).

3 Rain clouds travel over the plains of Bangladesh before reaching the Khasi Hills.
   - **a** Why does the change in relief lead to heavy rain?
   - **b** What hazards could such heavy rainfall bring to this area of steep relief?

4 Look at graphs B and C. Calculate:
   - **a** the total annual precipitation in each location by adding the amounts for each month
   - **b** the mean monthly precipitation for each location by dividing the annual precipitation by 12
   - **c** the range for precipitation in each location by taking the lowest monthly precipitation figure away from the highest monthly precipitation figure
   - **d** the percentage of the annual rainfall that falls in June by dividing the June figure by the annual figure and multiplying by 100.

Practice question

Describe the pattern of rainfall on map A. Use evidence from the map to support your answer. (3 marks)

Tropical storms and droughts

On this spread you will investigate where tropical storms and droughts are found and how this may have changed over time.

Tropical storms and where they occur

Tropical storms are powerful, rotating storms also known as **hurricanes**, **cyclones** and **typhoons**, depending on where they occur. They form over tropical oceans (see spread 1.6) and move from east to west during the summer and autumn, steered by the trade winds (map C). With wind speeds over 120 km/h, the storms can be up to 650 km across and bring heavy rain (up to 250 mm in one day). Each storm may last 6–14 days, bringing devastation to any land in its path.

A *Hurricane Lili, Cuba, 2011*

Droughts and where they occur

Droughts are periods when there is much less precipitation over a specific time than is usual for the area, leading to water shortages. But a drought in the UK is very different to a drought in the Sahel Desert in North Africa. Whilst droughts are inconvenient in the UK, they can be a real hazard in other places. They can lead to food shortages, hunger and death (map C).

Drought is linked to long periods of high pressure. This means few clouds and so less precipitation. Where there are also high temperatures, more water **evaporates** from plants, lakes and streams (**evapotranspiration**). So more water evaporates than falls as rain.

A drought is a natural hazard, but its impact is worse if people are using too much water or if water is being wasted, e.g. through leaky pipelines.

B *Nothing can be done with earth as dry as this*

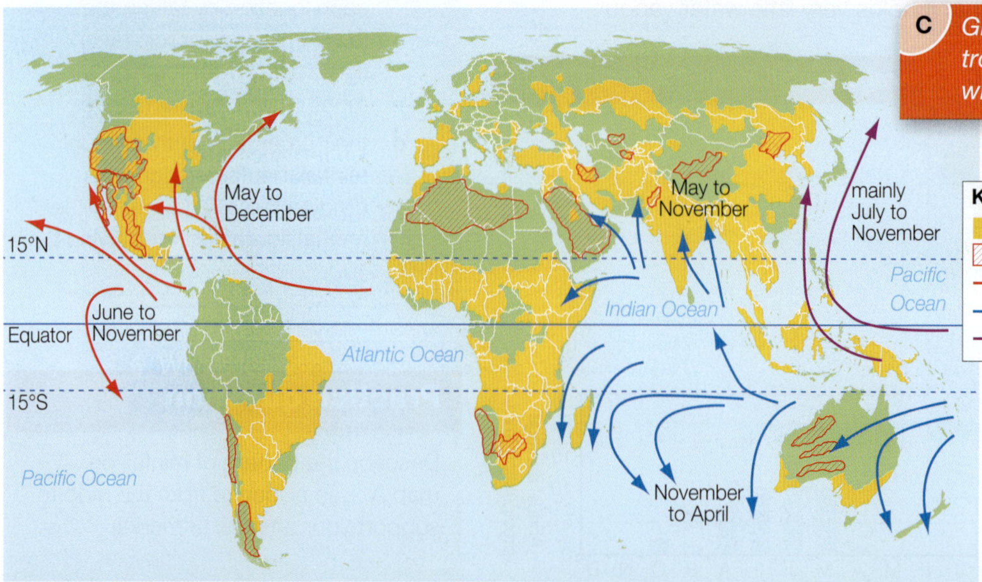

C *Global distribution of tropical storms and areas where drought is a hazard*

Is the pattern of tropical storms changing over time?

Tropical storms are complex and it is difficult to see a clear pattern of change over time. However, many scientists believe that global warming (see Chapter 2) is having an impact on these storms by increasing the temperature of the oceans. Scientists in the USA have identified periods of time with more and less storm activity (graph D). The period since 1991 has been an era of high storm activity, but is it higher than would have been expected anyway? We cannot be sure.

Is the pattern of drought changing over time?

It is clearer that the pattern of **drought** around the world is changing over time. Graph E shows that the severity of drought around the world has increased since 1940. This may be due to changing rainfall and evaporation patterns related to long-term climate change (see Chapter 2).

D *Changes in Atlantic hurricanes over time, 1950–2010*

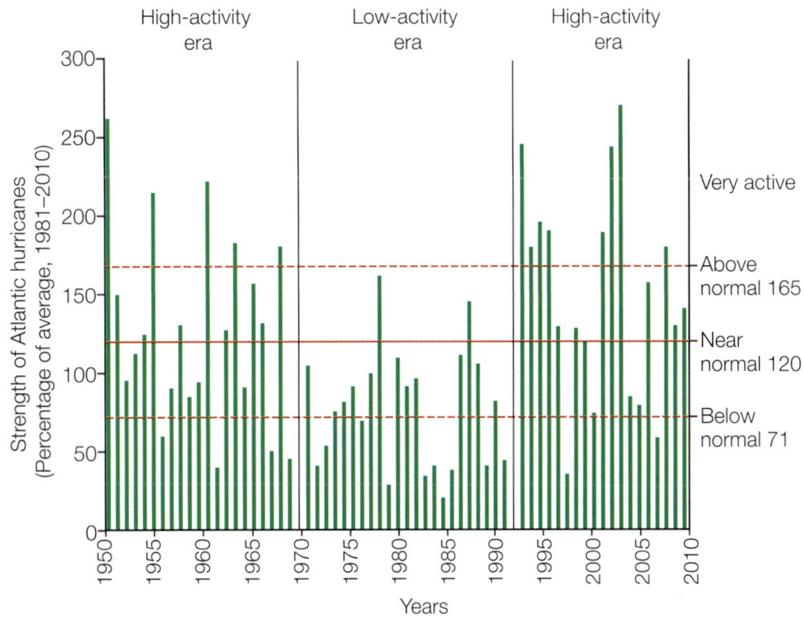

E *Changes in drought over time, 1900–2010*

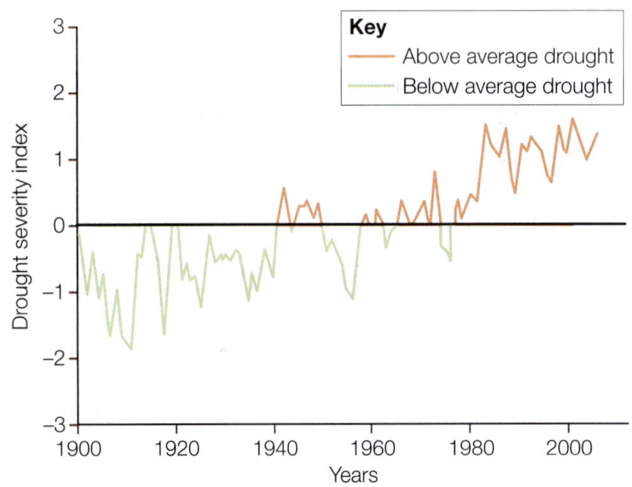

Activities

1. Look at map C. Identify three regions that frequently experience tropical storms.
2. Look at map C. Identify where in the world tropical storms are known as:
   - **a** hurricanes
   - **b** cyclones
   - **c** typhoons.
3. Write an article for the charity Water Aid's magazine. Include:
   - **a** a definition of drought
   - **b** examples of places that suffer regular droughts
   - **c** problems that drought can bring for people.
4. Look at map C here, an atlas and diagram B on spread 1.1.
   - **a** Identify areas where drought is a hazard.
   - **b** How can the Hadley cell help to explain why these areas are prone to drought?
5. Study graphs D and E. What does:
   - **a** graph D tell us about changes in the frequency of strong hurricanes over time?
   - **b** graph E tell us about changes in the severity of drought over time?

Go further

Research where droughts and tropical storms have happened in the last year. You could use the BBC website and follow links to weather news.

Practice question

Name two areas of the world that often suffer droughts. (2 marks)

1.6 Tropical storms and extreme weather

On this spread you will explore the causes of extreme weather conditions related to tropical storms.

Causes of tropical storms

Storms gathering over tropical oceans become tropical storms under certain conditions. The temperature of the ocean must be at least 26.5°C and the water must be at least 60 metres deep. So tropical storms occur in late summer and early autumn, when the oceans are at their warmest. The moist warm air over the ocean gives the storm more energy.

Tropical storms form between 5° and 15° N and S of the Equator, where the trade winds meet. They need to be at least 5° from the Equator so that the **Coriolis effect** starts the storm spinning (photo A). The Coriolis effect is the way the spinning of Earth makes winds veer to the right in the northern hemisphere and to the left in the southern hemisphere.

Tropical storms only form if winds blow at a constant speed at ground level. This allows the storm clouds to rise up in a column (diagram B).

Because a tropical storm depends on a warm ocean for its energy, it loses energy and eventually dies when it reaches land.

A *Typhoon Haiyan, 2013*

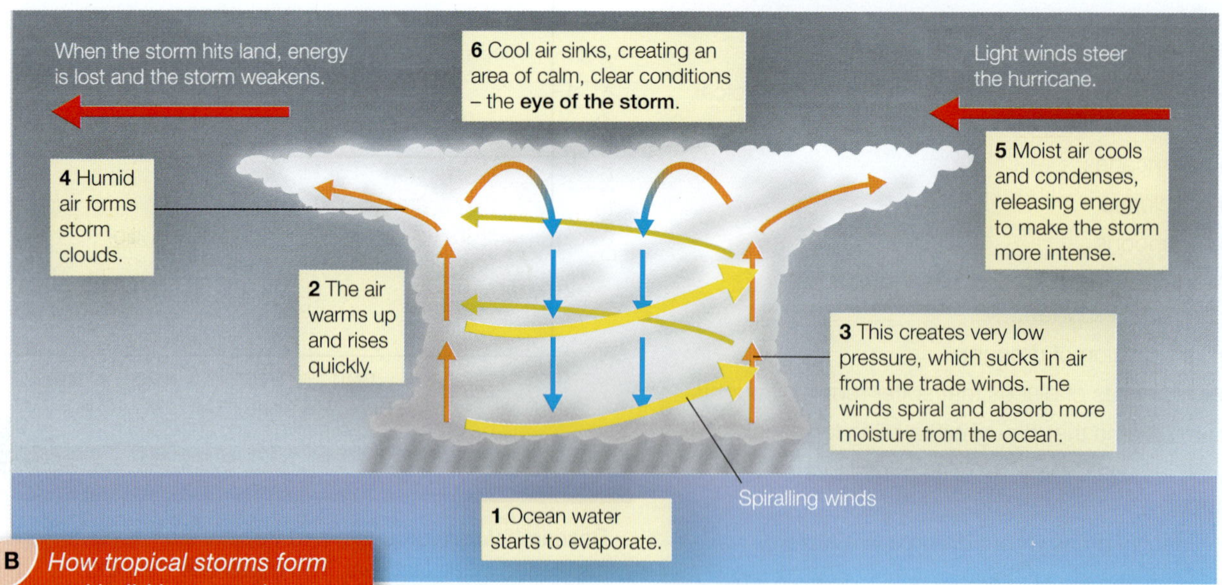

B *How tropical storms form and build in strength*

Measuring tropical storms

The wind speed of tropical storms is measured using the **Saffir–Simpson scale**. It ranges from Category 1, with wind speeds of 118–153 km/h, to Category 5, with wind speeds over 249 km/h.

Extreme weather

Tropical storms are associated with strong winds, heavy rainfall and coastal flooding, bringing danger to all in their path (photo C).

- Strong winds can blow the roofs off buildings, tear down trees and crops, and damage communication and transport networks.
- Heavy rain can cause short-term **flash flooding**, especially in urban areas, mudslides and **landslides**. It also flows into rivers, which can cause flooding downstream.
- The low air pressure raises the level of the ocean, especially as the storm nears land, leading to coastal flooding. These abnormal rises in sea level are known as **storm surges**, which cause most deaths during tropical storms.
- Industries such as fishing, shipping, transport and tourism are vulnerable to tropical storms so millions of people living along coastlines are affected.

These conditions combine to cause high death tolls and lots of destruction (table D).

C *Storm hazards and their consequences for people*

	Typhoon Haiyan, Asia	Hurricane Sandy, USA	Hurricane Katrina, USA
Year	2013	2012	2005
Category	5	3	5
Winds (km/h)	315	185	282
Storm surge (m)	6	4	8.5
Death toll	6300	285	1836
Diameter (km)	800	1520	650
Economic loss ($)	13 billion	65 billion	250 billion

D *How do recent tropical storms compare?*

Activities

1 Look at photo A.
- **a** Identify the eye of the storm.
- **b** Which hemisphere is the storm in? How do you know?

2 a Study diagram B. Draw a sketch to show the different parts of a tropical storm and the conditions needed for it to form. Add labels to show:
- the main parts of the storm, including the spiralling clouds and the eye
- the temperature and depth of the ocean
- the speed of the winds
- how much rain can fall in one day
- how wide storms can be
- where they form in relation to the Equator.

b Write a paragraph to explain how a tropical storm forms.

3 Write a statement to be read out over the radio to warn tourists what extreme weather conditions to expect from tropical storms. If you can, add advice on how to keep safe during the storm.

4 Study table D.
- **a** Compare the death tolls of Hurricane Katrina and Hurricane Sandy, which both affected the USA.
- **b** Suggest reasons for the difference in the death tolls for these two hurricanes.
- **c** Typhoon Haiyan affected parts of Asia including the Philippines. Suggest why its death toll was higher than that of Hurricane Katrina whilst its economic loss was lower.

Practice question

Outline the causes of the extreme weather conditions associated with tropical storms. (6 marks)

1.7 Drought, El Niño and La Niña

On this spread you will find out how El Niño and La Niña are linked to drought.

El Niño and La Niña

In some years there are fewer droughts than in others. The places affected by drought also vary over time. Some of this variation can be linked to **El Niño** and **La Niña** events, which are caused by changes in the sea surface temperature in a band across the Pacific Ocean (maps A and B). When the trade winds that blow from east to west are weaker, the surface temperature increases and causes the El Niño effect. When the trade winds are stronger, the surface temperature decreases and causes the La Niña effect.

Not all El Niño and La Niña events are the same. Some are stronger than others and can affect weather patterns around the world. El Niño events happen every three to seven years and usually last around nine months. La Niña events happen half as often. There are normal years in between.

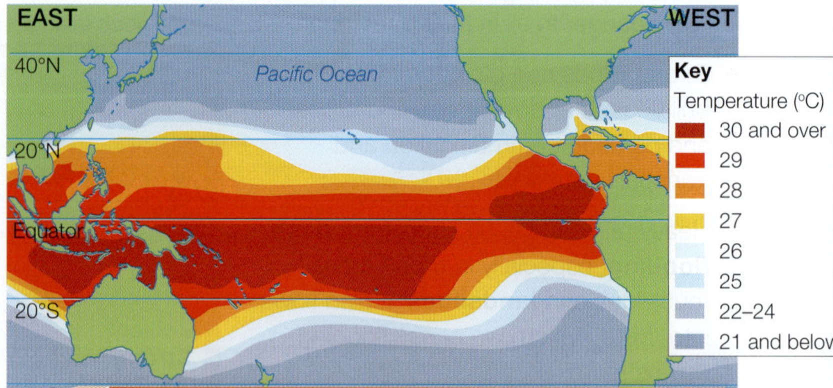

A Ocean surface temperatures during an El Niño event

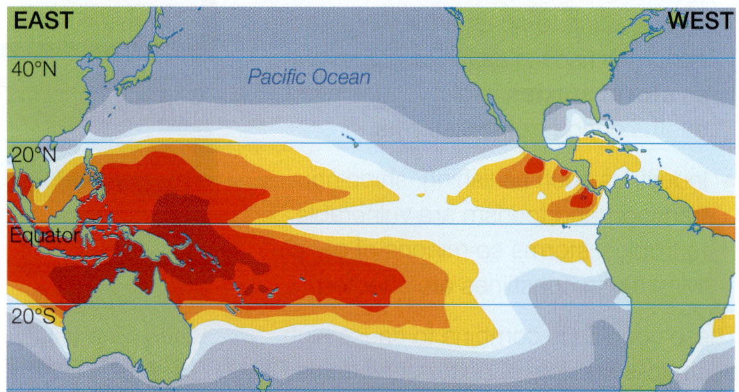

B Ocean surface temperatures during a La Niña event

Impact on rainfall patterns

El Niño and La Niña events change rainfall patterns. An El Niño event, with weak trade winds, results in less rainfall on the Australian/Asian edge of the Pacific Ocean and more on the North and South American edge (diagrams D and E). A La Niña event, with strong trade winds, reverses this.

The influence of an El Niño or La Niña event goes beyond the Pacific Ocean, causing changes to rainfall patterns around the world. The events also seem to influence the frequency and distribution of tropical storms in the Atlantic as well as the Pacific Ocean.

	El Niño	La Niña
Pacific Ocean	More storms	Storms are weaker than usual
Atlantic Ocean	Fewer storms	Storms are stronger than usual

C Typical impacts of El Niño and La Niña on tropical storms

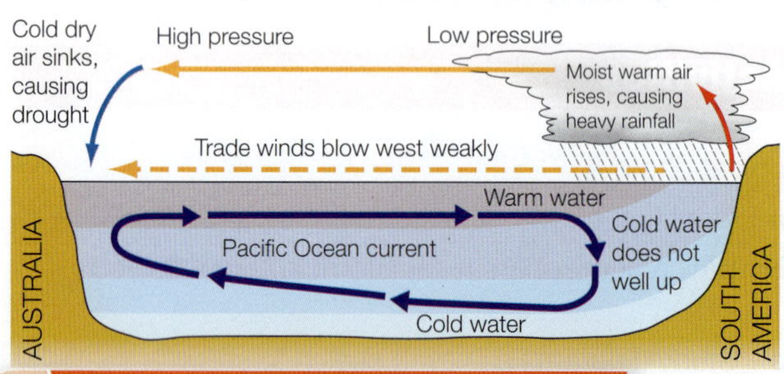

D Impact of El Niño on rainfall patterns in the Pacific Ocean

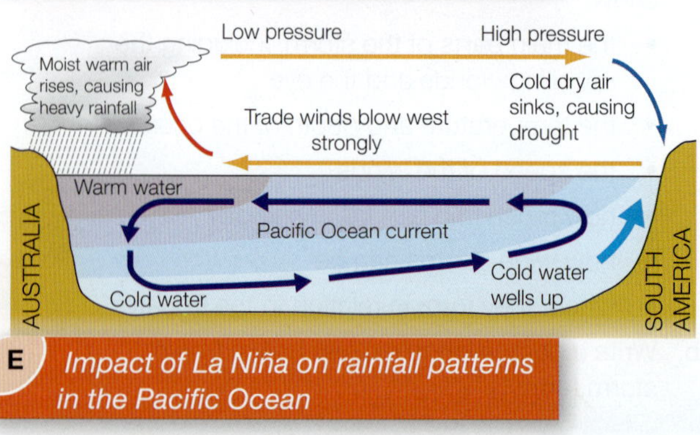

E Impact of La Niña on rainfall patterns in the Pacific Ocean

El Niño and drought

Scientists have found a strong connection between El Niño events and droughts around the world. The effects are felt most strongly on the Equator and in tropical areas where ocean temperatures are higher. The severity of the droughts and the season in which they happen vary with each El Niño event, but the strongest effects are fairly consistent (map F).

Asia is often hit hard during El Niño years, with droughts becoming more common. They cause:

- shortages of food and water
- bush fires and poor air quality
- power cuts in countries like Vietnam, Myanmar and Nepal, which rely heavily on hydro-electric power.

During the 2015 El Niño event, Thailand experienced one of its worst droughts, deepening the poverty of farmers, forcing food prices up and affecting the economy of the whole country.

However, El Niño events bring increased rain to other areas, where they sometimes end a drought. California has suffered drought in recent years (see spread 9.3), but often experiences higher levels of rainfall in El Niño years.

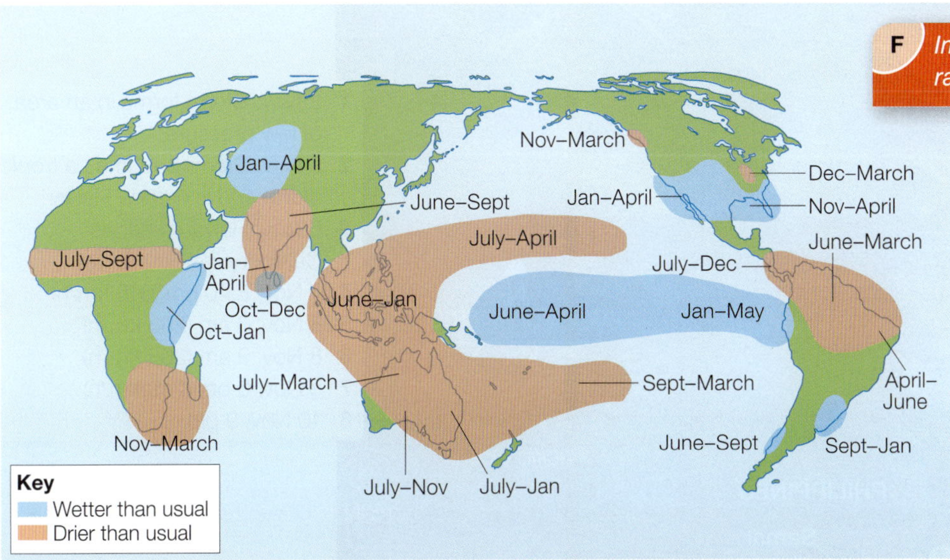

F *Impact of El Niño on world rainfall patterns*

La Niña and drought

La Niña events can also cause droughts, but in different places to El Niño. The areas affected include Peru and Chile in South America.

Activities

1 Look at diagrams D and E, and table C. Copy the paragraph below twice. Using words from the box, complete one paragraph for El Niño and the other for La Niña.

stronger	Asia	warmer	El Niño
Peru	cooler	Australia	
La Niña	weaker	drought	California

_____ events are related to _____ trade winds and lead to _____ than usual sea surface temperatures in the Pacific Ocean. They can cause _____ in such areas as _____ and _____.

2 Look at maps A and B.

a What is the surface temperature of the ocean off the coast of Peru in an El Niño year? How is this different in a La Niña year?

b What is the surface temperature of the ocean off the coast of Vietnam in an El Niño year? How is this different in a La Niña year?

3 Look at map F.

a Describe the distribution of drought associated with an El Niño year.

b Identify at least three countries that may suffer from drought in an El Niño year.

c Explain why El Niño brings drought to these countries.

d Identify three countries that may experience wetter weather than usual in an El Niño year.

e Suggest reasons for the different impacts of El Niño on the eastern and western sides of the Pacific Ocean.

Practice question

Outline the causes of the extreme weather conditions of El Niño/La Niña leading to drought. (6 marks)

Typhoon Haiyan – a tropical storm

On this spread you will investigate the extreme weather conditions that led to Typhoon Haiyan in 2013.

Case study

Typhoon Haiyan tragedy in the Philippines

When Typhoon Haiyan hit the Philippines in November 2013, tragedy was inevitable. The typhoon reached Category 5 on the Saffir–Simpson scale. It was the strongest storm ever recorded when it first reached land in Eastern Samar. It killed over 6300 people in the Philippines and affected about 11 million people in total.

The storm was small when it began in the Pacific Ocean but developed into a tropical storm as it moved west over the warm ocean (satellite image A).

A *Path of Typhoon Haiyan*

1. 2 Nov: Storm forms in an area of low pressure.
2. 5 Nov: Storm becomes a tropical storm and is named Haiyan.
3. 6 Nov: Haiyan reaches Category 5.
4. 7 Nov, 8.40 pm (315 km/h)
5. 8 Nov, 3 am (298 km/h)
6. 8 Nov, 9 am (265 km/h)
7. 8 Nov, 3 pm (249 km/h)
8. 10 Nov, 9 pm

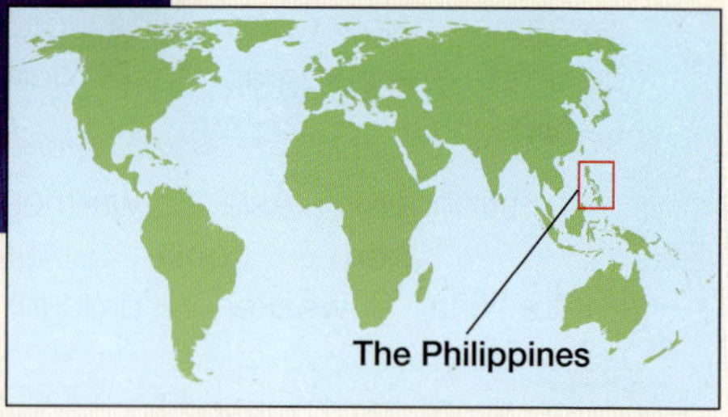

Causes of Typhoon Haiyan

Typhoon Haiyan formed in the Pacific Ocean at a latitude of around 7°N on 2 November. The sea temperature was over 26.5°C and the ocean was well over 60 metres deep. The Coriolis effect started the storm spinning (see spread 1.6). As the storm moved westwards, it passed little land to slow it down and grew as it gained energy from the warm ocean.

When Typhoon Haiyan reached the Philippines on 7 November, it battered the islands with winds up to 315 km/h and drenched them with heavy rain. The low air pressure of the typhoon meant that storm surges reached up to 6 metres in places, causing coastal flooding that extended 1 km inland around the coastal city of Tacloban.

The Philippines – a victim with little protection

The Philippines government started issuing warnings that Typhoon Haiyan was on its way on 6 November. People in areas at risk of landslides or flooding were advised to evacuate. But where could they go?

The Philippines is an **emerging and developing country (EDC)**. In terms of development (see Chapter 6), it is ranked 117 out of 187 countries in the world (factfile B). The area hit by the typhoon was also still recovering from the damage caused by the Bohol earthquake in October, just one month before.

Typhoon Haiyan crossed densely populated areas in the Philippines. Large numbers of people were affected by the wind, rain and storm surge (table C). Many lived in simple wooden houses, which offered little protection from the storm (photo D).

Distance from typhoon path	Population
Within 50 km	3 169 332
50–100 km	1 900 840
100–150 km	1 793 977
150–200 km	4 038 015
0–200 km	10 902 164

C *Population along the path of Typhoon Haiyan through the Philippines*

D *Typhoon Haiyan destroyed many homes*

Total population	99.14 million
Population in rural areas	50.64%
GNI/capita (see spread 6.3)	$3470
Percentage below the poverty line	25.2%
Average life expectancy	69 years
Number of islands	7107

B *Philippines factfile*

Activities

1. Study satellite image A.
   - **a** Describe the location of the Philippines.
   - **b** Describe the path taken by Typhoon Haiyan.
   - **c** Describe how its strength increased along its path.
2. Using factfile B, write one or two sentences summarising the main characteristics of the Philippines.
3. Look back at diagram B on spread 1.6. Write a paragraph explaining the physical factors that led to Typhoon Haiyan being such a powerful storm.
4. Produce a factfile for Typhoon Haiyan using the information on this spread. Think about the five Ws – what, where, when, why, who?
5. Study tables B and C. Why is this information important in explaining why the typhoon caused so much damage?
6. Do you think poverty plays a role in making people more vulnerable to hazards such as typhoons? Imagine you are the child in photo D. What would you say if a TV reporter asked you that question?

Practice question

Describe and explain the causes of a natural weather hazard that you have studied [Typhoon Haiyan]. (6 marks)

1.9 Typhoon Haiyan's path of destruction

On this spread you will explore the consequences of and responses to Typhoon Haiyan.

Case study

Deadly consequences

The most deadly hazard Haiyan brought was the storm surge that flooded coastal areas. In Tacloban, it was 5.2 metres high. It destroyed 90% of the city, including the convention centre, which was being used as an evacuation shelter. There were other **primary consequences** during the storm, followed by **secondary consequences** in the following days and weeks (see panels, right).

The consequences of the storm can also be categorised in a different way (photo A).

- **Social consequences** affect people.
- **Economic consequences** relate to money.
- **Environmental consequences** relate affect the natural environment.

Primary consequences

- Heavy rainfall led to flash flooding and landslides.
- Strong winds caused damage to buildings.
- People were drowned and killed or injured by falling debris.
- Infrastructure, trees and crops were damaged.

Secondary consequences

- People found themselves without shelter, clean water or electricity.
- Roads were blocked by trees and 453 flights were cancelled.
- People were devastated by the loss of at least one family member.

A *Damage caused by Typhoon Haiyan*

Social consequences
- Over half a million families were made homeless.
- Over 6300 deaths left families devastated.
- Damage to 571 health clinics and hospitals.

Economic consequences
- $2.86 billion of damage was caused.
- 77% of farmers and 74% of fishermen lost their main source of income.

Environmental consequences
- Crops were damaged.
- Trees were uprooted.
- A barge was punctured and 85 000 litres of oil leaked into the sea.

Keeping people informed

The Philippines is often affected by tropical storms. A government agency is responsible for monitoring storms and releasing warnings to make sure people know the level of the threat and how they should respond. The warning system has four levels in different colours (panel B). People can find out about storm warnings by watching the television, listening to the radio or using social media.

B *The warning system used in the Philippines*

RED WARNING	More than 30 mm rain in 1 hour and likely to continue for next 2 hours	Serious flooding is expected in low-lying areas	EVACUATION
ORANGE WARNING	15–30 mm rain (intense) in 1 hour and likely to continue for next 2 hours	Flood is threatening	ALERT for possible evacuation
YELLOW WARNING	7.5–15 mm (heavy) rain in 1 hour and likely to continue for next 2 hours	Flooding is possible	MONITOR weather conditions
GREEN STATUS	Less than 7.5 mm rain in 1 hour	No threat	No action

Emergency responses to Typhoon Haiyan

By 11 November the Philippines government had put seven provinces under a 'state of national calamity', controlling the price of vital goods and beginning the relief effort. **Emergency (short-term) aid** to provide food, water, shelter and medical supplies was the first priority.

- The United Nations (UN) appealed for £190 million for emergency aid.
- The World Health Organisation helped to manage the healthcare workers who poured in from around the world.
- Many countries sent aid (table C). For example, Australia put together a $28-million package including medical staff, materials for building shelters, water containers and hygiene kits.

The emergency aid effort in areas such as Tacloban faced many difficulties as the airport terminal had been destroyed and there was looting and crime. But social media was used to ask for aid and volunteers, and hashtags were created for survivors to call for rescue or to search for missing people.

The Red Cross warned people to be careful when returning to their homes. They needed to make sure buildings were stable and watch out for dangers such as hidden snakes and live electricity cables.

Long-term responses to Typhoon Haiyan

Once the initial emergency had been tackled, the government and aid agencies refocused on **long-term aid**, recovery and development. By April 2014, services had been restored to 560 schools, 220 rural health centres and 30 hospitals. In addition, 1500 km of roads and 1100 km of drainage canals had been repaired. Schemes were also set up to give cash for training (to support small businesses) or cash for work, which involved:

- clearing debris and managing waste
- repairing **infrastructure** (e.g. water supply systems, road and communication networks, schools, hospitals)
- replanting mangroves around the coasts to provide a buffer against future storm surges.

Activities

1. Study the text and photo A. Identify:
   - **a** three primary consequences of the typhoon
   - **b** three secondary consequences of the typhoon.
2. Look at source B. Write a radio advert to make Filipinos aware of the warning system. Make sure they know how important it is to move to an evacuation centre or high land when a red warning is given.
3. Study the text and table C. Draw a spider diagram of aid given after the typhoon. Colour it to show short- and long-term aid.
4. Why did mobile phones and social media prove so useful during and after the disaster?

Aid given	Who benefited
Food assistance	4 million people
Seeds	44 000 families
Fertiliser	80 000 families
Educational materials	420 000 children
Basic emergency shelter materials	500 000 households
Tools and materials to rebuild homes	55 000 people
Cash for materials and labour	40 000 people
Solar lanterns	50 000 families

C *Some of the aid delivered in the first 100 days after Typhoon Haiyan*

D *Repair was well underway by April 2014*

Go further

Find out how people are preparing for future typhoons in the Philippines by searching newspaper websites. Try using 'Philippines preparing for typhoon' in your search. What improvements have been made since Typhoon Haiyan?

Practice question

Discuss the responses to a natural weather hazard you have studied [Typhoon Haiyan]. (6 marks)

Drought in the UK

On this spread you will consider the causes, consequences and responses to the drought that hit the UK in 2012.

Case study

Drought strikes!

People often complain of relentless rain in the UK, but much of England and Wales suffered from drought in 2012. It was one of the worst droughts on record in southern England, leading to very low levels in reservoirs and water shortages (photo A).

From April 2010 to May 2012 most of England received less than 85% of average rainfall, with some areas receiving less than 75%. The extent of the drought is shown on map B.

A *Not so dangerous in 2012!*

B *Areas of England and Wales in drought in 2012*

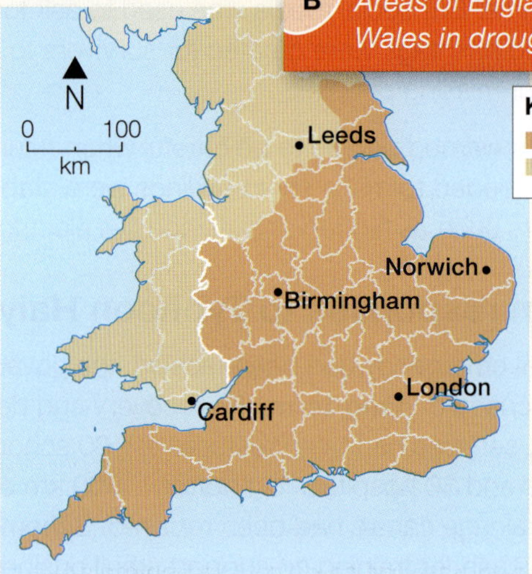

Causes of the drought

Droughts develop over a period of time where rainfall is below average. Reserves of water in soils, lakes and rivers are 'used up', resulting in water shortages. Several factors contributed to the drought in 2012:

- **Less rain:** The area affected by drought received only 55–95% of usual rainfall between April 2010 and May 2012 (map C). Unusual wind patterns had brought dry winds from Europe in the east rather than the usual wet winds blowing from the Atlantic in the west.
- **Warmer temperatures:** The weather was also warmer than usual, so more water evaporated from reservoirs than usual and soils dried out.
- **Dry soils:** It was difficult for the rain that did fall to soak into the ground as the soil was dry and baked hard.
- **High water usage:** People in the UK are in the habit of using large amounts of water in their everyday lives (chart D). Massive amounts are also used in industry and farming. This leads to around 1.7 billion litres of water being used every day. Water is also wasted through leaking pipes.

C *Rainfall in the UK, April 2010 to March 2012*

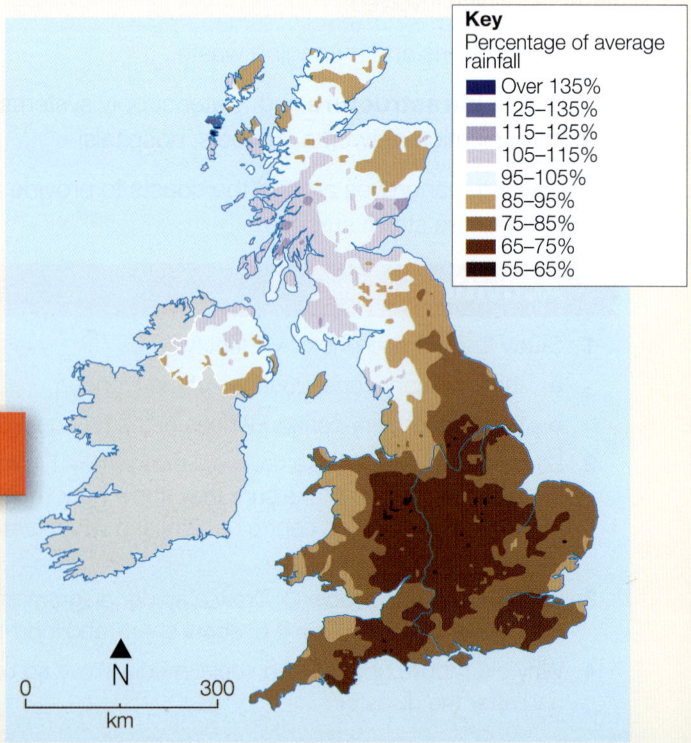

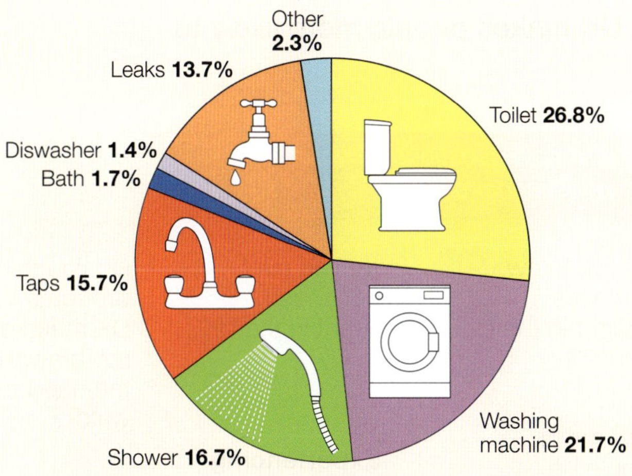

D *How do we use water in our homes?*

Consequences of the drought

- **Problems for farming:** Water shortages made it difficult to find water for crops and livestock. The dry ground also made it difficult to harvest crops in the autumn of 2011.
- **Environmental damage:** Dry areas of moorland caught fire easily, with **wild fires** raging across parts of South Wales, Surrey and the Scottish Borders. River water was used to boost the water supply in some areas, reducing river levels and causing damage to plants and animals.
- **Restrictions in homes:** Homeowners were still able to turn on the tap for water. But hosepipe bans meant that around 20 million people weren't allowed to use hose pipes to water their gardens or wash their cars.

Responses to the drought

As the seriousness of the drought became clear the water companies took action to keep water flowing through the pipes of the UK.

- Permits were granted to allow the water companies to extract water from rivers.
- They issued hosepipe bans to conserve the water that was available.
- There were campaigns to get ordinary people to use less water in their homes. Even something as simple as turning off the tap when you brush your teeth can save 6 litres of water per minute!

Activities

1. Look at map B. Describe the area of the UK affected by drought in 2012.
2. Look at map C.
   - **a** Name an area that received more than the average rainfall. (Clue: look at the map key.)
   - **b** Name an area with less than average rainfall.
   - **c** How does map C help to explain the severity of the drought in 2012?
3. Describe the consequences the drought might have for:
   - **a** families living in the south of England
   - **b** businesses that need large amounts of water to make products
   - **c** wildlife in the rivers being used to top up water supplies.
4. Look at chart D.
   - **a** Does the chart reflect how water is used in your house? Identify similarities and differences.
   - **b** What could you do to use less water? Design a poster to remind your family how to conserve water.

Your enquiry

How can weather be hazardous?

Which do you think will be the greater challenge to the world in future – droughts or tropical storms?

a Consider the opinions of:
- farmers • fishermen/women • city dwellers
- medical workers.

Use a table like this:

		Affected by drought	Score	Affected by tropical storms	Score
Farmers	Q1				
	Q2				
	Q3				

For each type of people, make notes on how they might respond to these three questions:

1 What risks do you face from this hazard?
2 What consequences would this hazard have for you?
3 How quickly would you recover from these consequences?

You could add more questions if you wish.

b Give a score of 1–5 for each answer, with 1 for slightly affected to 5 for very badly affected. Add up the scores for each hazard.

c Write a conclusion on which hazard will present the greater challenge. Justify your opinion.

1.11 Hazards fieldwork

On this spread you will investigate if drought in the UK makes people more likely to conserve water.

Carrying out fieldwork to test a hypothesis

Fieldwork will help you to answer a question or test a **hypothesis**. A hypothesis is a statement that you try to prove to be correct or incorrect. Or you might find that the hypothesis needs to be changed. The hypothesis you are going to work on is *'Experiencing drought in the UK has raised awareness of the need to conserve water and has led to people changing their habits.'*

Designing a questionnaire

A good way to test this hypothesis is to do a survey, using a questionnaire. You could find out what people have experienced and if they have changed their habits as a result of their experiences.

You need to design questions to find out:

- if people know there have been droughts in the UK
- if droughts make people more aware of the need to conserve water
- if and how people change their habits to conserve water.

You could use the questions in diagram A or design your own. Use two basic kinds of question so you can gather different types of data.

- **Closed questions:** These can be answered with one word or phrase. They could be multiple-choice. They are quick and easy to answer, so they make good opening questions. They can give facts and the results are easy to put on graphs.
- **Open questions:** Answers are usually lengthy and allow respondents to explain themselves. They draw out opinions, reasons and feelings. The results are not easy to put on graphs.

Avoid bias in your questions. You want to find out what people think, not suggest particular answers to them.

You could ask basic information about respondents, such as age and gender, at the beginning of the questionnaire. This would allow you to analyse if these factors influence people's views and actions.

A *Example questions for your questionnaire*

Experiencing the 2012 drought in the UK has raised awareness of the need to conserve water and has led to people changing their habits.

- Did you know there was a drought in the UK in 2012?
- Have you ever experienced drought in the UK?
- Do you think droughts in the UK make people more aware of the need to save water?
- How could people be encouraged to save more water?
- Do you think people in the UK should do more to save water?
- Do you use any of these ideas to save water...?
- What changes have you made to your habits to use less water?
- Which of these reasons is the most important for saving water..?
- Have you tried to save more water since 2012?

Key

Closed questions | Open questions

Improving your questionnaire

Once designed, you should pilot your questionnaire, testing it out by getting one or two people to answer it. Find out if the questions are easy to understand and if they get the sort of information you need. Then make your improvements.

Conducting your survey

Make sure the sample size for your survey is as large as possible given the time and resources available. The larger the sample size, the wider the range of views you can record, making them more representative of the whole population. Your data are then less likely to be biased.

You could work in a group and share your findings so you have more data to analyse. You could use a free online survey tool so that you can e-mail your questionnaire to a larger number of people. If you conduct your survey in public, stay safe! Keep with your group, supervised by your teacher.

Analysing your data

Once you have collected the data, you need to study them carefully in relation to your hypothesis. You can arrange and analyse the data in different ways, including:

- **Graphs:** You could plot the answers to closed questions, making it easier to pick out any patterns (diagram A).
- **Statistics** are useful for analysing data, e.g. you could work out what percentage of people responded in a particular way.
- **Tables:** You could put the answers to open questions in categories or tables to make them easier to analyse.

In your analysis, draw together your key findings and explain what they mean. A good way to do this is to write short notes under each graph or table before pulling everything together in your conclusion.

B *Drawing a bar graph*

Concluding and evaluating your research

In your conclusion, say if your hypothesis is correct, incorrect or if it needs changing. Base this on your research findings, not just on your own views.

An example of a change to the hypothesis could be *'Experiencing drought in the UK has raised awareness of the need to conserve water but most people have not made significant changes to their habits.'*

Finally, evaluate the research and your methods, thinking about what went well and what could be improved. Also reflect on who might be interested in your findings and how you could extend the investigation.

Activities

1 Plan your data collection.
- **a** Working in small groups, design a short questionnaire to test the hypothesis.
- **b** Pilot, and then improve, your questionnaire.

2 Collect your data.
- **a** Each ask at least three people to complete your questionnaire.
- **b** Share your findings with your group.

3 Analyse your data.
- **a** Produce at least one graph to show how people answered one of your closed questions.
- **b** Compile a table to summarise the answers to your open questions.

4 Draw conclusions from your data.
- **a** What experiences did the respondents have of drought?
- **b** Did experiencing drought raise their awareness of the need to conserve water?
- **c** Did they change their habits to conserve water?
- **d** Are you going to accept, reject or change your hypothesis?

5 Evaluate your investigation.
- **a** How successful was your use of the questionnaire?
- **b** How successful have you been in testing your hypothesis?
- **c** If you did this investigation again, how would you improve it?
- **d** Who might be interested in your findings?
- **e** How could you take this investigation further?

Plate tectonics – shaping our world

On this spread you will explore Earth's structure, tectonic plate boundaries and the distribution of earthquakes and volcanoes.

Earth's structure

The outer layer of Earth, including the land we live on, is known as Earth's crust. What would we find if we could drill through this hard outer shell? It depends on whether we drilled through **continental crust** on land or through **oceanic crust** under the ocean (diagram A).

- If we started on land, we would drill through about 35 km of crust until we reached the **mantle**.
- Under the ocean we would only need to drill through about 7 km of crust before reaching the mantle.

The crust and the very top of the mantle make up the **lithosphere** (diagram A).

We couldn't explore very far because we would rapidly get too hot! Near the crust, the rock in the mantle gets extremely hot but does not melt because it is under extreme pressure. Further below the crust, when temperatures reach 1300°C, the mantle gets close to melting point and flows slowly.

We would need to drill through 2890 km of mantle to reach the **core**. The outer core is liquid, but the inner core is solid rock.

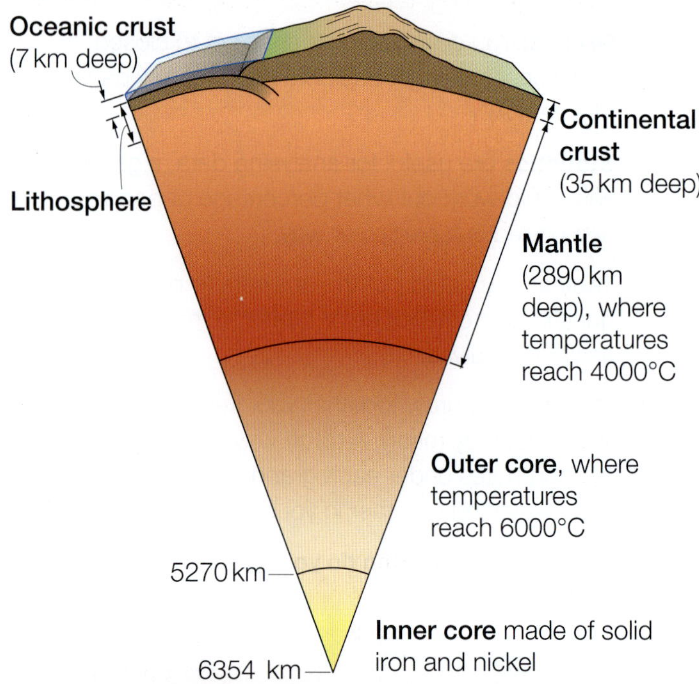

A *A section through Earth's structure*

Tectonic plates and their boundaries

Earth's crust is broken into pieces called **tectonic plates**, which move slowly in different directions (map B). The places where two or more plates meet are called **plate boundaries**.

B *Tectonic plates, earthquakes and volcanoes*

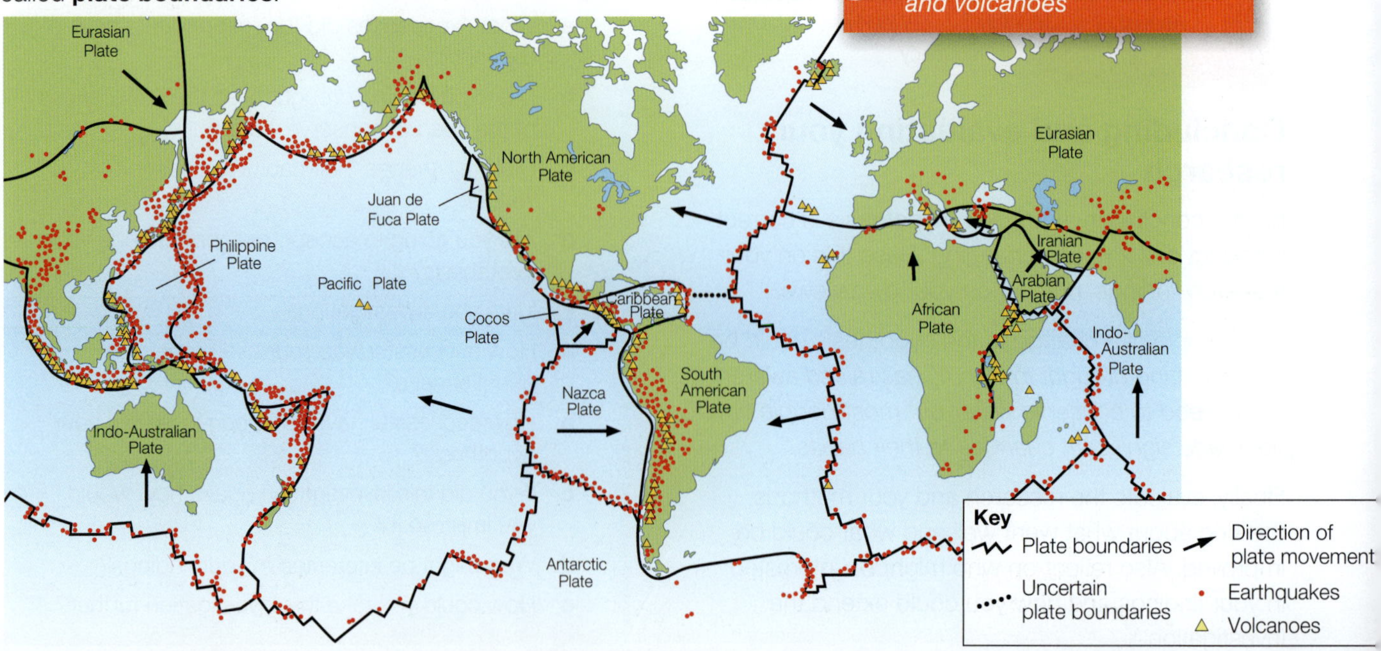

Different types of plate boundary

- **Constructive boundary:** plates are moving apart and creating new land.
- **Destructive boundary:** plates are moving towards each other with one being forced under the other.
- **Collision boundary:** plates are moving towards each other and creating mountains.
- **Conservative boundary:** plates are moving alongside each other with no land lost or gained.

Moving plates

Scientists once thought that the heat of Earth's core causes convection currents, vast slow flows of heat in the mantle, which pull the lithosphere. But they cannot find evidence of currents strong enough to do that.

Now they suggest that **ridge push** and **slab pull** might be a more important process in causing the plates to move (diagram C). However, more evidence is needed to find out how important each of these processes is.

Plate movement causes **earthquakes** and **volcanoes**, where **lava** (melted rock), solid rock, hot vapour and gas erupt (see spreads 1.13–1.19).

C *Why plates move*

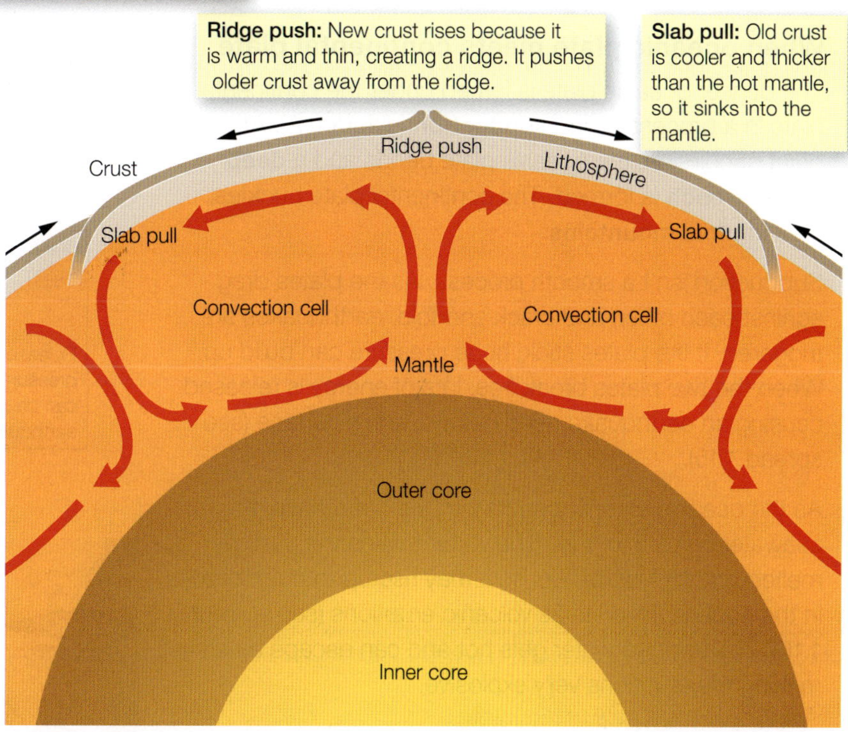

Activities

1. Study diagram A. Annotate a larger version with all you have learned about Earth's structure.
2. Look at map B.
   - **a** Describe the distribution of earthquakes and volcanoes. How does it relate to the plate boundaries?
   - **b** Do all earthquakes and volcanoes occur on plate boundaries? Using an atlas, refer to specific areas.
   - **c** Identify two plates that are moving apart and two plates that are moving together.
3. Study diagram C.
   - **a** Using items from your pencil case and school bag, make a model to show how tectonic plates move. Use it to explain plate movement to a partner.
   - **b** Draw a sketch of your model. Annotate it to record what happens.
   - **c** Use your sketch to explain to a friend or a parent how these processes work together. Which processes did they find difficult to understand? Improve your annotations.

Go further

Carry out research to find three earthquakes or volcanic eruptions that have happened in the last year. You could use http://earthquakes.volcanodiscovery.com/. Where did they happen? How near to a plate boundary were they?

Your enquiry

How do plate tectonics shape our world?

As you work through this section, make notes about anything that will help you to answer the question. You will bring your ideas together to do an activity at the end of spread 1.19.

Destructive and collision plate boundaries

On this spread you will learn about the processes at work at destructive and collision plate boundaries.

When plates push together they form a destructive plate boundary or a collision plate boundary.

Destructive plate boundaries

When oceanic plate meets continental plate

The denser oceanic plate is forced under the continental plate in a process called **subduction** (diagram A). At the point of collision, the oceanic plate bends and a **deep ocean trench** is formed. The continental plate buckles, creating **fold mountains**.

Subduction isn't a smooth process. As the plates drag against each other, and crack and fold, earthquakes are triggered. If the plates stick, huge pressure can build up. When the two plates break free, lots of energy is released, causing an earthquake deep below Earth's surface (see spread 1.15).

As the oceanic plate subducts into the hot mantle, the seawater helps it to melt. The water makes the partially melted mantle less dense, so it may rise up through cracks in the continental plate as volcanic eruptions (see spread 1.16). The trapped water gets hot and can escape as steam, making the eruptions very explosive.

A *Processes at a destructive boundary*

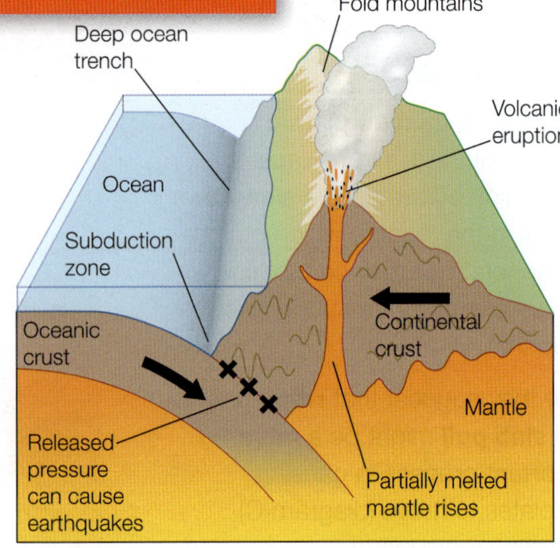

The Andes fold mountains

The oceanic Nazca Plate is being subducted under the South American Plate (map B). The area experiences earthquakes when the plates drag and stick against each other. A large earthquake hit the Maule region of Chile in 2010, killing at least 214 people and destroying buildings, bridges and roads.

The South American Plate has folded to create the Andes. This mountain range includes active volcanoes that have explosive eruptions. The Calbuco volcano in Chile erupted three times in April/May 2015 and 2500 people had to be evacuated from the area.

B *The Andes – fold mountains on a destructive boundary*

> **When two oceanic plates meet**
>
> The denser plate subducts under the other one, forming a deep ocean trench. Volcanoes can erupt under the ocean and may eventually form islands, such as the Marianas Islands in the Western Pacific.

Collision plate boundaries

When two continental plates push against each other they collide (diagram C). Continental plates are less dense than the mantle beneath them. So there is no subduction and therefore no volcanic activity. But the continental crust thickens and the surface lifts into fold mountains. The pressure built up by the plates pushing against each other also causes the crust to crack into **faults** which can trigger earthquakes fairly close to Earth's surface.

The Himalayan fold mountains

The Himalayan mountain range was formed by the collision of the Indian Plate and the Eurasian Plate (map D). There used to be an ocean between them, but it closed up as the plates moved together. The sedimentary rock (see spread 3.3) under the ocean folded to form the mountains. The mountains are still being folded upwards at a rate of more than 10 mm a year. Devastating earthquakes also still occur in this region, including the one that hit Nepal in April 2015 (see spreads 1.17–1.18).

C *Processes at a collision boundary*

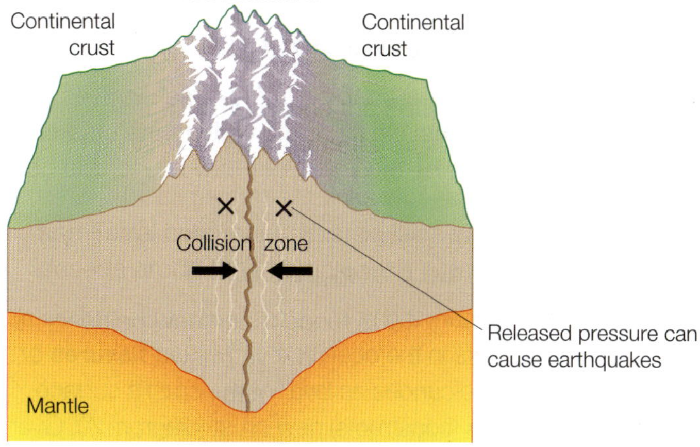

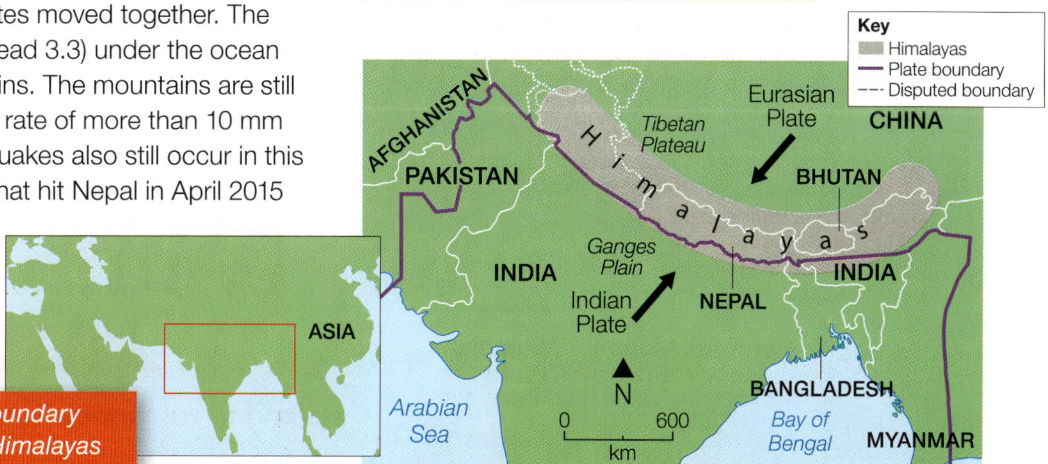

D *The collision plate boundary associated with the Himalayas*

Activities

1 Copy and complete the following paragraph, using the words in the box.

earthquakes fold mountains oceanic continental subduction

When an oceanic and a continental plate meet, the denser ______ plate is forced down and under the ______ plate. This is called ______. The plates don't move smoothly, triggering ______ and forming ______.

2 Write a short warning from the UK Foreign Office to advise tourists why Chile experiences volcanic and earthquake hazards.

3 Start a summary table about plate boundaries. You will complete it on spread 1.14.

Plate boundary	Key processes	Earthquakes	Volcanoes	Example
Destructive				
Collision				

4 a At which types of plate boundary do fold mountains form?

b Why are there volcanoes in the Andes but not in the Himalayas?

c Study diagram C and map D. Draw a diagram to show how the Himalayas were formed. Remember to label the plates.

Go further

Search for the plate tectonics resource on the geolsoc.org.uk website. Visit the plate margins section on the resource to watch animations of how the plates move at different types of boundary.

Practice question

Draw a diagram to show the processes that take place at destructive plate boundaries. (4 marks)

1.14 Constructive and conservative plate boundaries

On this spread you will learn about the processes at work at constructive and conservative plate boundaries.

Constructive plate boundaries

When plates pull apart partially melted mantle rises. Under the ocean this forms **mid-ocean ridges**. On land, **rift valleys** are created. As the plates pull apart, small earthquakes may be triggered, but they cause little damage.

When oceanic plates pull apart

Under the ocean, the partially melted mantle comes up through cracks in the oceanic crust called **fissures** or erupts from volcanoes as lava. At the crust's surface, the lava forms (constructs) new crust (diagram A). Over time, this produces chains of mid-ocean ridges under water. The Mid-Atlantic Ridge has formed in this way because two pairs of plates – the Eurasian and North American plates and the African and South American plates – are moving apart. Mid-ocean ridges can rise up thousands of metres above the ocean floor and some become islands, such as Iceland.

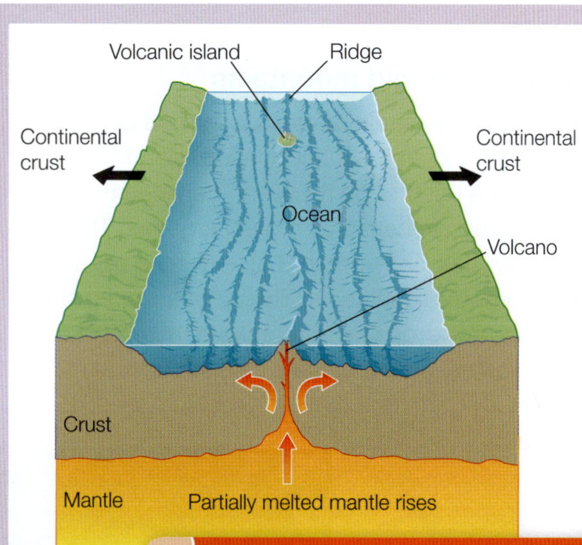

A *Processes at a constructive boundary*

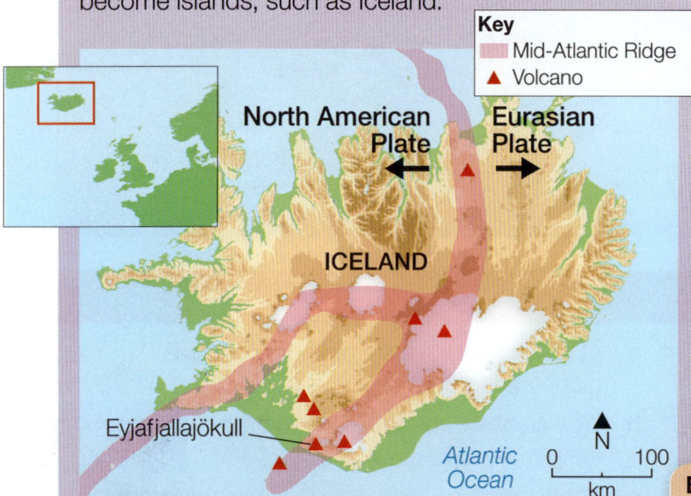

B *Iceland and the Mid-Atlantic Ridge*

Island of volcanoes

Iceland lies on the Mid-Atlantic Ridge (map B). There the plates are still moving apart. As the hot rock in the mantle rises to the surface, the pressure decreases and the rock starts to melt, producing volcanoes that are still active. Eyjafjallajökull erupted in 2010, sending an ash cloud into the air that disrupted flights across Europe.

When continental plates pull apart

On land, when the partially melted mantle rises, it heats the plates above, causing them to expand and bulge. This forms highlands. As the plates stretch, they fracture along fault lines. This leads to fallen blocks of lowland, which form rift valleys. It is thought that these processes are happening where the Arabian Plate is pulling away from the African Plate. This is forming the Great African Rift Valley (map C), which runs for thousands of kilometres. If the stretching continues, the lowland areas will eventually drop below sea level. Then a new ocean could form, splitting eastern Africa from the rest of the continent.

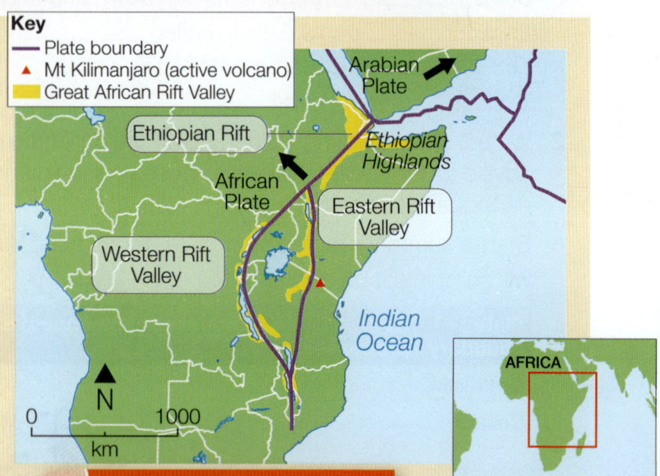

C *Great African Rift Valley*

Conservative plate boundaries

Tectonic plates move past each other at conservative boundaries. Along some boundaries they move in different directions, but along others they move in the same direction but at different speeds. A famous example is the San Andreas Fault (photo D and map E).

The plates move in jerks. They can be locked together for years, building up pressure. When they break free suddenly, the release of stored energy causes earthquakes. The biggest earthquakes can be very powerful and destructive because they are triggered near Earth's surface. They are rare, but small earthquakes happen frequently. There are no volcanoes at conservative plate boundaries because there is no subduction.

D *The San Andreas Fault runs for 1300 km, mostly through California*

The Parkfield earthquake

On 28 September 2004 an earthquake shook the small town of Parkfield, which lies on the San Andreas Fault. Earthquakes happen in that area quite regularly, on average about 22 years apart. They seem to occur because Parkfield is halfway between a locked section of the fault to the south and a moving section to the north.

The 2004 earthquake was caused by the San Andreas Fault suddenly moving about 500 mm. It was a relatively small earthquake of magnitude 6.0 on the Richter scale (see spread 1.15). The damage it caused was slight, but roads cracked and bridges needed to be rebuilt. If a large earthquake hit the megacity of Los Angeles, the damage would be much greater (see spread 9.4).

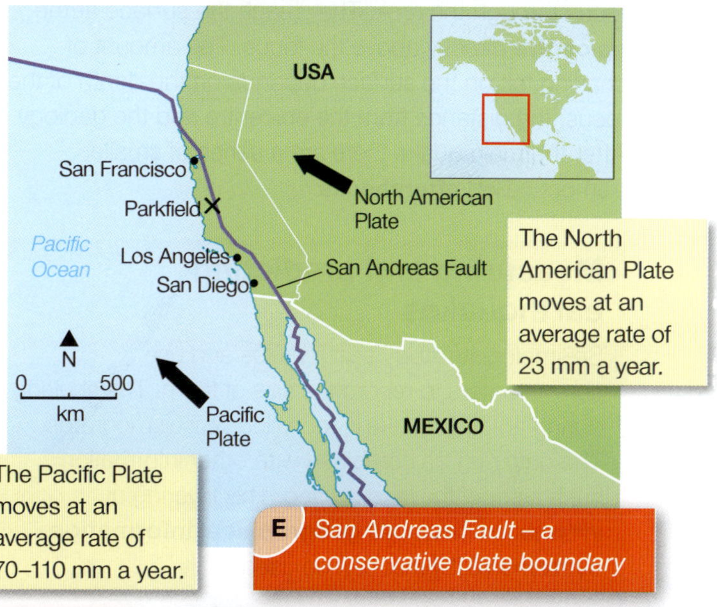

E *San Andreas Fault – a conservative plate boundary*

Activities

1. Study maps B, C and E.
   - **a** Identify the plates that are pulling apart.
   - **b** Identify the plates that are moving past each other.
2. Using diagram A and map B, describe the process that creates volcanoes like Eyjafjallajökull at constructive plate boundaries.
3. Describe what is happening at the San Andreas Fault. Remember to mention the direction and speed of the plates.
4. Study map E and the text. Draw a poster to educate visitors to Los Angeles about the earthquake threat. Include:
   - **a** a diagram showing the role of the San Andreas Fault in causing earthquakes
   - **b** a brief explanation of why these earthquakes can be so destructive. (Clue: think about physical and human factors.)
5. Complete the summary table you started on spread 1.13 by adding constructive and conservative plate boundaries to it.

Go further

Tsunamis are often associated with earthquakes.

- **a** Research what a tsunami is.
- **b** Then explain why a large tsunami is unlikely to happen because of the San Andreas Fault.

You may find it useful to look for tsunami on the ngkids.co.uk website or use your browser to search for 'california tsunami risk' and select the link for www.time.com.

Practice question

Outline the processes that take place at constructive plate boundaries. (4 marks)

How earthquakes happen

On this spread you will find out how movement of tectonic plates causes shallow and deep focus earthquakes.

Causes of earthquakes

Earthquakes can be both fascinating and terrifying. The more powerful ones happen where plates are pushing together at destructive or collision plate boundaries, or moving past each other at conservative plate boundaries.

If the plates stick, huge pressure can build up. When the plates break free or one plate gives way, the energy released causes an earthquake beneath Earth's surface at the **focus** (diagram A). Fast waves of energy called **seismic (shock) waves** spread out from the focus through the rock. They reach the surface at the **epicentre** directly above the focus. The amount of shaking felt on the surface depends on the depth of the focus, the distance from the epicentre and the geology. After the main quake there are a series of smaller tremors called **aftershocks**.

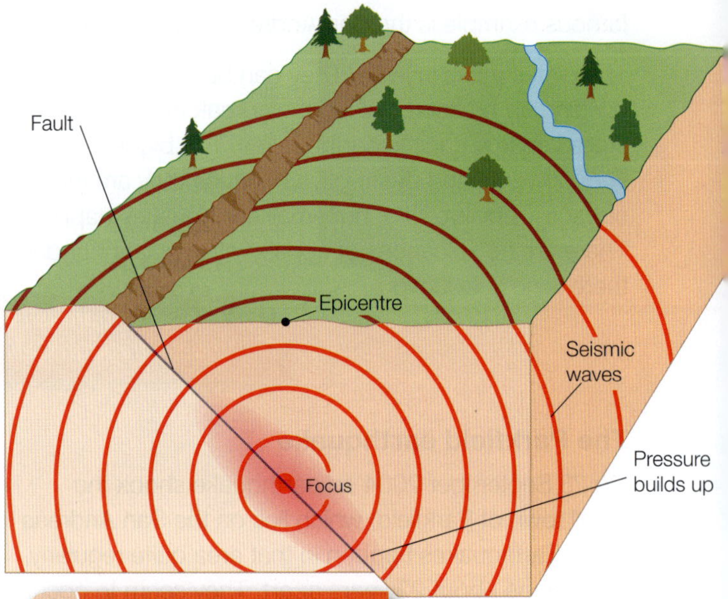

A *An earthquake on a fault*

Measuring and predicting earthquakes

Seismologists study earthquakes and use seismometers to record the size of them. The shaking during an earthquake is recorded on seismograms. The data can be combined with other information, such as how the land is used. This layering of information produces **geographical information systems (GIS)** that can be used to:

- predict the timing and severity of earthquakes
- calculate the risk of earthquake damage
- help to minimise risk during future earthquakes.

Seismologists often know where pressure is building up, but can't predict exactly when and where earthquakes will happen.

Richter scale

The amount of energy released in an earthquake is known as its **magnitude**. This is measured using the **Richter scale**. It is a logarithmic scale so, for example, an increase of 1 means the quake is ten times more powerful, with just over 30 times more energy, than the previous level. So a magnitude 7.0 earthquake is ten times more powerful than a magnitude 6.0 earthquake and 100 times more powerful than one measuring 5.0.

Mercalli scale

The intensity (strength) of an earthquake and the effects and damage caused can be measured using the **Mercalli scale**. This ranges from I to XII. Level I can only be recorded on seismograms and cannot be felt. At level XII everything would be destroyed and the ground would move in waves.

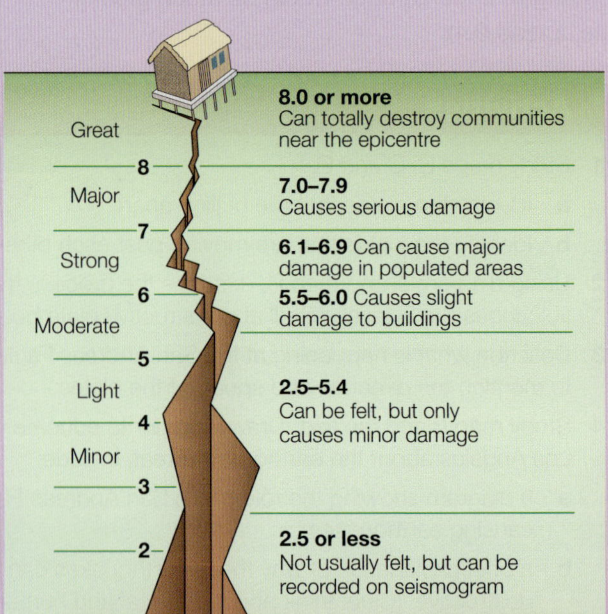

B *The scale of damage caused by earthquakes of different magnitudes on the Richter scale*

Deep focus earthquakes

Deep focus earthquakes start 70–700 km below Earth's surface (diagram C). They are triggered where an oceanic plate is subducted under a continental plate, so only occur at destructive plate boundaries. They can be very powerful, but the seismic waves spread out vertically so a fairly small area suffers the damage.

The 2013 Okhotsk Sea earthquake was 609 km deep, not the deepest earthquake ever recorded. However its magnitude was 8.3, making it the strongest deep focus earthquake on record. It mainly affected Russia, where the shaking lasted five minutes in some places.

Shallow focus earthquakes

Shallow focus earthquakes start within 70 km of Earth's surface (diagram C). They occur at conservative boundaries and in the upper part of destructive plate boundaries and collision boundaries. They are usually small and most cannot be felt. But because the seismic waves spread out horizontally they can cause damage over a large area.

The 2015 Nepal earthquake (see spread 1.17) is an example of a shallow focus earthquake as its focus was only 15 km deep. However, with a magnitude of 7.8, it was unusually strong.

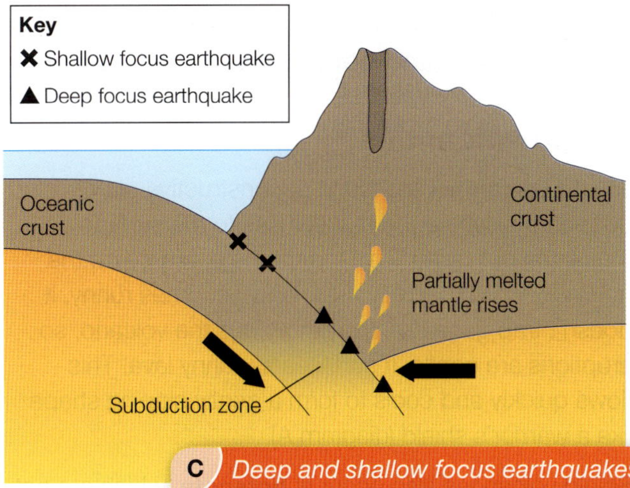

C *Deep and shallow focus earthquakes at a destructive boundary*

	Shallow focus earthquakes	Deep focus earthquakes
Depth	0–70 km	70–700 km
Frequency	Common	Rare
Average magnitude	1.0–5.0 on Richter scale	6.0–8.0 on Richter scale
Seismic waves	Radiate horizontally	Radiate vertically
Damage	Widespread	Localised

D *Characteristics of shallow and deep focus earthquakes*

Activities

1. Chose five words that you think are particularly important in understanding earthquakes. Define these terms in your own words.
2. Study diagram C and look back at spread 1.13.
   - **a** Why are shallow and deep focus earthquakes found at destructive plate boundaries?
   - **b** Shallow focus earthquakes are triggered by movement at the San Andreas Fault (see spread 1.14 for conservative plate boundary). How is that likely to affect the damage caused? (Clue: think about the likely magnitude of the earthquake and how large an area would be affected.)
3. Compare the characteristics in table D.
   - **a** Which type of earthquake do you think is most dangerous – shallow or deep focus?
   - **b** Justify your decision, referring to the likely magnitude of the earthquake and the area affected.

Go further

The epicentre of the 2013 Okhotsk Sea earthquake was just off the coast of the Kamchatka Peninsula in Russia, but the earthquake was felt as far away as the cities of Tokyo, Nanjing and Moscow.

- **a** Use an atlas to find out how far away these cities are from the epicentre.
- **b** Suggest reasons for the shaking being felt over such a large area.

Practice question

Explain how the movement of tectonic plates causes shallow focus earthquakes. (4 marks)

Why volcanoes erupt

On this spread you will learn how the movement of tectonic plates causes shield and composite volcanoes.

Shield volcanoes

Shield volcanoes are found at constructive plate boundaries, where the mantle rises to the surface as the plates pull apart. This lowers the pressure on the hot rock, which partially melts and becomes runny. It doesn't trap gases and nothing plugs the volcano, so eruptions are gentle. It surfaces as runny lava. This flows quickly and cools to form a gently sloping shape like a warrior's shield (diagram A).

Shield volcanoes also occur at **hotspots**, which are not on plate boundaries but where the oceanic crust is very thin.

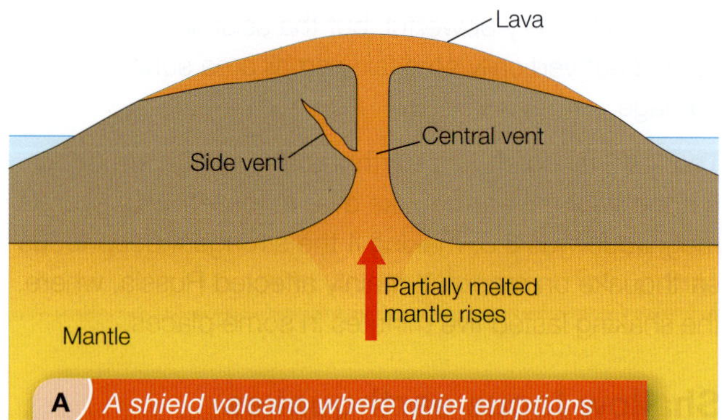

A *A shield volcano where quiet eruptions gradually form a gently sloping mountain*

Volcanoes on hotspots

The nearest plate boundary to the Hawaiian Islands is 3200 km away, so why are volcanoes found there (photo B)? Some geologists think these volcanoes have formed because a **mantle plume** rises there, forming a hotspot. The rock in a plume is hotter than the mantle around it, so it rises from the outer core over millions of years. The thin oceanic plate above heats up and the plume burns through the crust, spilling over to form a shield volcano.

The eight volcanic islands of Hawaii were formed in this way. The island of Hawaii has two active volcanoes, Kilauea and Mauna Loa. The Pacific Plate is moving in a north-westerly direction, so each volcano created over the hotspot is eventually carried away from it. The extinct volcanoes that are no longer active cool, are eroded and gradually sink into the ocean. At the same time the new piece of oceanic plate over the hotspot starts to heat up (diagram C).

The hotspot is now 35 km south-east of Hawaii. A new island called Lo'ihi Seamount is already forming. It is 975 metres below sea level, but is expected to surface in 10 000–100 000 years' time.

B *Kilauea – a shield volcano on Hawaii*

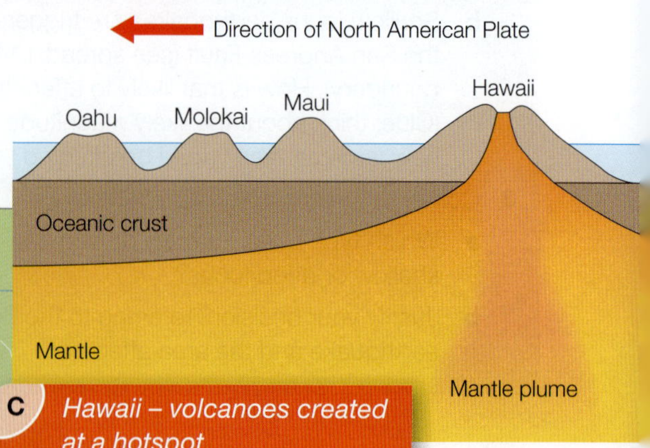

C *Hawaii – volcanoes created at a hotspot*

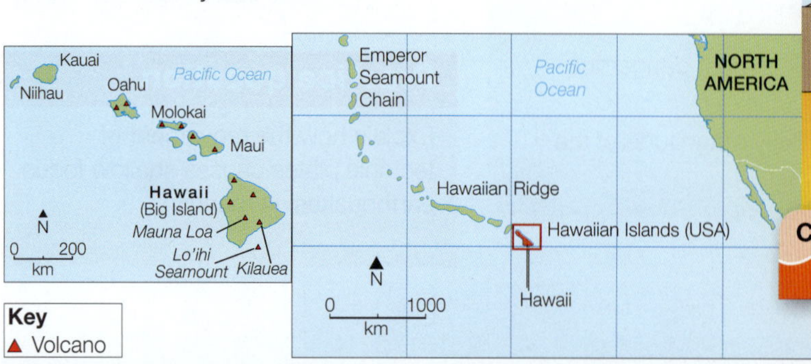

Composite volcanoes

Destructive boundaries can be dangerous places, where subduction can lead to volcanic activity as well as earthquakes. The partially melted mantle that rises up is viscous (sticky). It contains trapped seawater and hot gases. This builds up pressure. When it is released suddenly, steam and dangerous gases erupt violently from the volcano. The sticky lava cools quickly so it flows slowly, creating steep slopes. Ash also escapes from the volcano and lands on the slopes. This alternate layering of lava and ash creates **composite volcanoes** (diagram D).

Sometimes cooled rock forms a plug at the top of the main vent. This leads to an enormous build-up of pressure before the next eruption, which will be even more explosive.

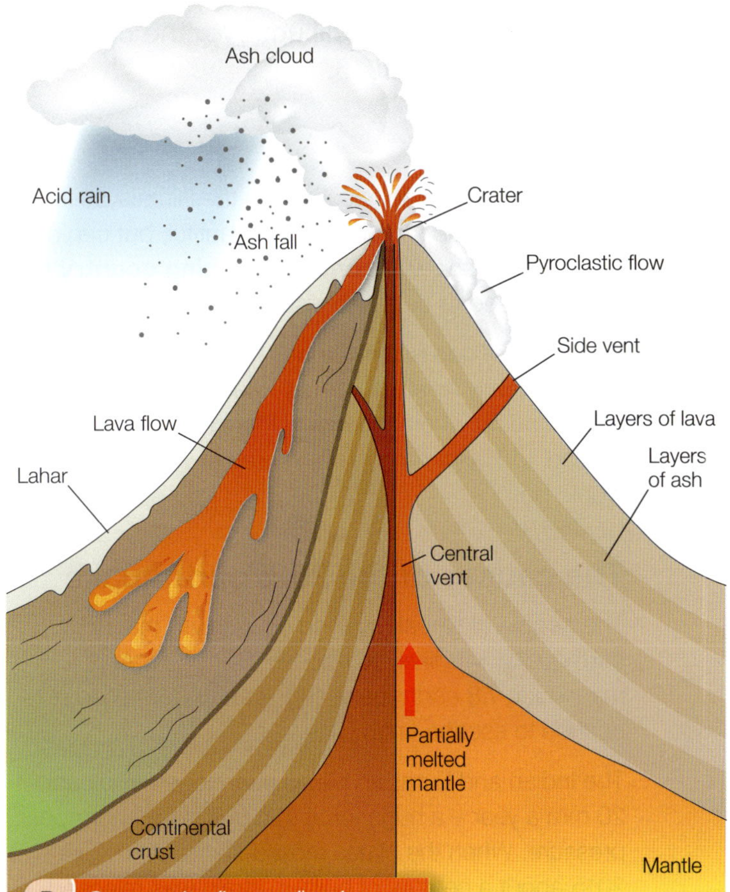

D *Composite (layered) volcano*

Popocatépetl is active again

Popocatépetl is an active composite volcano in Mexico (photo E). There is evidence of three major eruptions in the distant past. Torrents of hot ash, rocks, gases and steam formed fast-moving **pyroclastic flows**. Melting glaciers released mudflows of melted snow and volcanic ash called **lahars**.

The volcano fell dormant (asleep) for a while, but constant smoke has been rising from the crater since 1993. An eruption in 2013 began with ash rising 3 km into the sky. Then explosions sent volcanic rock up to 700 metres from the crater.

Popocatépetl is only 70 km from the megacity of Mexico City and there are around 25 million people living in the region around the volcano. A major eruption could lead to catastrophe.

E *Popocatépetl in Mexico – a composite volcano*

Activities

1. **a** Identify three ways in which shield and composite volcanoes are similar.
   b Identify three ways in which they are different.
2. Study diagram C and the text.
   a Draw a map and identify:
   - the islands that are extinct volcanoes
   - where the next two volcanoes will appear.

   b Copy diagram C and add annotations to explain how the islands were formed.
3. Study photos B and E. Explain the reasons for the shape of each volcano.
4. Create a guide to volcanoes, including their:
   - location • shape • formation
   - eruptions • danger.

   Use maps and diagrams to illustrate your guide.

Practice question

Explain how the movement of tectonic plates causes different types of volcano. (6 marks)

Earthquake disaster in Nepal – 1

On this spread you will uncover the causes of the earthquake that hit Nepal in 2015.

Case study

Nepal in 2015

Nepal is a mountainous country in the Himalayas (see map D on spread 1.13). The mountains are beautiful, but can also be deadly. Nepal is also a **low-income developing country (LIDC)**, so it is not well equipped to deal with disaster (factfile A).

A meeting of geo-scientists was held in Kathmandu, the capital of Nepal, in early April 2015. They knew pressure was building up along the boundary where the Indian Plate and Eurasian Plate collide (see spread 1.13). But no one could predict when the next earthquake would be.

Total population	30 million
Population of Kathmandu	1.183 million
Population in rural areas	81.4%
GNI/capita (see spread 6.3)	$730
Percentage below the poverty line	25.2%
Average life expectancy	68 years
Literacy rate	64%

A *Nepal factfile*

The causes of the 2015 earthquake

The people of Nepal received no warning that an earthquake was likely to happen. But on 25 April a magnitude 7.8 earthquake struck and people had no chance to escape (map B).

The Indian and Eurasian plates push together by about 20 mm a year – a huge force that builds up enormous pressure. When the plates moved suddenly in April 2015 about a century of built-up strain was released.

The focus of the earthquake was 15 km below the surface. This shallow focus caused extremely strong shaking at the surface and a great deal of damage. The intensity measured VIII (severe) on the Mercalli scale. The crust moved 3 metres in some places. The fault line ran right underneath the densely populated city of Kathmandu, putting many people and their property at risk.

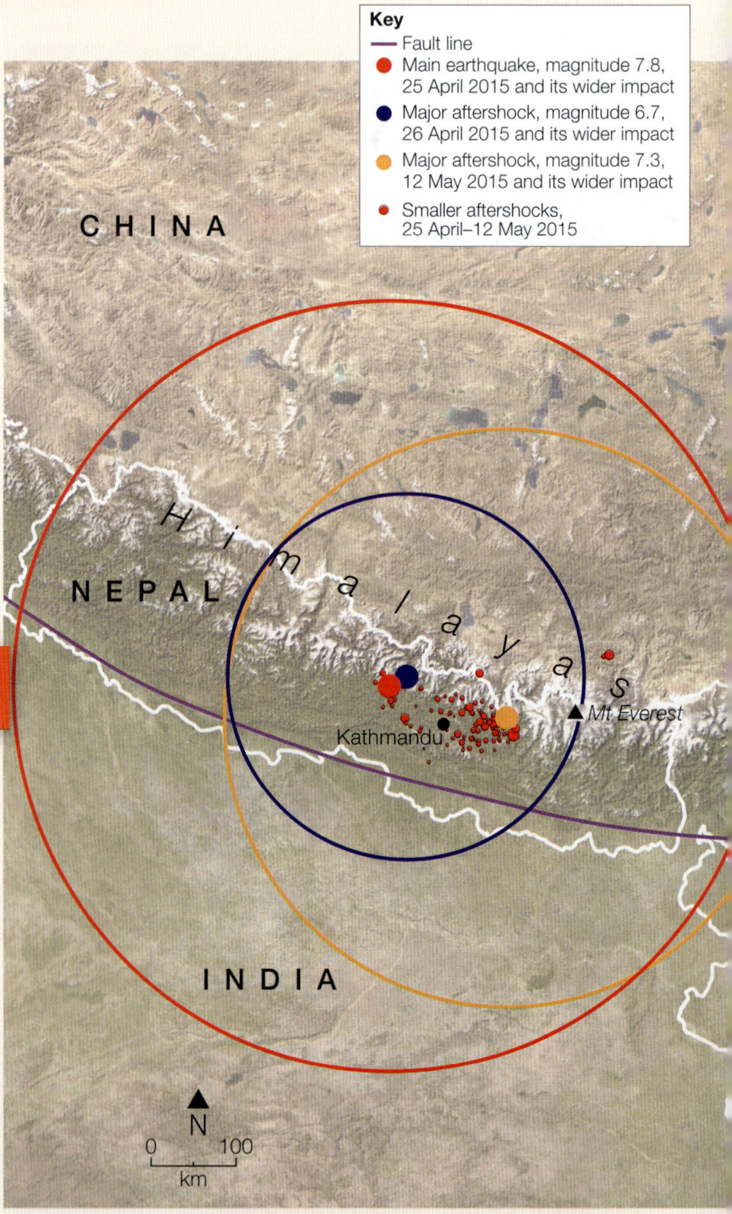

B *Epicentre and aftershocks of the 2015 earthquake in Nepal*

C *Rural Nepal before the earthquake*

Aftershocks and landslides

The earthquake was followed by a large number of aftershocks, including one measuring 6.7 on 26 April and a worse one measuring 7.3 on 12 May. It also triggered deadly avalanches on Mt Everest.

The mountainous landscape of Nepal added to the dangers. Steep slopes can become unstable when the earth shakes, leading to landslides. These can happen during the earthquake or the shaking can loosen soil, ready to collapse weeks later. For example, a landslide on 15 May flattened most buildings in the village of Singati, killing at least 12 people. The landslide also affected main roads and vehicles carrying relief materials. In all, 547 landslides were linked to the 2015 earthquake.

D *A landslide hit the village of Singati on 15 May 2015*

Activities

1. What type of plate boundary does Nepal lie on? Explain why it is that type of boundary.
2. Look at factfile A. Now make a factfile for the 2015 earthquake in Nepal. Include the following facts and any others you think are relevant:
   - date
   - location
   - magnitude (Richter scale)
   - depth
   - intensity (Mercalli scale)
   - aftershocks.
3. Look at map B here and map D on spread 1.13.
   - **a** Describe the location of the earthquake on 25 April 2015.
   - **b** Describe the distribution and size of the aftershocks up to 12 May.
4. Scientists were asked to explain the causes of the earthquake. Prepare a statement to explain how the movement of plates at this collision boundary led to a shallow focus earthquake. Include a diagram.
5. Compare photos C and D. You are asleep in your room in Singati when you are woken by horrendous noises. You can hear things breaking and someone calling your name. Write about what happens next.

Go further

In 2011 the Tohuku earthquake hit Japan.

- **a** Research development indicators for Japan like the ones for Nepal in factfile A.
- **b** Compare the two factfiles.
- **c** How do you think the consequences of the two earthquakes would differ, given the level of development in each country?
- **d** Why is it important for someone studying earthquakes to consider the level of development of the countries affected?

Practice question

Explain the specific causes of a tectonic event that you have studied [Nepal earthquake in 2015] . (6 marks)

Earthquake disaster in Nepal – 2

On this spread you will explore the consequences of the earthquake and how people responded to it.

Case study

Consequences of the earthquake

The combination of a large earthquake and an LIDC is disastrous. Many buildings were destroyed by the severe shaking in Nepal because they were poorly built. And people got trapped and killed in falling buildings. The US Geological Survey calculated that an earthquake of magnitude 7.8 in California would be likely to cause the deaths of 10–30 people per million residents, compared to 1000 or more per million in Nepal.

The primary consequences happen whilst the ground is still shaking. In Nepal they included buildings falling down, landslides, death and injury. Once the shaking stopped the people of Nepal found themselves facing secondary consequences such as homelessness, lack of health facilities and difficulty reaching remote areas.

A *Damage caused in the countryside by the earthquake*

Over 180 buildings were reduced to rubble in Kathmandu.

Historical and religious buildings were destroyed.

Thousands of homeless people had to sleep outside.

19 009 people were injured.

Schools, health facilities and government offices had to close.

8635 people were killed.

People in Kathmandu struggled to return to their families in rural areas.

Landslides in the mountains buried houses and cut off roads.

$10 billion of damage was caused.

Aftershocks threatened to destroy weakened buildings.

B *Primary and secondary consequences of the earthquake*

Responses to the earthquake

The earthquake was instantly recorded on seismometers. Many early reports focused on deaths caused by avalanches on Mt Everest. Reports soon followed from Kathmandu, but it took much longer to find out what had happened in remote rural areas because many were cut off by landslides.

Emergency aid

Countries quickly pledged emergency (short-term) aid to help Nepal's emergency response. **Non-governmental organisations (NGOs)** such as the Red Cross and Oxfam helped to support the injured and homeless. India was in a prime position to provide emergency aid quickly, including:

- 10 tonnes of blankets
- 50 tonnes of water and 22 tonnes of food
- doctors and 2 tonnes of medical supplies
- an engineering task force.

However, the aid effort was hampered by damaged infrastructure – the airport shut at one point and many roads were in poor condition (photo C).

C *Damage to roads made it hard to reach some areas to distribute aid*

Long-term aid

After the first emergencies, long-term aid was needed to help Nepal repair and rebuild. 'Cash for work' projects were set up: survivors were paid to work on rebuilding projects in their own communities.

Less than 10% of the money spent on relief has come from overseas. I would like future international donations to be managed by my government. *(Nepal's Finance Minister)*

We have provided food for 1.8 million people. This is one of our most complex operations because of the difficult terrain. *(World Food Programme)*

We have banded together to provide food and help build shelters. *(Local volunteer)*

We have provided relief materials, services and goods. We don't give money to governments as they may be corrupt. *(International aid worker)*

There is a danger that children could be taken by traffickers. We need to keep an eye on children. *(Save the Children)*

The British Government has pledged £33m to help Nepal. £10m of this will be spent on primary health care and rebuilding hospitals. *(UK politician)*

D *One month after the earthquake people reflected on the responses*

Activities

1. Study the consequences around photo B. Sort them into primary and secondary consequences. You could use a table.
2. Look at photos A and B.
   a. What consequences did the earthquake have in the two places?
   b. Which are similar and which are different?
3. Read the different views in panel D.
   a. Who was focusing on emergency aid?
   b. What were they providing?
   c. Who was focused on long-term aid?
   d. What were they doing?
   e. What disagreement had arisen about how aid was provided?

Go further

Carry out your own research to find out how Nepal has recovered from the earthquake. The www.dec.org.uk website is a good place to start.

Practice question

With reference to a tectonic event you have studied [Nepal earthquake in 2015], discuss the consequences of the event. (6 marks)

Saving lives in earthquake zones

On this spread you will find out about how technology can help to reduce earthquake damage.

Building to survive

It is sometimes said that earthquakes don't kill people, but buildings do. But if buildings are carefully designed the people inside will survive an earthquake even if the buildings fall down around them. Even simple buildings can be made safer. The design in diagram B has secure foundations to hold the building together and a light roof that won't crush people if it falls. Designs like this can be very useful in LIDCs as they are simple and cheap to build.

Advanced countries (ACs) have more money to spend on designing buildings to withstand earthquakes. A modern 'earthquake-proof' building uses strong building materials, shock absorbers and a damper in the roof to minimize shaking and damage (diagram C).

These buildings are designed to reduce the damage likely to be caused by an earthquake. This is known as **mitigation**.

A *Buildings can injure and kill people*

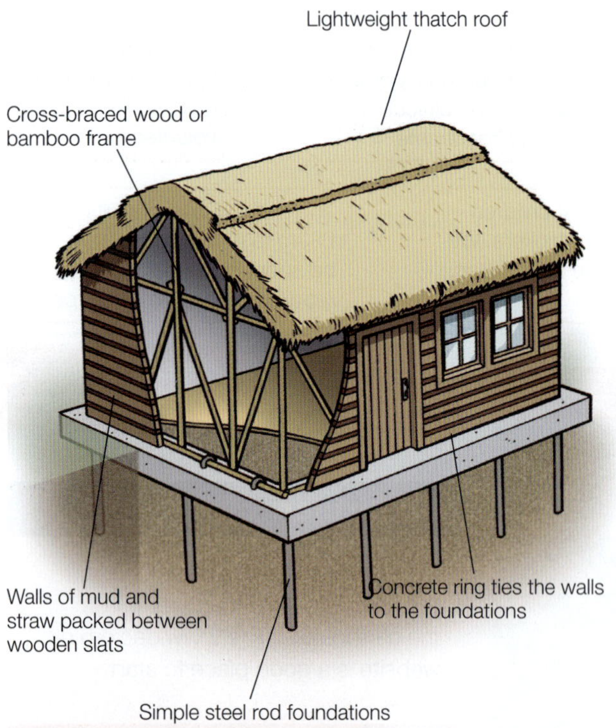

B *Simple building designs can make a big difference*

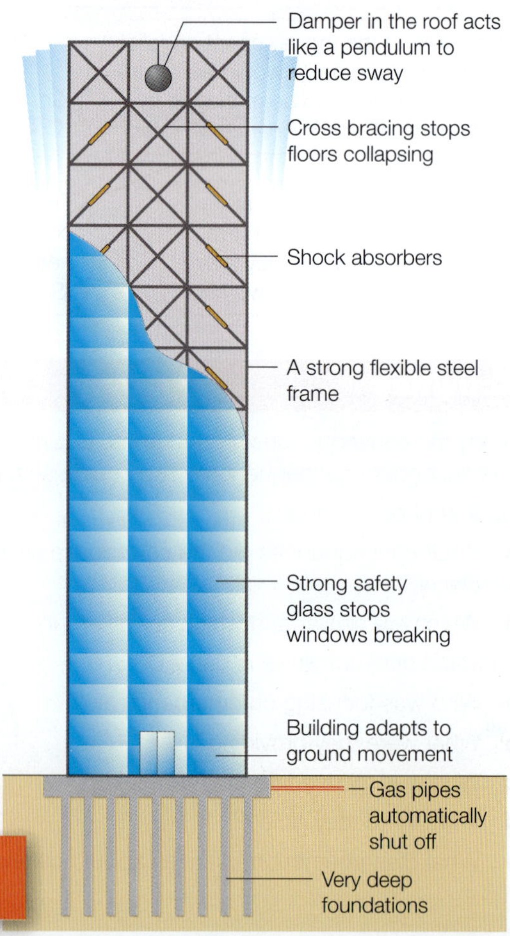

C *Building to survive future quakes*

Prediction and warning systems

There is no way to predict earthquakes accurately. However, data recording the movement of plates are collected from satellites and seismometers. This information can be used in GIS to produce hazard maps and warnings can be given.

One way to identify areas that could experience earthquakes is to look for **seismic gaps**. These are areas along known faults that haven't experienced earthquake activity for a long time. They are likely to be where pressure is building.

Once seismic waves start, more warnings can be sent out. When the Tohuku earthquake hit Japan in 2011 residents in Tokyo received a 30-second warning on their mobile phones. Smartphones are also being developed to measure the ground shaking and warn people when an earthquake is about to happen.

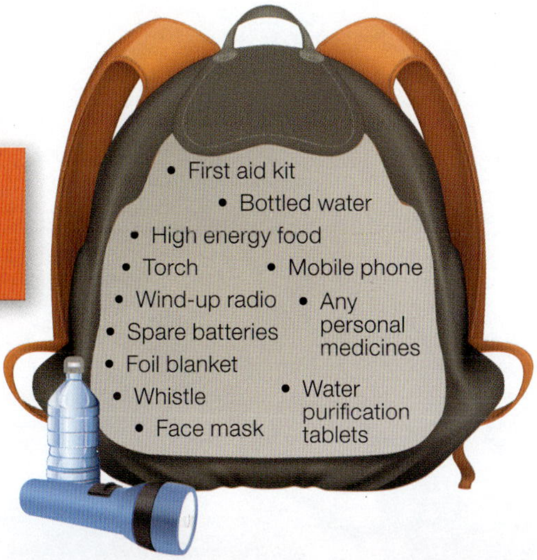

E A 'go bag' contains a range of survival equipment

Activities

1 Study the photos and diagrams.
   a How would you prepare for possible earthquakes if you lived in an earthquake zone?
   b Look at the list of items found in a 'go bag'. Would you add anything else? Why?

2 a Produce a mind-map to show the different methods used to mitigate the effects of earthquakes.
   b Choose three methods from your mind-map. How useful would they be to mitigate earthquake damage in Nepal in the future? (Clue: look at the strengths and weaknesses of each technique, then decide how much help it would be.)

Be prepared!

If people in areas prone to earthquakes have enough money, they can make sure they are prepared. Workers and school children in Japan do regular drills (photo D). Ideas such as a 'go bag' (drawing E), can also help people survive if an earthquake strikes.

D How Japan prepares for earthquakes

Your enquiry

How do plate tectonics shape our world?

Look back at your work on the 2015 earthquake in Nepal and how plate tectonics shape the physical and human geography of the country. Produce a report, including a map and diagrams, answering the following questions:

- How have plate tectonics shaped the physical geography of Nepal? Mention fold mountains and earthquakes.
- What impact has the earthquake hazard had on the human geography of Nepal? Mention:
  - damage to the capital city, Kathmandu
  - impact on services such as education and healthcare
  - damage to the infrastructure
  - knock-on effects on Nepal's development.
- What can Nepal do to manage the earthquake threat in future?
  - Should Kathmandu be rebuilt with 'earthquake-proof' buildings?
  - Should the capital city be moved to a different location?
  - Will the release of pressure on the fault in 2015 mean that Kathmandu can relax because it is unlikely to have another earthquake for a long time?

2 Changing Climate

What is this topic about?

Climate change is one of the most important issues facing us. Since 2000, Earth has experienced at least nine of the warmest years since accurate records of temperatures began in 1880. The warmer temperatures are having a huge range of impacts, such as flooding and species extinction, but not everyone is affected in the same way.

Natural processes have always caused the climate to change, but most scientists agree that human activity is also having a huge impact on climate change today.

Your enquiry

In this chapter you will investigate the specification's key question.

What evidence is there to suggest climate change is a natural process? You will:

- Understand that we are currently living in an ice age.
- Describe how Earth's climate has changed over the last 2.6 million years.
- Compare evidence of climate change from a variety of sources.
- Explain the natural causes of climate change.
- Understand how the greenhouse effect works.
- Evaluate whether human activity has enhanced the natural greenhouse effect.
- Explain and classify the global impacts of climate change.
- Describe and explain the consequences of climate change for the UK.

Look at the photo opposite.

- Which region of the world do you think this is?
- What do you think the women are doing?
- People didn't always have to do this. Why are they doing it now?
- What impact do you think climate change is having on these people?

2.1 Earth's changing climate

On this spread you will discover how Earth's climate has changed over the last 2.6 million years.

You are living in an ice age!

Surprising, isn't it? Your image of an **ice age** might be one of woolly mammoths wandering about, but scientists use the term 'ice age' to mean any period of time when Earth has permanent ice sheets such as those in Antarctica.

A *Penguins brave the cold in Antarctica*

The Quaternary Ice Age

Earth is 4.5 billion years old. Scientists divide Earth's history into different time periods and we are currently in the **Quaternary period**, which began about 2.6 million years ago. Scientists also divide the Quaternary period into two epochs. The **Pleistocene epoch** lasted from 2.6 million years ago to 11 700 years ago. The current **Holocene epoch** began 11 700 years ago (diagram B).

Period	Quaternary	
Epoch	Pleistocene	Holocene

B *The Quaternary period*

Fluctuating temperatures

Although the Quaternary period is an ice age, the temperature has always fluctuated, moving up and down, sometimes warmer but often colder than it is today (graph C). The colder periods normally last about 100 000 years and are called **glacial periods**. The warmer periods normally last about 10 000 years and are called **interglacial periods** (see graph D).

In the last glacial period huge sheets of ice covered most of Britain. Temperatures reached their coldest about 22 000 years ago when 30% of Earth was covered in ice. Temperatures then increased, many glaciers melted and Earth began a warmer interglacial period. We are currently living in the interglacial period that began 11 700 years ago. Today only 10% of Earth is covered in ice.

Opinion depends on time

The graphs on this spread show Earth's temperature change over a number of different time periods. Graph C looks at the last 2.6 million years, but graphs D–F look at shorter periods of time. It is possible to draw different conclusions about Earth's changing temperature depending on the length of time you are looking at.

Even in the current warmer interglacial period there have been big changes in the temperature (graph E). For example, there was the Medieval warm period from AD 950 to 1250 and the Little Ice Age from 1300 to 1870.

C *Earth's temperature in the Quaternary period*

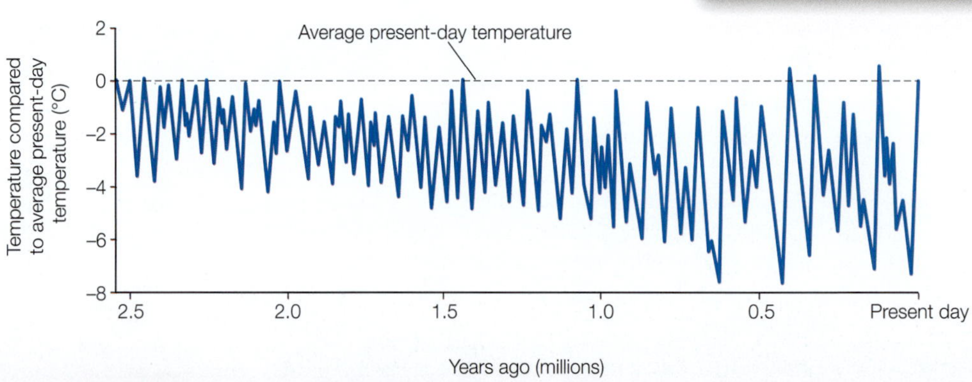

Recent changes

Most scientists believe the period since about 1880 is very significant because the rise in temperature is so fast (graph F). It is also a period when human activity might have had the biggest effect on Earth's climate.

D *Glacial and interglacial periods in the last 450 000 years*

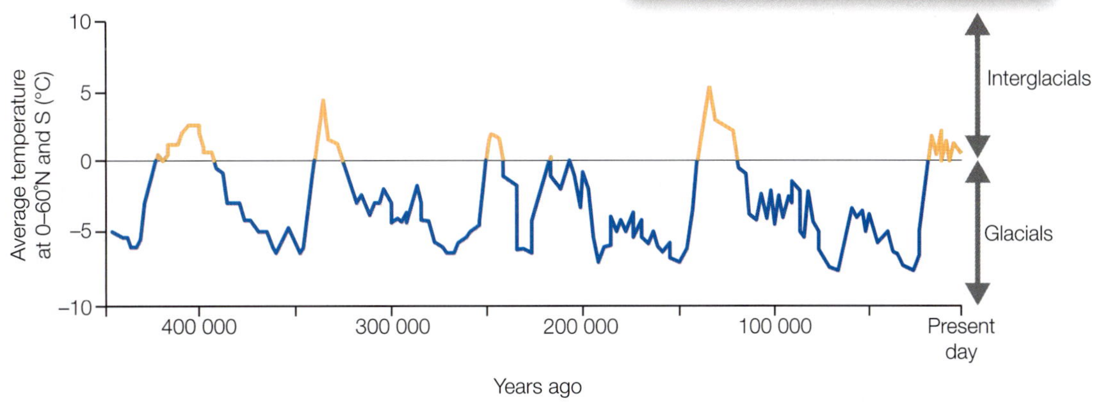

E *Earth's temperature in the last 1500 years*

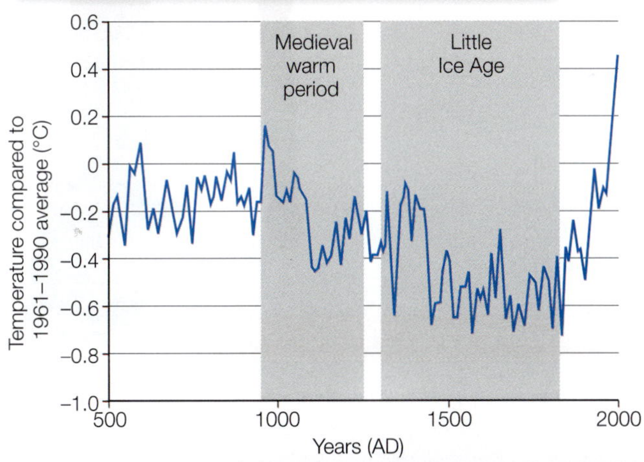

F *Earth's temperature since 1880*

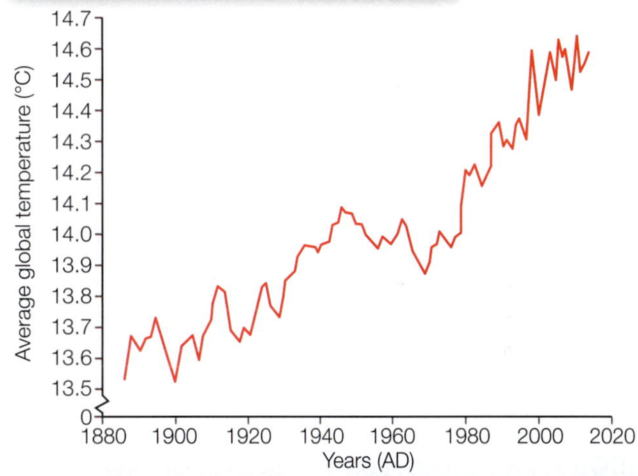

Activities

1 a What is the Quaternary period?

b Why do scientists say that we are currently living in an ice age?

c Look at graph C. What is the general trend in Earth's temperature over the last 2.6 million years?

2 a What is meant by the terms 'glacial period' and 'interglacial period'?

b What happened to Earth's temperature about 11 700 years ago?

c Look at graph D. Write four or five sentences to describe how Earth's temperature has changed in the last 450 000 years. Try to use some of these words in your description: increase, decrease, cycle, fluctuate, glacial, interglacial, highest, lowest.

3 a Using graphs C–F, write a paragraph arguing that the current warm temperatures on Earth are not unusual.

b Using graphs C–F, write a paragraph arguing that the current temperatures on Earth are unusually high.

Your enquiry

What evidence is there to suggest climate change is a natural process?

As you work through this section, make notes anything that will help you to answer the question. You will bring your ideas together to do an activity at the end of spread 2.11.

Evidence of climate change – 1

On this spread you will explore how studying ice cores and sea ice positions gives evidence of climate change.

As you know, Earth's temperature has changed over the last 2.6 million years. But how do scientists know what the temperature was like so long ago? They start drilling (photo A)!

Ice cores

Every year when it snows, more snow and ice are added to the sheets of ice in the Arctic and Antarctic. Scientists can drill down into this ice and remove tubes of ice called **ice cores**. They can drill so deep they are able to study ice laid down 800 000 years ago.

You can see the different layers in the ice in photo B. The lighter coloured layers show snow laid down in summer and the darker coloured layers show winter snow. Although not much snow falls in winter, strong winds blow the snow about and mix dust into it, making it darker. Thick summer layers means there was lots of new snow.

A A scientist drilling an ice core

Evidence in the snow

Ice cores show how much snow fell, but scientists can also study what is in the snow. By looking at gases trapped in each layer, they can learn more about Earth's climate.

Ice cores contain small bubbles of gas that give us evidence of what the air was like thousands of years ago. The bubbles can contain heavy or light types of oxygen (O) called **isotopes**. One type is ^{16}O, which is lighter than ^{18}O. By looking at the type of oxygen in different layers in the ice, scientists can tell whether the climate was warmer or colder. Lots of ^{16}O and Earth was colder, lots of ^{18}O and Earth was warmer.

Scientists also look at the amount of **carbon dioxide (CO_2)** in the bubbles. Lots of CO_2 suggests warmer temperatures. The evidence from oxygen isotopes and CO_2 levels can be used to show Earth's fluctuating temperature (see spread 2.1). They also tell us that Earth's current temperature is high compared to the past.

Matching data

We now have data from ice cores that include ice that is thousands of years old. Scientists have also looked at much younger ice, such as from the 1950s, to see whether that matches other data collected in the last few decades from sources like weather stations (see spread 2.3). Matching in this way has proved that data from ice cores are accurate and reliable.

B An ice core showing the different layers of ice

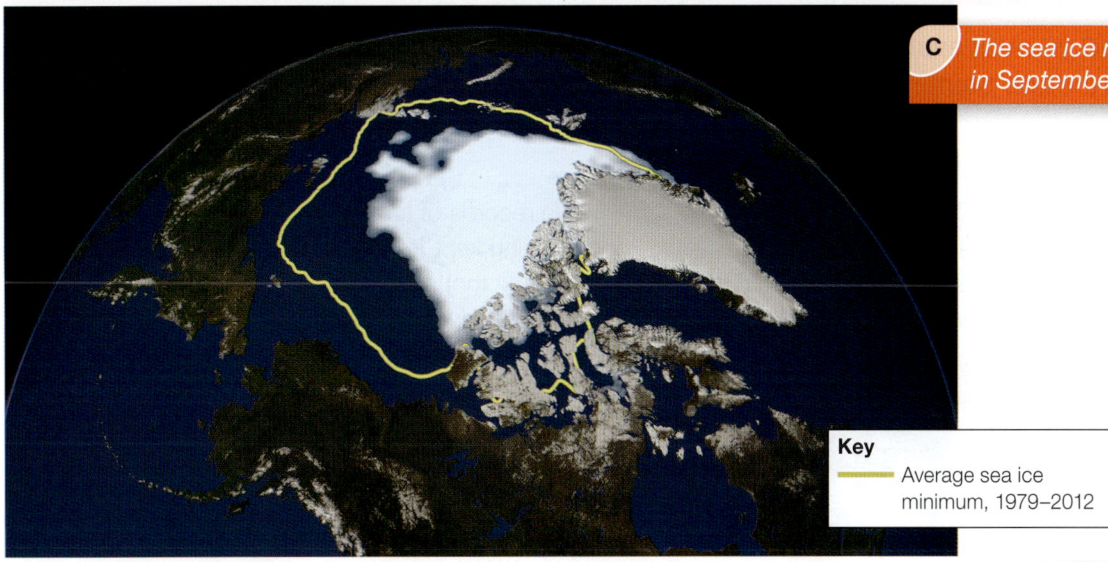

C *The sea ice minimum in September 2012*

Sea ice positions

The expanse of ice in the Arctic is another piece of evidence for climate change. Since 1979 scientists have been using satellite photos to study the **sea ice maximum** and **sea ice minimum** (image C). The sea ice maximum is the maximum area in the Arctic covered in ice in any year. This occurs in winter and it normally reaches its maximum in February or March. As the Arctic warms up in summer, the sea ice melts and reaches its minimum in September.

The size of the sea ice maximum and minimum have both been getting smaller. On average the area covered by sea ice in the Arctic has become 13.3% smaller every ten years since 1979 (graph D). Scientists believe this gives evidence of **global warming** – an increase in the average temperature of Earth's atmosphere and oceans.

D *The area in the Arctic covered by ice in September, 1979–2014*

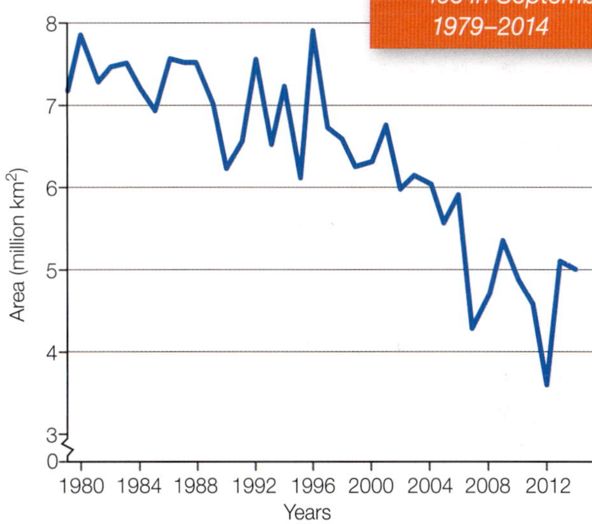

Activities

1. Is the scientist in photo A doing an important job? Explain why or why not.
2. **a** In your own words, explain what ice cores are.
   b Look at photo B and draw a sketch of an ice core. Label the summer and winter layers, and name the trapped gases. Add annotations to explain how the gases can provide evidence of climate change.
   c Are data from ice cores reliable? Explain why or why not.
3. **a** What is meant by the terms 'sea ice maximum' and 'sea ice minimum'?
   b Draw a rough sketch map from satellite image C. Using an atlas, label the Arctic Ocean, Greenland, Canada and Russia.
   c Using graph D, describe the changes in the area of September sea ice between 1979 and 2014. Remember to comment on the overall trend and give specific figures in your description.
   d Using graph D, make a prediction of when you think there will be no Arctic sea ice in September. Explain your prediction.

Go further

Find out the size of the sea ice minimum in the Arctic this year. Is it bigger or smaller than in 2014?

Practice question

Explain how ice cores can be used to find evidence of climate change. (4 marks)

Evidence of climate change – 2

On this spread you will evaluate the evidence of climate change found in temperature data, paintings and diaries.

Global temperature data

In 1659 the Central England Temperature Record began recording monthly average temperatures for the Midlands in England. This is the oldest record of temperature measurements in the world.

Reliable records of temperatures across the world began being kept from about 1880. Today there are more than a thousand weather stations recording daily temperatures. New weather stations are added each year, which makes the data even more reliable.

A *Land and ocean temperatures, April 2015*

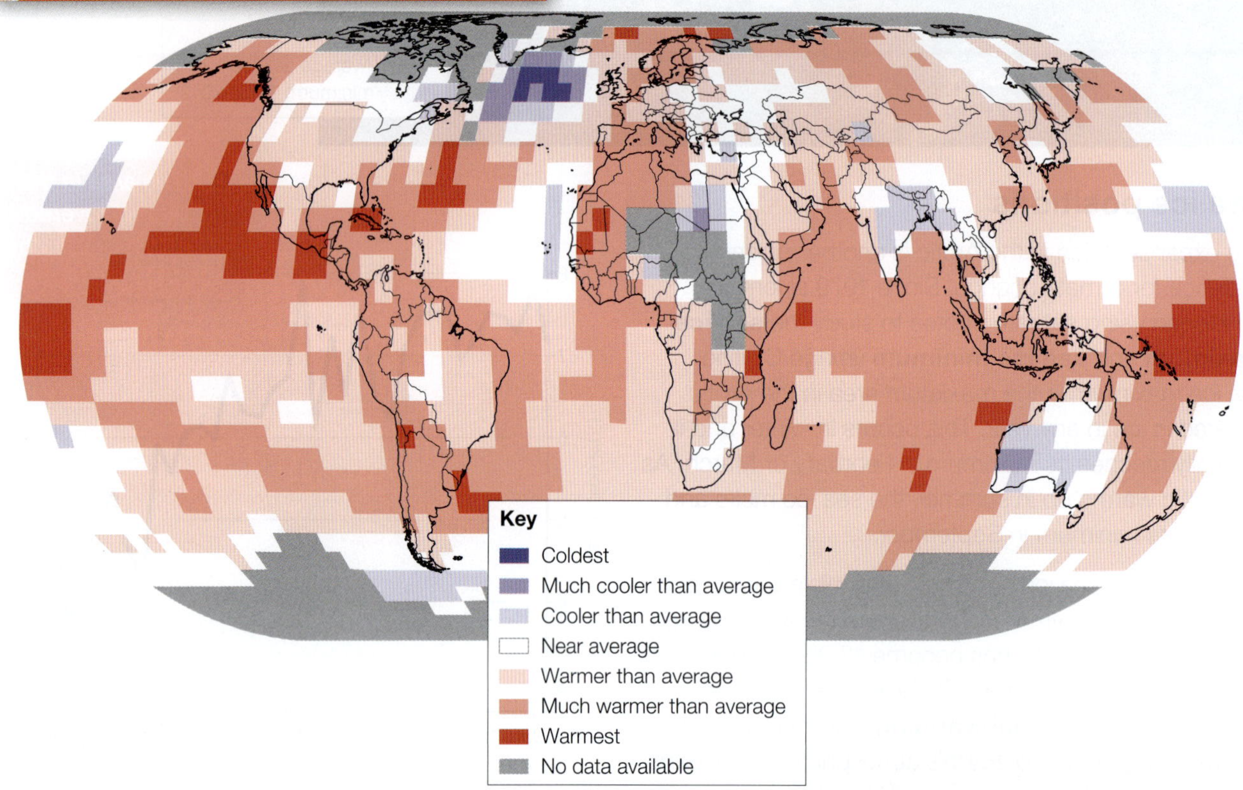

Using the data

Scientists use these records to study Earth's temperature (map A) and compare it in different time periods. They have also been able to use temperature data to work out the global average temperature each year. They can then develop lists of the hottest years (table B) and the coldest years. 2014, for example, was 0.69°C warmer than the average of all the years 1880–2014.

Criticising the data

Some critics say that there aren't enough weather stations to give an accurate picture of Earth's temperature. Others are concerned that data might not always be accurate. Recording instruments might be placed in positions that don't give a fair record of an area's real temperature. For example, choosing to collect data from the top or bottom of a hill would give quite different results.

Rank	Year	°C warmer than the average of all years
1	2014	0.69
2	2010	0.65
2	2005	0.65
4	1998	0.63
5	2013	0.62
5	2003	0.62
7	2002	0.61
8	2006	0.60
9	2009	0.59
9	2007	0.59

B *Top ten hottest years, 1880–2014*

Evidence in paintings and diaries

Finding out about Earth's temperature before accurate daily temperature records were kept can be difficult. Studying things like old paintings and people's diaries do not give scientists mathematical data, but they can still help to show what the temperature was like in the past.

We know from paintings that between 1309 and 1814 the River Thames froze at least 23 times, sometimes for months at a time. On five occasions the ice was thick enough for Londoners to set up market stalls, pubs, shops and ice rinks for a frost fair (picture C). These were rowdy events and there are even reports of an elephant walking on the frozen river!

We also know about the frost fairs from the diaries of people like Samuel Pepys and John Evelyn (diary D). Poems and written reports tell us about this cold period too. This sort of evidence has led to the period 1300–1870 being called the Little Ice Age.

C A Frost Fair on the Thames at Temple Stairs, *painted by Abraham Hondius, 1684*

Coaches plied [carried paying passengers] from Westminster to the Temple, and from several other staires to and fro ... bull-baiting, horse and coach races, puppet plays ... so that it seemed a carnival on the water ... and the very seas so lock'd up with ice, that no vessels [ships and boats] could stir out or come in.

D *Diarist John Evelyn describes the frost fair in 1683*

Activities

1 a Write five sentences describing what map A shows about global temperatures in April 2015. Use an atlas to help you name regions of the world in your description.

b Why do you think the grey area on the map has no data available?

2 a Look at table B. How many of the top ten hottest years from 1880–2014 occurred since the year 2000?

b What does this tell us about recent climate change?

3 Imagine you are visiting a frost fair. Use painting C and diary D to help you write a Tweet describing what you can see and do, and what the temperature is like. You can only use 140 characters.

4 a Write a diary entry describing what you did on a hot day last summer. Write five sentences.

b What do you think a person from the 22nd century might conclude about that summer?

c Do you think diaries provide evidence of weather or of climate (see spread 1.1)? Give reasons for your answer.

5 Using the information here and on spread 2.2, draw a table like this, summarising the evidence for climate change. One example has been started for you.

Evidence	How does it tell us about the past?	What does it tell us about climate change?	How reliable is it?
Ice cores	Looks at the type of oxygen and other gases trapped in ice to learn about the climate.	• Earth's past temperatures have fluctuated. • There is a link between CO_2 and high temperatures.	Comparing recent ice cores with other data has shown this is a reliable source of evidence.

Go further

Use a search engine to research how tree rings provide evidence of climate change.

Practice question

Evaluate how reliable sources like diaries and paintings are in providing evidence of climate change. (4 marks)

2.4 Natural causes of climate change

On this spread you will find out about the natural causes of climate change.

Humans were not around for most of the many periods of climate change throughout Earth's history. If humans weren't there, there must be some natural reasons for Earth's climate changes. There are various theories, which involve volcanic eruptions, sun spots and Milankovitch cycles.

Milankovitch cycles

In the 1930s a Serbian mathematician called Milutin Milankovitch argued that climate change was linked to the way that Earth moves around (orbits) the Sun, and how it wobbles and tilts as it does this. The three ideas he linked are:

- **Eccentricity:** changes in the shape of Earth's orbit.
- **Obliquity:** changes in how Earth tilts on its axis.
- **Precession:** the amount Earth wobbles on its axis.

The effects of these three changes combine and can be linked to many big changes in Earth's climate.

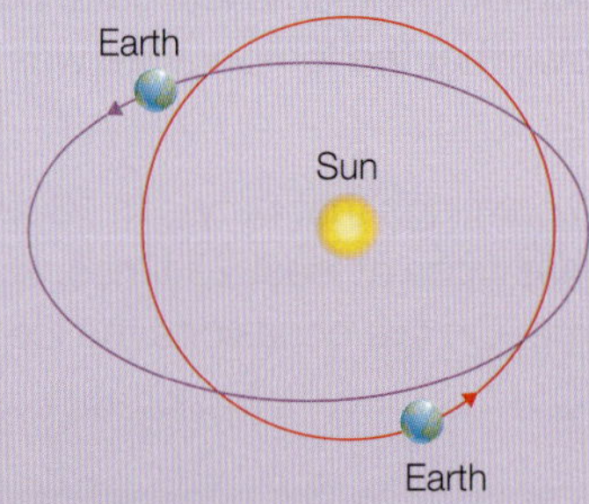

Eccentricity (orbit)

In the 21st century Earth is orbiting the Sun in a nearly perfect circle. However, over a period of approximately 100 000 years the orbit changes slowly from a circle to an oval and back again. When the shape is oval, Earth is much closer to the Sun at some times of the year and further away at other times of the year. This has most effect on our climate when Earth is further from the Sun during northern hemisphere summers.

Precession (wobble)

As Earth spins on its axis it wobbles. The wobble is caused by the force of gravity from the Sun and the moon. It makes the axis trace a circular shape, slowly changing the direction it points in. The axis wobbles from one extreme to the other over a period of approximately 23 000 years. This can change the severity of the seasons and which hemisphere is closest to the Sun at different times of the year.

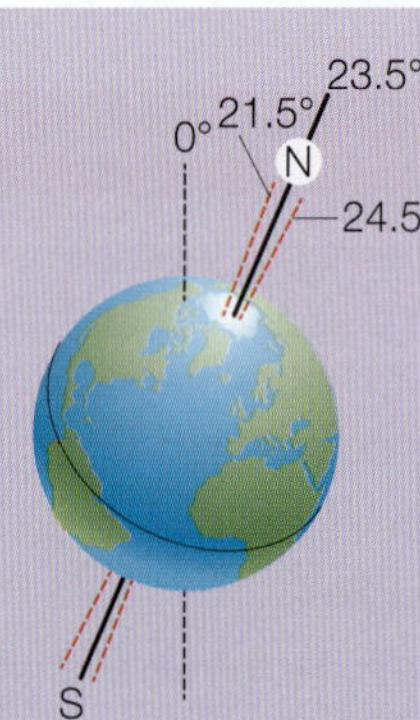

Obliquity (tilt)

Earth is tilted on its axis. Over roughly 40 000 years the angle of tilt changes slowly from 21.5° to 24.5°. We get our summer when Earth is tilted towards the Sun in its annual orbit and winter when it is tilted away from the Sun. With less tilt our summers are cooler and winters are milder. In the 21st century the angle of tilt is 23.5 degrees.

A *The effects of Milankovitch cycles*

How Milankovitch cycles affect Earth's climate

Milankovitch thought the most important reason for climate change was the temperature during northern hemisphere summers. If it was cool, the winter ice would not melt as much and would build up – eventually triggering an ice age. Cooler northern hemisphere summers are more likely when:

- Earth's orbit means Earth is furthest from the Sun during northern hemisphere summers
- Earth has less tilt
- Earth's wobble means the northern hemisphere has cooler and shorter summers.

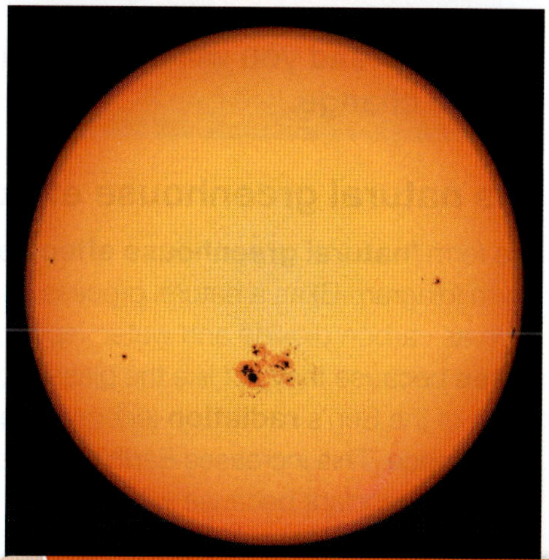

B *Sun spots on the surface of the Sun*

Sun spots

There are dark spots on the Sun's surface called **sun spots**. These are temporary and are caused by magnetic storms. They increase the amount of energy Earth receives from the Sun, causing it to warm up.

There is an 11-year cycle of sun-spot activity, although some cycles are longer. The Little Ice Age (1300–1870) is linked to a time when there were few sun spots. The Medieval warm period has been linked to more sun spots.

Volcanic eruptions

When volcanoes erupt they release huge amounts of dust (containing ash, sulphur dioxide, carbon dioxide and water vapour) into the atmosphere. This dust blocks sunlight, which results in cooler temperatures on Earth.

On 15 June 1991 Mt Pinatubo erupted, putting 20 million tonnes of sulphur dioxide and ash into the atmosphere. Over the next two years Earth's average temperature fell by 1°C.

Looking at past changes in climate, during the Medieval Warm period (950–1250) there was less volcanic activity and during the Little Ice Age (1300–1870) there was more volcanic activity.

C *The volcano Mount Pinatubo erupting*

Activities

1. Look at photo B. Explain why sun spots increase Earth's temperature.
2. Explain why temperatures often reduce after large volcanic eruptions.
3. **a** Study panel A. In pairs, use one ball (or an apple) to represent the Sun and another to represent Earth. Put a skewer (or a pencil) through the middle of the ball representing Earth. Move the two balls to imitate each of the Milankovitch cycles to help you visualise more clearly how they might affect Earth's climate.
   b Work out a commentary for your demonstration of each cycle. Then make a short video of your complete demonstration.
   c Using three diagrams, explain how eccentricity, obliquity and precession can change Earth's temperature.
4. Write a paragraph arguing that the current warming of our climate might be happening because of natural causes.

Go further

Type 'feedback loops and climate change' into a search engine to research how small changes in temperature can lead to much bigger changes.

Practice question

Explain one natural cause of climate change. (4 marks)

Human causes of climate change – 1

On this spread you will begin to consider whether human activity is causing climate change.

The natural greenhouse effect

The term **'natural greenhouse effect'** describes how Earth is kept warm (diagram B) as a natural process. Gases in the atmosphere, such as carbon dioxide and water vapour, are called **greenhouse gases** because they act like the glass roof on a greenhouse. They trap some of the Sun's **radiation** (energy), preventing it from escaping into outer space. This increases Earth's temperature. Greenhouse gases are very important because without them it would be too cold for life to exist on Earth. The average temperature would be –18°C rather than +15°C.

A A greenhouse traps the Sun's heat

The enhanced greenhouse effect

Most scientists believe that the increase in Earth's temperature over the last 250 years is because of the **enhanced greenhouse effect** (diagram C). Humans have put more greenhouse gases into the atmosphere than occur naturally. This makes the layer of greenhouse gases thicker and means more of the Sun's energy is trapped. This increases Earth's temperature and is causing climate change.

B Natural greenhouse effect – before human intervention

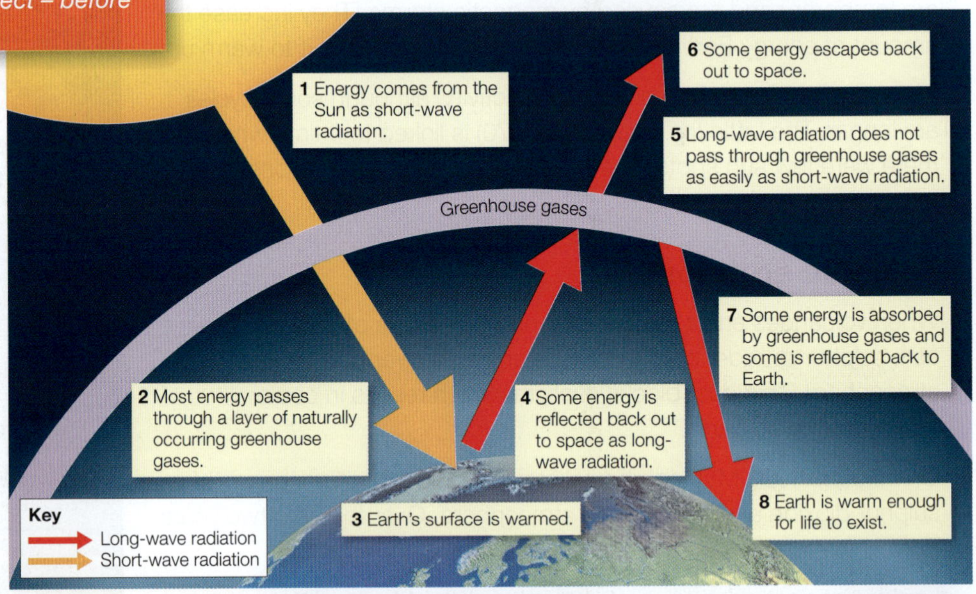

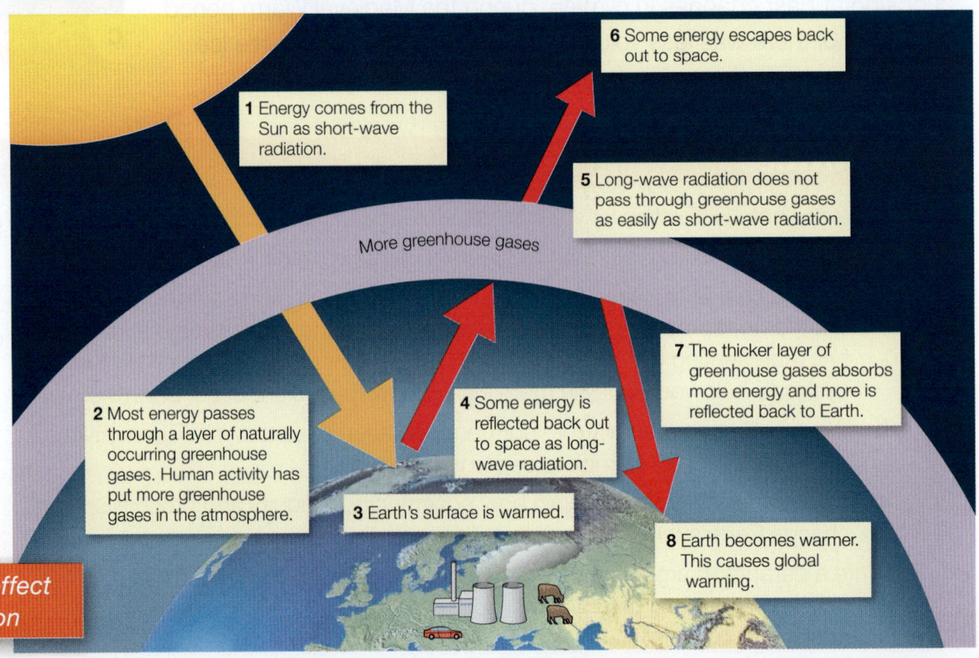

C Enhanced greenhouse effect – after human intervention

Greenhouse gases

Most greenhouse gases occur naturally.

Water vapour has the biggest effect because there is much more of it than any other greenhouse gas. However, humans do not influence the amount of water vapour in the atmosphere. So scientists do not normally include it when they consider how much different greenhouse gases contribute to global warming.

It is also important to consider the **global warming potential** of gases. The same quantities of different gases trap and absorb different amounts of radiation.

Carbon dioxide (CO_2) is the most significant greenhouse gas. It accounts for 60% of the enhanced greenhouse effect. It is produced by burning fossil fuels like coal and gas to make electricity. Industry, cars and **deforestation** all release CO_2 into the atmosphere.

Methane accounts for 15% of the enhanced greenhouse effect. The actual amount of methane in the atmosphere is quite small but it is 25 times more efficient than CO_2 at trapping radiation. Its main sources are landfill sites, rice production and farm animals like cows.

Halocarbons, mainly CFCs (chlorofluorocarbons), are the only human-made greenhouse gases and make up only a tiny proportion of all greenhouse gases. But they can be 15 000 times more efficient at trapping radiation than CO_2 so they account for 15% of the overall effect. They come from air-conditioning, refrigerators, aerosols and foam packaging.

Nitrous oxide is 250 times more efficient than CO_2 at trapping radiation and accounts for 6% of the enhanced greenhouse effect. The main sources are agricultural fertilisers and car exhausts.

D *Sources of greenhouse gases*

Activities

1 Study panel D. Which greenhouse gases are emitted from things in each photo?
2 Study diagrams B and C. Explain why Earth's temperature is increasing, using the phrases 'natural greenhouse effect' and 'enhanced greenhouse effect'.
3 Explain what is meant by a gas's global warming potential, giving some examples.
4 Draw a pie chart showing the different contributions of the main greenhouse gases to the enhanced greenhouse effect. Don't forget to include a segment for 'Other'. Does the pie chart help your understanding? Why or why not?

Practice question

Name two greenhouse gases that contribute to the enhanced greenhouse effect. (2 marks)

Human causes of climate change – 2

On this spread you will consider whether your lifestyle is to blame for climate change.

Carbon dioxide emissions and Earth's temperature

There is a strong link between increases in carbon dioxide (CO_2) in the atmosphere and increases in Earth's temperature (graph B). 97% of climatologists agree that this increase is caused by human activity that is emitting more greenhouse gas into the atmosphere. The lifestyle of people who live in **advanced countries (ACs)** like the UK is the reason that so much CO_2 is being put into the atmosphere. This lifestyle uses lots of fuel and energy in things like cars, aeroplanes, food, phones and other electrical products.

A *Who is to blame for the enhanced greenhouse effect?*

B *Changes in Earth's temperature and CO_2 levels over the last 1000 years*

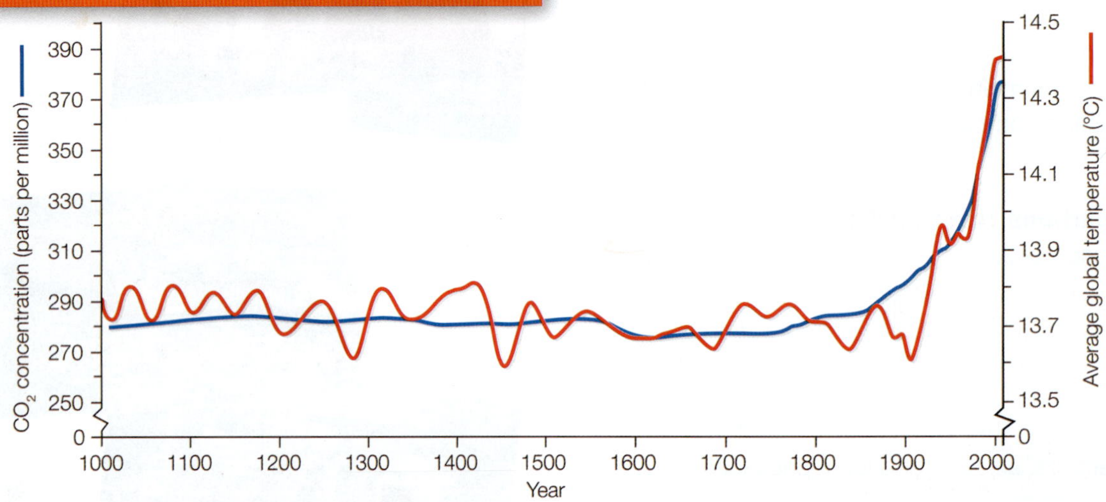

The geography of carbon dioxide

Not all countries emit the same amount of CO_2 (chart C). **Low-income developing countries (LIDCs)**, including most countries in Africa, emit very low levels of CO_2, but ACs and **emerging and developing countries (EDCs)** emit much more. China now emits more CO_2 than any other country (see spread 12.2). EDCs like China and India are emitting more CO_2 as they develop industry and their people start to want a lifestyle similar to that enjoyed by people in ACs.

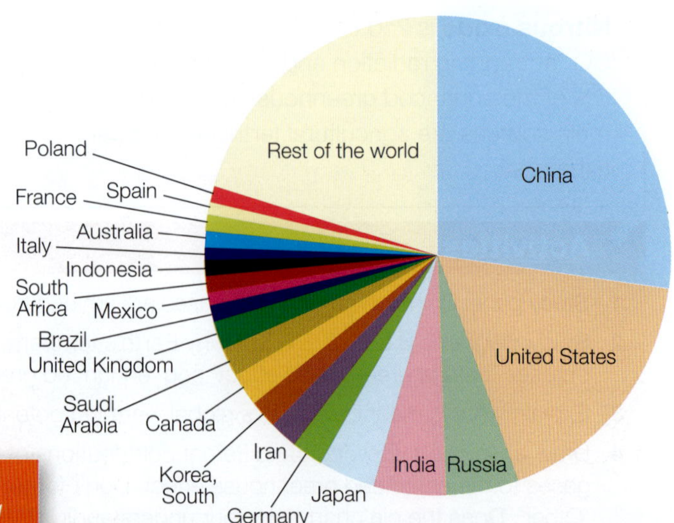

C *Each country's share of total CO_2 emissions, 2011*

Chart C suggests that China is mostly to blame for high CO_2 emissions but the Chinese would argue that just looking at total CO_2 emissions is misleading (see spread 12.2). There are 1.36 billion people living in China compared to 320 million in the USA. If we want to think about which countries are causing climate change, it is important to look at CO_2 emissions per capita (per person) (map D).

D CO_2 emissions per capita in each country

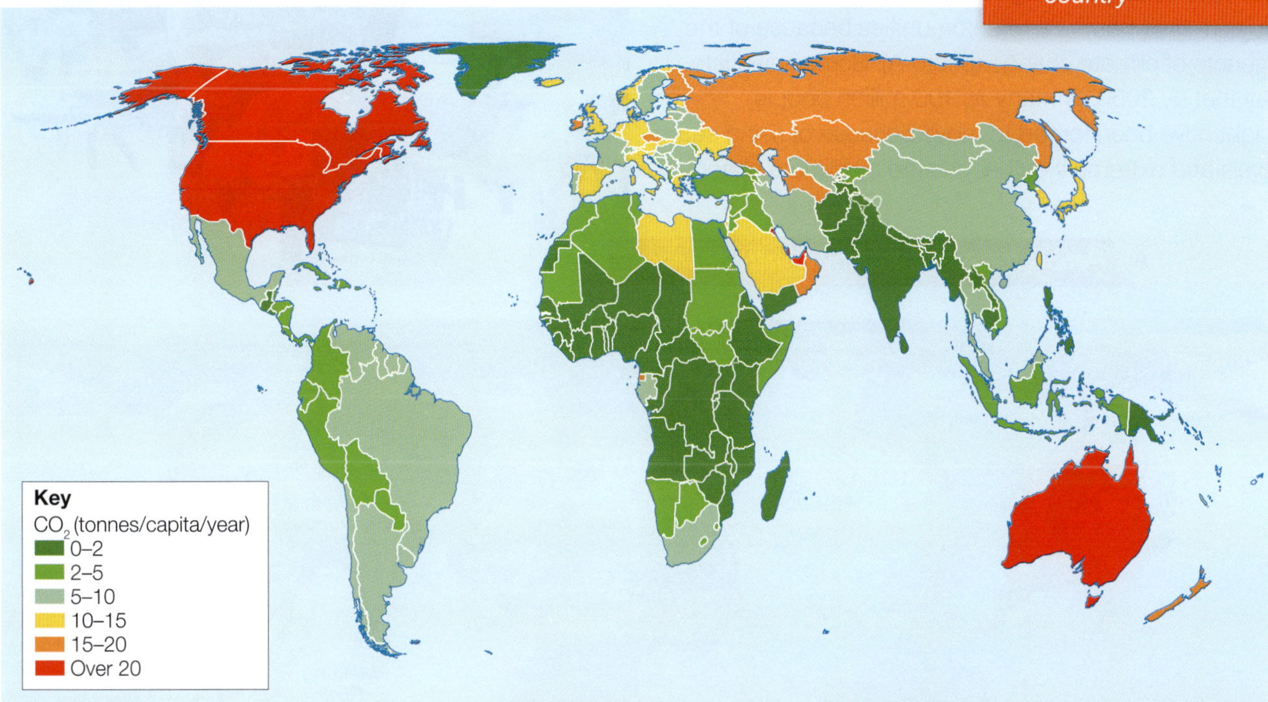

The carbon footprint of countries

Even looking at CO_2 per capita can be misleading. It does not take into account the fact that countries like China produce a lot of CO_2 making products like cement, clothes and electronic tablets, which are used by people like us in ACs. There is growing debate about whether the CO_2 emitted in making these products should count in the country where they are made or in the country where they are bought and used.

It might be more sensible to think about the **carbon footprint** of countries. This takes into account the fact that the lifestyle of people in ACs depends on carbon emissions in EDCs and LIDCs. If this is taken into account, countries like the USA, Japan and the UK are responsible for significantly more emissions than EDCs like China.

Activities

1 Describe the relationship between CO_2 concentrations and Earth's temperature. Use graph B to include some data.

2 **a** Look at map D. Name three countries with CO_2 emissions per capita per year above 20 tonnes a year.

b What are the CO_2 emissions per capita per year in China?

3 **a** Look at chart C. Name the three countries that emit the most CO_2.

b Explain why EDCs would say that ACs are more to blame for climate change.

4 Explain why looking at a country's carbon footprint might be a fairer way of measuring CO_2 emissions.

5 Discuss cartoon A. Who do you think is most to blame for the enhanced greenhouse effect?

6 Explain what might happen to Earth's temperature if cars became more affordable in EDCs and LIDCs.

Go further

Calculate your own carbon footprint with one of the online calculators. The one on the WWF (World Wide Fund for Nature) website is a good one to use.

Practice question

Explain why the lifestyle of people in ACs might be causing climate change. (6 marks)

Global impacts of climate change – 1

On this spread you will learn about the impacts of rising sea levels and extreme weather.

Climate change refugees

Climate change refugees are people who are forced to leave their homes and communities because of the impacts of climate change (photo A). Some scientists say that by 2050 as many as 100 million people might have been forced to move because of rising sea levels and extreme weather caused by climate change.

A *People forced to move by climate change*

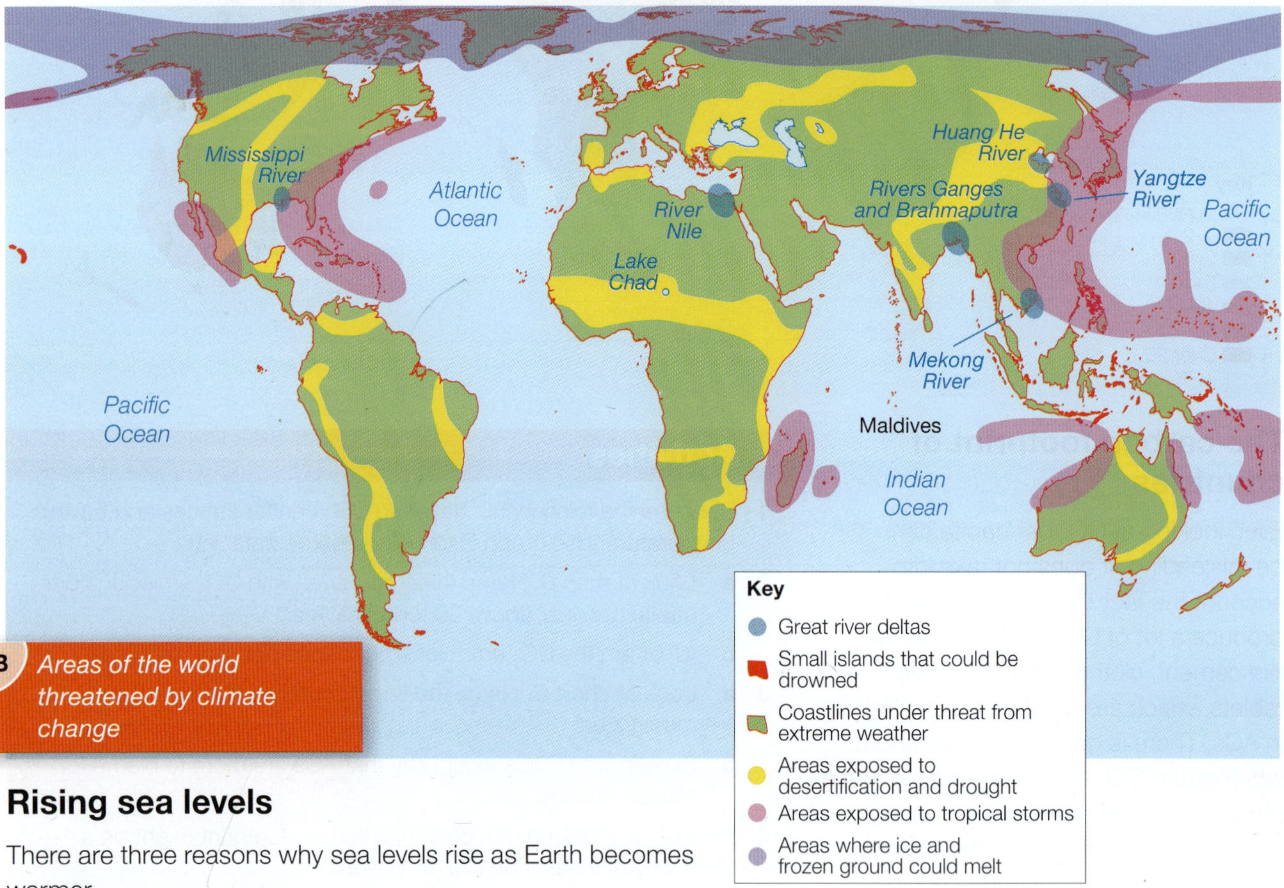

B *Areas of the world threatened by climate change*

Rising sea levels

There are three reasons why sea levels rise as Earth becomes warmer.

- **Thermal expansion:** As the temperature of water in the oceans increases, the water expands making sea levels rise.
- **Melting glaciers:** As glaciers melt, water that was stored as ice is added to the world's oceans.
- **Melting ice caps:** As the huge expanses of ice in the Arctic and Antarctic melt, more water is added to the oceans.

Since 1901 sea levels have risen by about 200 mm. They are predicted to rise by between 500 mm and 1 metre by 2100. This will have severe effects on some countries and communities as the land they once lived on disappears under water (map B).

Disappearing islands

Rising sea levels threaten to flood many small islands and low-lying coastal areas. The Maldives, for example, is made up of a small group of islands in the Indian Ocean (map B). In 2009 the Maldivian government held an underwater meeting to draw attention to the impact of rising sea levels (photo C).

The Maldives relies on tourism for a lot of its income. As sea levels rise there is more coastal erosion and flooding – and tourist resorts will shut down. As the ocean around the islands warms up it absorbs more CO_2. This makes the water more acidic, which reduces the numbers of fish, which the local people rely on. Because of the flooding it is likely that in 100 years no one will be able to live there. In 2008 the President said that the country was saving money so it could buy a new country!

C *The Maldivian government made a protest by meeting under water in 2009*

Extreme weather

Climate change is also causing more extreme weather. Warmer oceans seem to be making tropical storms more frequent. For example, Typhoon Haiyan (see spreads 1.8 and 1.9) was one of the most damaging tropical storms in history, creating 4 million refugees. High temperatures are also making heatwaves more common and more extreme. So droughts are more severe and last longer (see spread 9.3).

Heatwaves in Europe

In the summer of 2003, countries across Europe suffered from record high temperatures. It was the hottest summer for 500 years.

- Over 20 000 people died from the heat, with old people particularly at risk.
- Hospitals were overflowing and France had to set up temporary places to store dead bodies.
- The farming industry lost £10 billion because crops and animals died.
- Railway lines buckled in the heat and roads melted.
- The London Eye closed for a day because it was too hot in the cabins.

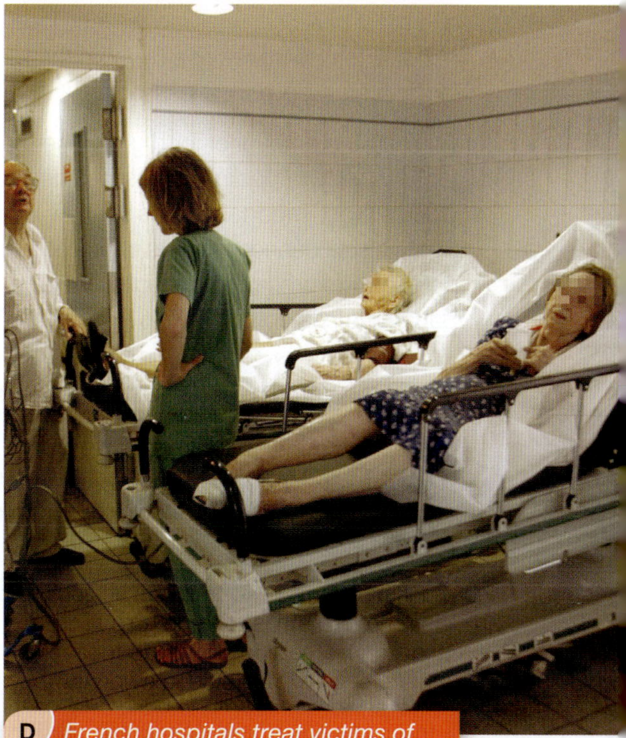

D *French hospitals treat victims of the heatwave of 2003*

Activities

1. Explain why climate change is causing sea levels to rise.
2. Study map B. Using map B and an atlas:
   - **a** identify three regions of the world affected by climate change. Then identify the threat to each region.
   - **b** describe the effects of climate change on South-East Asia. Try to name individual countries and different types of impacts in your description.
3. Imagine you live in the Maldives (photo C). Write a letter explaining the impact that climate change is having on your country.
4. Create a spider diagram showing the impacts of the 2003 heatwave in Europe. Try to classify these as social, economic or environmental impacts.

Practice question

The rise in sea levels is an environmental impact of climate change. Describe a social impact and an economic impact that result from rising sea level. (4 marks)

Global impacts of climate change – 2

On this spread you will learn how climate change is affecting the world's food and water supplies.

Food and water are essential for life on Earth. As Earth's temperature increases there is likely to be a decrease in the amount of food and water available.

Food supply

The supply of food is affected by warmer temperatures, changing rainfall patterns and extreme weather. In general, northern parts of the world will benefit from warmer temperatures and so be able to produce more food. Other parts of the world, like the Mediterranean and most of Africa, will produce less food as they get hotter. If temperatures rise by more than 1°C, a much larger area will produce less food.

Globally, the amount of food grown per hectare (the **yield**) is already declining. This includes some of the world's most important crops such as wheat and rice. In 2011 Russia banned exports of wheat after a heatwave damaged its crops. It is thought that crop yields in sub-Saharan Africa could be 22% less by 2050.

Rice in Vietnam

About half of Vietnam's rice is grown in the Mekong Delta. With sea levels rising, there are more floods, which makes the water too salty to grow rice in. By 2050 the decrease in rice production could affect 1 million people and cost $17 billion.

A *Rice production is threatened*

Water, water – no longer everywhere

Already, many people in the world do not have enough water. They need freshwater to drink and stay alive. Seawater is too salty to drink and it is very expensive to take the salt out of it. A recent scientific report estimated that 1 billion more people will not have enough water by 2025 due to the impact of climate change.

Freshwater is also needed to keep other living things alive and for farming, industry and energy production.

Warmer temperatures affect water supply in four ways (map B). In general, areas of the world most likely to suffer from decreased water supply are the dry semi-arid and arid regions such as the Mediterranean, northern and southern Africa, and north-east Brazil.

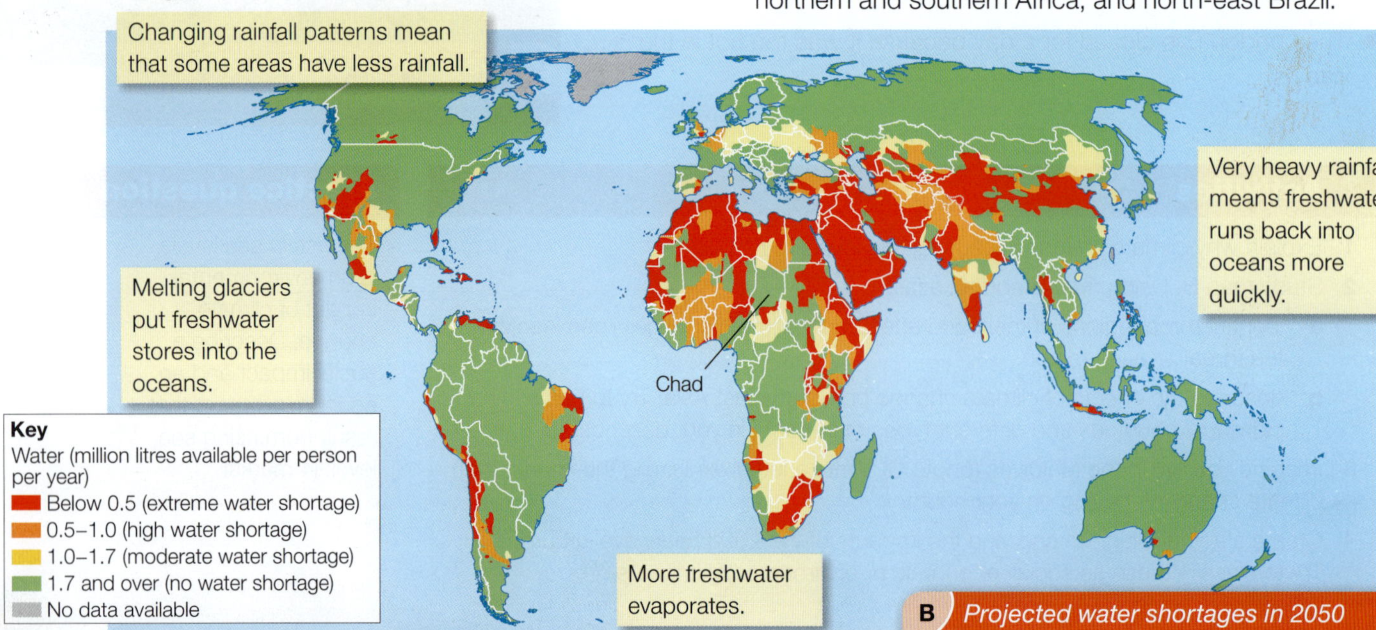

B *Projected water shortages in 2050*

C Lake Chad in 1972

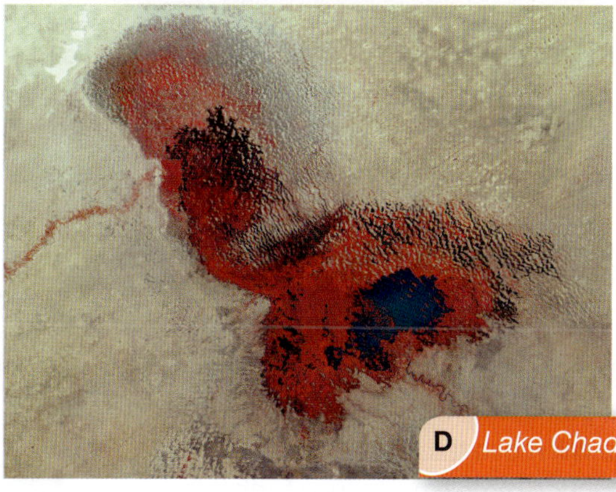

D Lake Chad in 2007

Lake Chad is disappearing!

Lake Chad in central Africa was one of the world's great lakes (photo C). There are a number of reasons why the lake is shrinking:

- A series of droughts meant water lost through increased evaporation was not replaced by rainfall or from the rivers than run into the lake.
- Rainfall in the whole region has reduced by 5–10 mm a year.
- Less rainfall means that people take more water out of the lake to water their land and grow crops.

The water in the lake is used by 30 million people in Chad, Nigeria, Niger and Cameroon to provide drinking water, food and an income. The lake is used for fishing and provides water for the crops and animals.

The disappearing lake causes many conflicts in the region. Fishermen want farmers to stop taking water for their crops and different countries argue over who should control the use of water from the lake.

Activities

1 Look at map B. Describe the distribution of countries that are likely to suffer from extreme water shortages. Using an atlas, include the names of continents and countries.

2 Explain how global food production is affected by climate change.

3 a Look at photos C and D. Estimate how much water had disappeared from Lake Chad in 2007 as a percentage of its extent in 1972.

b Look at map B and an atlas. Which countries have suffered most from the disappearing water?

4 a Imagine you are interviewing a farmer in Nigeria about the effects of the disappearance of Lake Chad. Think of five questions you would ask, e.g.

- Are you worried that the lake is shrinking?
- Will you still be living here in 20 years' time?

b Ask these questions to a partner who can play the role of the farmer.

5 a Study this spread and spread 2.7. Draw a Venn diagram of climate change impacts like the one opposite. Put each of the impacts in the list opposite on your diagram.

- People starve.
- Vietnam exports less rice.
- The price of rice increases.
- Extreme rainfall floods crops.
- Drought is more common.
- People are forced to move.
- The ocean becomes more acidic.
- Sea levels rise.
- The London Eye is temporarily closed.
- Coastal towns are flooded.

b Add three more statements to your diagram.

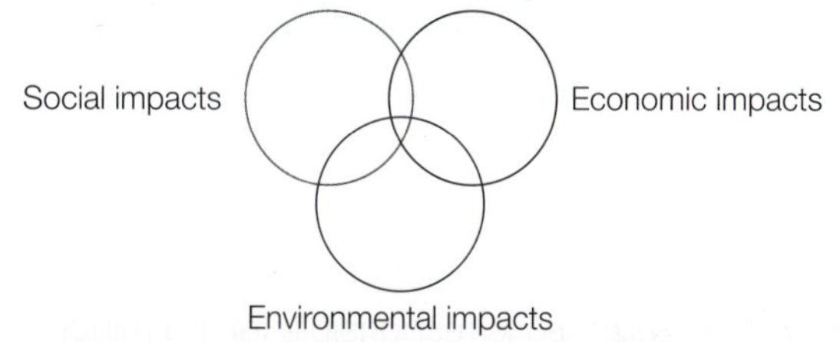

Practice question

Explain two ways in which people are affected by climate change. (4 marks)

2.9 Global impacts of climate change – 3

On this spread you will investigate threats to plants and animals, and summarise the global impacts of climate change.

Plants and animals

Up to a quarter of all plant and animal species on Earth could be extinct by 2050. They are threatened because warmer temperatures, different rainfall patterns and dry spells change the conditions they need to survive. Even if they do not die out, there will be **species refugees** because species will move to different parts of the world. Butterflies, for example, will move further north as their normal environments become too hot.

Threats in the Arctic

The ice in the Arctic, and the ecosystems it supports (Chapter 4), is particularly threatened by warmer temperatures.

In the Arctic (see spreads 4.9–4.14), polar bears are already under threat (photo A). Polar bears eat mainly seals, but hunt for their food from the ice. When there is less ice they have to swim further to find food, which takes more energy. They also need to rest on the ice so they can save energy, especially in winter when there is less food.

The population of seals is also declining, which makes it more difficult for polar bears to find food. As the Arctic ice shrinks, the seals have less ice on which to mate, rear their pups and rest. When the ice breaks up, some pups are separated from their mothers and die of starvation or exposure.

If polar bears don't get enough food they become weaker and stop reproducing. Lack of food is the main reason why polar bear cubs die.

A *Scientists found this polar bear starved to death in the Arctic*

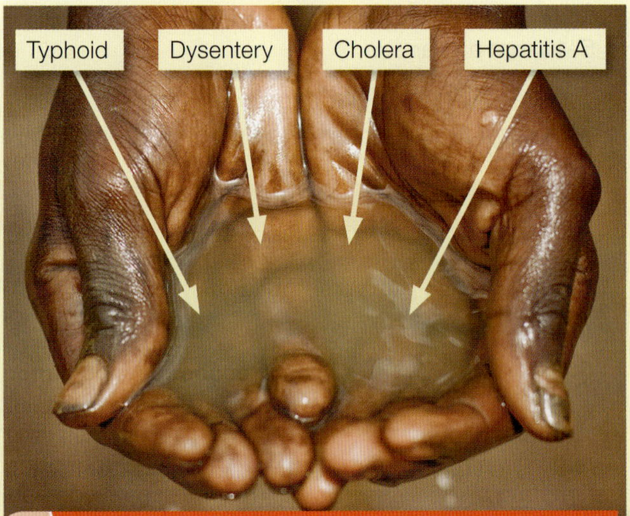

B *Flooding leads to water becoming contaminated (made diseased) by human and industrial waste*

Disease and health

Climate change is also impacting on animal and human health by increasing infectious diseases. Warmer temperatures mean that insects like mosquitoes live longer and survive in parts of the world that used to be too cold for them. This means that diseases like malaria are threatening more parts of the world such as Europe.

Many diseases spread in infected water (photo B). For example, diarrhoea is responsible for 1.5 million children under 5 years of age dying each year. Because climate change is causing more floods, the spread of waterborne diseases is affecting more people.

Disease and food supply

Disease also affects plants and crops. Pests that attack crops are living longer and living in more areas. For example, coffee is very important to the economy in Ethiopia but the country's coffee crops are threatened by the Coffee berry borer beetle. This thrives in warmer temperatures and attacks coffee berries. 700 000 families in Ethiopia depend on coffee for their income, now threatened by climate change.

Global distribution of impacts

The impacts of climate change are different around the world. For example, some places are threatened by extreme heat but for others the problem might be flooding (table D).

Some parts of the world are more vulnerable to the impacts of climate change than others (map C). How much impact climate change will have, and how vulnerable countries are, can be linked to how rich a country is. ACs can afford to prepare for the impacts of climate change by building things like better flood defences. They are also more able to adapt to climate change by building new cities or growing different types of crops.

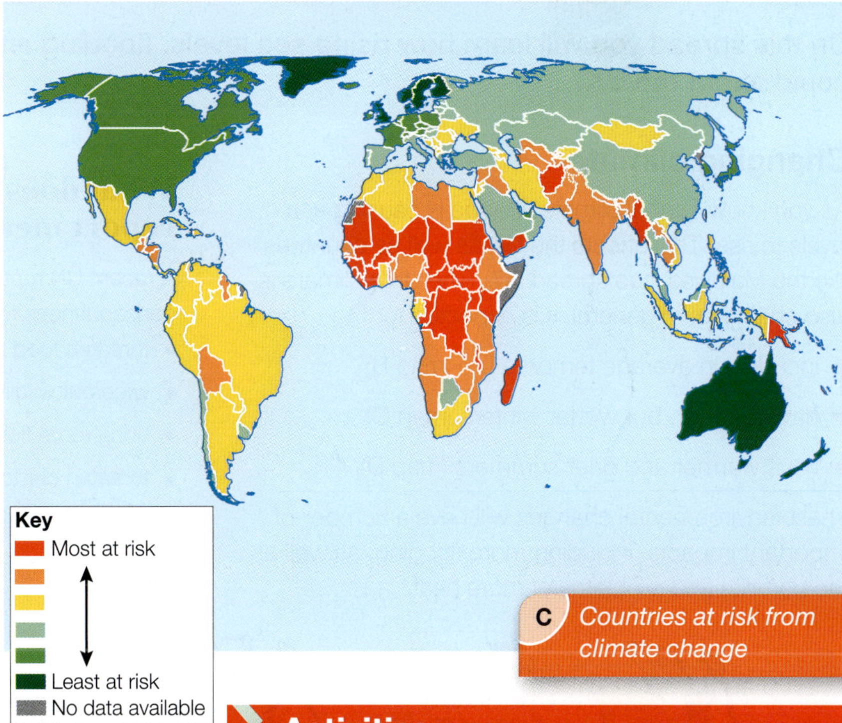

C *Countries at risk from climate change*

Region	Key impacts
Africa	Water shortages and drought
	Reduced crop production
	Increase in diseases
Europe	Economic losses due to flood damage
	Reduction in water availability especially in southern Europe
	Threat to human health
Asia	Coastal and urban flooding
	Threat to human health
	Water and food shortages
Australasia	Economic losses due to flood damage
	Damage to coral reef ecosystems
	Water shortages
North America	Property damage and death from wildfires
	Heat-related deaths
	Urban floods
Central and South America	Reduced water availability
	Flooding and landslides
	Decreased food production
Polar regions	Threat to ecosystems
	Threat to indigenous way of life

D *Regional impacts of climate change*

Activities

1 In your own words, explain what a species refugee is.

2 Explain why polar bears might be extinct in 100 years.

3 Give two reasons why climate change will increase the amount of disease.

4 Using the example of the Coffee berry borer beetle, explain how environmental changes have social and economic impacts.

5 Look at map C.

   a Name the continent most at risk from climate change and two continents least at risk.

   b Describe the pattern of countries most and least at risk from climate change. Give examples of specific countries in your answer and refer to ACs and LIDCs.

   c Suggest reasons for this pattern.

6 a Use map C, table D and the information on spreads 2.7–2.9 to produce a world map of the different impacts of climate change. Annotate the map using images and words.

   b Colour code your map to identify the social, economic and environmental impacts.

Practice question

Using examples, assess whether the social and economic impacts of climate change are worse than the environmental impacts. (8 marks)

Impacts of climate change for the UK – 1

On this spread you will learn how rising sea levels, flooding and extreme weather could affect the UK.

Changing climate

As you know, global climate change is causing sea levels to rise. This affects the UK as well as countries like the Maldives (see spread 2.7). The UK's climate is also changing. In general, it is expected to:

- increase in average temperature (map B)
- have warmer, but wetter winters (map C)
- have warmer and drier summers (map D).

These environmental changes will have a number of important impacts, including more flooding, as well as the benefits and problems of more heat.

What does UN climate change report mean for us?

A recent UN report on climate change predicts dire consequences for the UK. Climate change will:

- threaten food prices
- wipe out wildlife
- bring more floods
- threaten our forests
- cost more than we thought
- bring many climate change refugees.

A *Adapted from newspaper reactions to Climate* Change 2014, *from the UN's Intergovernmental Panel on Climate Change*

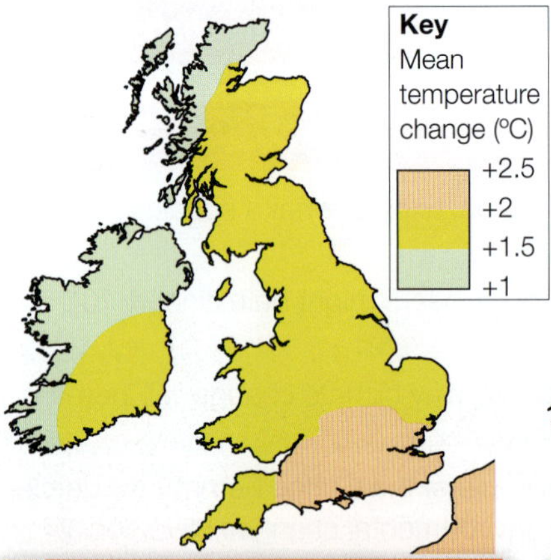

B *Possible temperature changes by 2050*

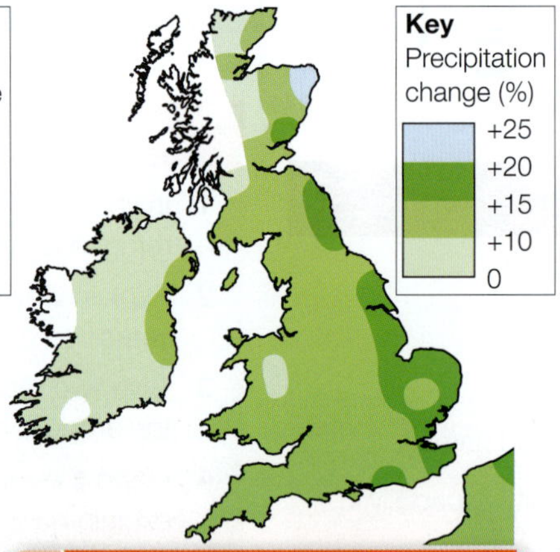

C *Possible winter precipitation changes by 2050*

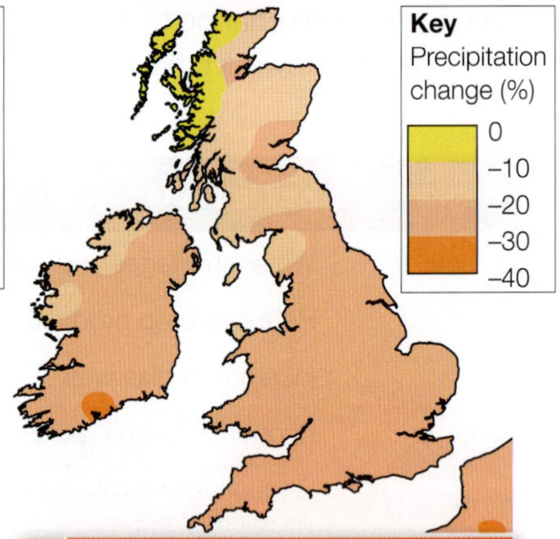

D *Possible summer precipitation changes by 2050*

Coastal flooding

About 30% of the UK's population lives within 10 miles of the coast. Rising sea levels threaten many parts of the UK (map E):

- Low-lying areas, particularly in the east of England, could flood. London could experience serious flooding with the Thames Flood Barrier unable to cope (see spread 3.17).
- Coastal regions contain lots of manufacturing, transport routes, power stations, oil and gas terminals, and agricultural land. £120 billion worth of infrastructure is at risk from coastal flooding.
- Erosion of coastal areas will increase and more will need to be spent on coastal defences (see spreads 3.10 and 3.11).
- Coastal towns tend to have high numbers of elderly people – the least able to cope with flood events (see spread 7.5).

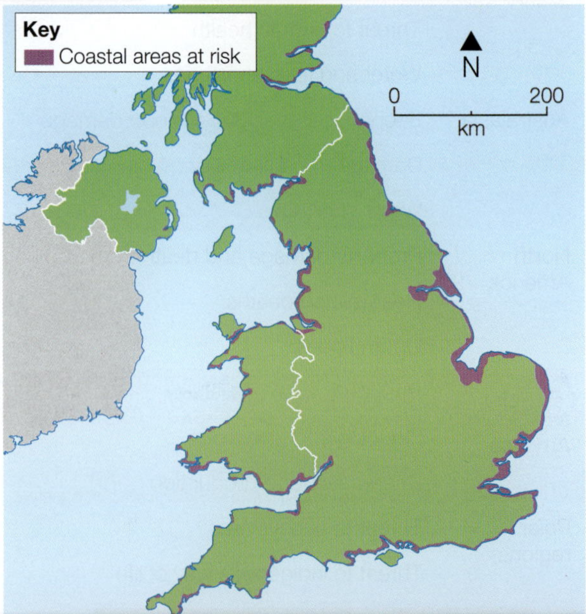

E *Flood risks in the UK if sea levels rise by 2 metres*

Extreme rainfall

The UK is likely to have more extreme rainfall, especially in winter. This will increase the chance of flooding from rivers. Heavy downpours in particular can cause **flash floods**, which happen very quickly and have devastating effects. River flooding also makes coastal flooding worse and puts even more properties at risk. Currently, damage caused by rivers and the sea flooding costs the UK about £1.3 billion a year but this could increase to up to £12 billion a year by 2080.

Flooding in 2007

In the summer of 2007 the UK suffered its wettest May and June since records began in 1776. There were floods across the country (map F) and widespread damage to homes, businesses and transport networks. The total cost to the UK's economy was £3.2 billion.

Some areas suffered more than others. Tewkesbury in Gloucestershire was badly affected (photo G). On 20 July 2007 two months' worth of rain fell in just 14 hours. Three people died, 1800 houses were completely cut off and 50 000 had no electricity. A water treatment works was flooded leaving 135 000 houses without water for 17 days. Birds like the Lapwing and Redshank suffered when breeding grounds were flooded.

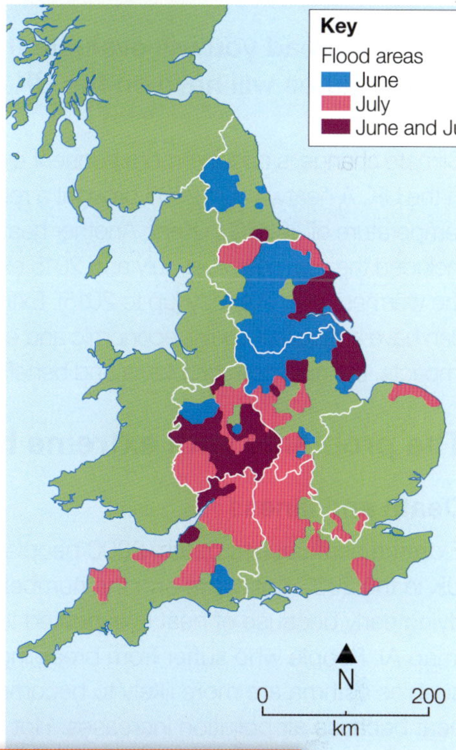

F *Areas affected by flooding in the summer of 2007*

Activities

1. Look at newspaper report A.
   - **a** Sort the points into social, economic and environmental impacts of climate change.
   - **b** Pick one impact that you think is the most worrying. Explain why you chose it.
2. Describe how the UK's climate will change in the future. Use information from maps B–D in your description.
3. Look at map E.
   - **a** Using an atlas, identify three towns or cities that are at risk from coastal flooding.
   - **b** Explain the impacts that coastal flooding will have for people and for the UK's economy.
4. **a** Look at photo G. List the social, economic and environmental impacts of flooding you can see.
   - **b** Add more impacts to your list that you can't see but can infer from the photo.

G *Flooding in Tewkesbury, 2007*

Go further

The UK also suffered from severe flooding in the winter of 2013/14. Research the impact of these or other recent floods for one place in the UK.

Practice question

Using examples, explain why climate change has social impacts in the UK. (6 marks)

2.11 Impacts of climate change for the UK – 2

On this spread you will evaluate the positive and negative impacts warmer temperatures will have on the UK.

Climate change is causing more frequent extreme heat in the UK. A heatwave in 2003 brought a record high UK temperature of 38.5°C in Kent. Another heatwave, in 2006, included the warmest-ever July and 2015 experienced the warmest-ever July day (up to 2015). Extreme heat can have a range of social, economic and environmental impacts, bringing both problems and benefits.

The problems with extreme heat

Death and illness

Extreme heat causes death. 2000 people died in the UK in the 2003 heatwave and the number of people dying early because of heat is expected to increase (map A). People who suffer from breathing difficulties such as asthma are more likely to become ill in extreme heat because air pollution increases. Hot weather also means more bacteria breeds on food. It is estimated that by 2050 there will be an extra 9000 cases of illness a year caused by salmonella bacteria.

Water shortages

High temperatures, combined with less summer rain, will lead to water shortages. London and the south-east of England will be worst affected. This is a particular problem because over a quarter of the UK's population live there. There might have to be restrictions on water use, such as hosepipe bans, or even cuts to domestic water supplies. Water shortages also affect industry. For example, power stations need large quantities of water and may be forced to shut down during droughts. Farmers might face water shortages, as well as more extreme weather conditions, making it difficult to grow some crops. Then the UK will have to rely more on imported food.

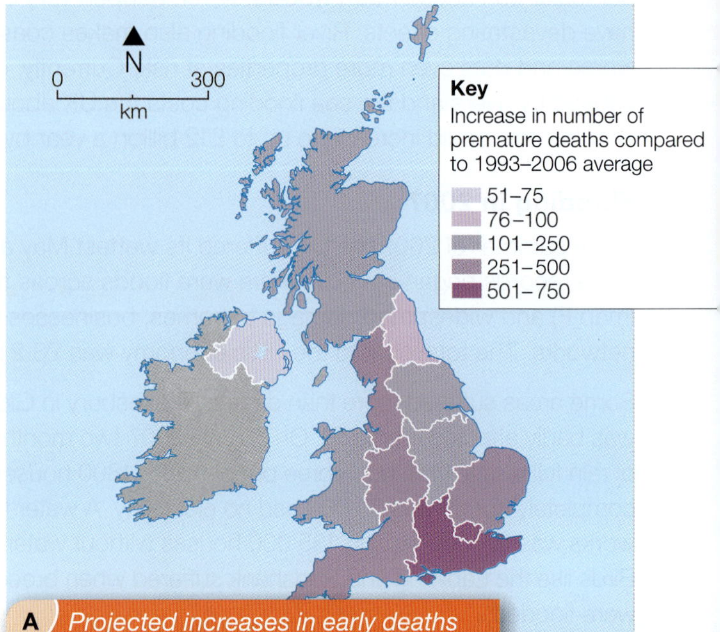

A *Projected increases in early deaths due to heat in 2050s*

B *Tourists enjoy Brighton Beach during the 2013 heatwave*

The benefits of warmer temperatures

The UK's climate is expected to be 1–2.5°C warmer by 2050. Warmer temperatures mean there are fewer deaths due to cold, and industries like tourism and farming benefit.

A better climate for tourism

Better weather will mean that more people will take holidays in the UK (photo B). Warmer temperatures will also make the tourist season longer. This will be good for the economy because people will spend more money on things like hotels, days out, ice creams, drinks and barbecue food.

The Tourism Comfort Index (TCI) describes how suitable the climate is for outdoor activities. The maps in panel C show the improvement that Cornwall, Devon and Somerset are predicted to have. This will encourage more people to visit and add more money to the economy.

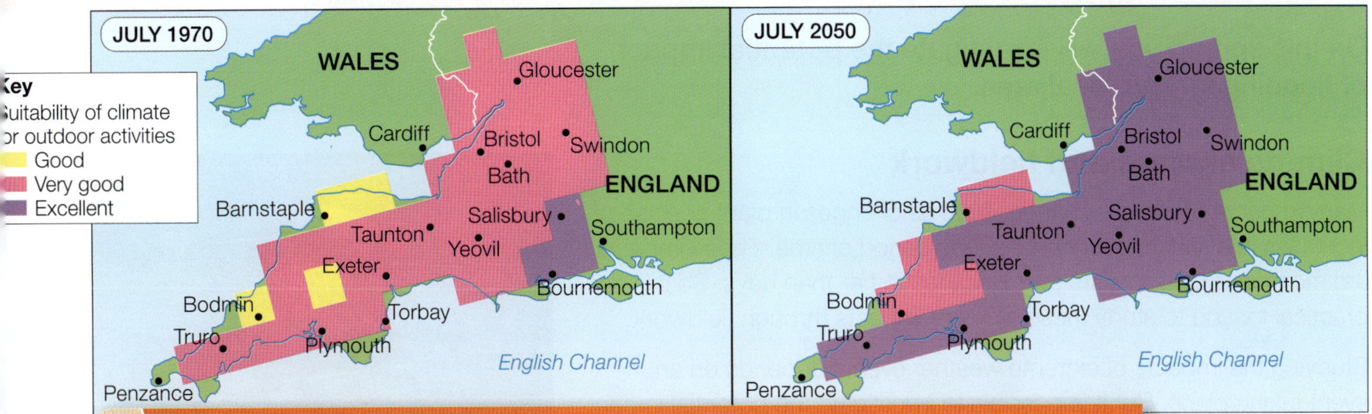

C *Projected changes in the TCI in south-west England between July 1970 and July 2050*

A better climate for farming

Warmer weather could also help the UK's farming industry. Growing seasons will be longer, so more food will be grown. There would also be opportunities for growing new types of food. Farmers in the south will be able to grow crops like sunflowers, olives and melons, which all need higher temperatures.

A corking success!

When Prince William married Catherine Middleton, English wine was served at their wedding. England has not been well known for the quality of its wine, but in the last 16 years English sparkling wines have won 14 international trophies – better than any other country!

Warmer temperatures now mean conditions in England are similar to those in France 20 years ago. This creates a better environment for vineyards. Land used for vineyards more than doubled from 2004 to 2014. In 2014 a record 6.4 million bottles were sold – a 43% increase on the previous record of 4.45 million bottles in 2013.

Extreme weather, however, can ruin farmers' crops. 2012 was a very bad year for wine. Only 1.03 million bottles were produced because floods wiped out many crops.

D *Chapel Down vineyard in Kent where the grapes for William and Kate's wine were grown*

Activities

1 a Look at map A. Describe how the impact of extreme heat on people will vary across the UK.
  b What reasons can you give for the different numbers of deaths?
2 Create a table showing the positive and negative impacts of warmer temperatures in the UK.
3 Do you think the potential benefits of climate change in the UK outweigh the negative impacts? Give reasons.
4 Using the information on this spread and spread 2.10, create an annotated map showing the impacts of climate change on different parts of the UK.

Your enquiry

What evidence is there to suggest climate change is a natural process?

There are a few climate change sceptics, who don't believe that humans are causing climate change. They argue that Earth's climate has been constantly changing for 4.5 billion years and that recent changes are just part of natural processes.

Write a letter to a sceptic explaining why you agree or disagree with them. Use evidence from this chapter in your letter. It will help if you break this down into shorter sections. You could consider:

- What caused climate change before humans?
- What is causing climate change today?
- Are the extreme impacts of climate change natural?
- Should we be worried about climate change?

Climate change fieldwork

On this spread you will investigate the potential impact of flooding in your local area.

Climate change and fieldwork

Many impacts of climate change, such as changes in plant species or increases in disease, occur over a long period of time. These impacts are monitored by scientists over decades rather than days. So it is not practical for you to study these types of impacts through fieldwork.

Studying the impacts of extreme weather events depends on an event taking place. But it is possible to investigate the potential impact of events like increased flooding. The activities here will help you investigate the fieldwork question, *'What impacts would a greater risk of flooding have on my area?'*

A *Places have the potential to flood*

Identifying areas at risk

Not everywhere will be impacted by sea level rise or flooding, but you probably live close to somewhere that might be. If you type 'Environment agency interactive map' into a search engine you will find a link to their flood risk maps. The interactive maps allow you to search by place name or postcode. The map will then show the flood risk for that place (map B).

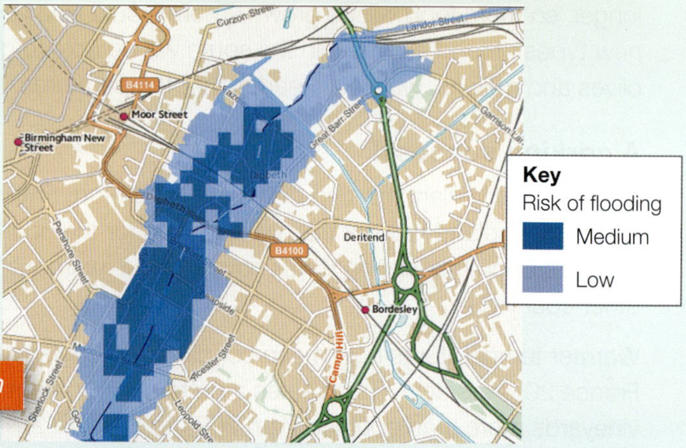

B *Flood risk in central Birmingham*

Using secondary data

- **Secondary data:** Use secondary data to gain an overview of the study area to help you understand the types of problems it might face. Census data can tell you how many people live and how many households there are in a street, the age profile of an area, unemployment statistics, and so on. There are many searchable data sets at www.nomisweb.co.uk/.
- **Economic impact survey:** Calculate an estimated economic cost of flooding for different streets (table C).
- **Social impact survey:** Identify any essential services like hospitals, fire stations, telephone exchanges or community centres that would be flooded.

Climate change fieldwork activities

- **Annotated sketch/photo:** Add notes about the potential impacts of flooding to a photo or sketch (photo D).
- **Photo evidence:** Take photos to show social, economic and environmental impacts.
- **Flood mapping:** Create an annotated map of the potential impacts of flooding.
- **Questionnaire:** Interview people about how they would be affected by flooding. You could ask questions such as:
  - Do you think you live in a high flood risk area?
  - Do you have insurance for flooding?
  - Would you have a place to stay if your house was flooded?
- **Role play:** Imagine how different people might be affected by flooding. For example, you could write about the different problems faced by people like these:
  - a shopkeeper
  - a young married couple with a baby
  - a pensioner who lives alone
  - a teenager.

Street name	(A) Number of small houses and ground-floor flats	(B) Number of medium houses	(C) Number of large houses	(D) Number of businesses	Estimated cost of flooding to houses (£)	Estimated cost of flooding to businesses (£)	Total cost of damage (£)
Church Street	90	10	5	5	1 800 000 + 300 000 + 200 000 = 2 300 000	150 000	2 450 000
Priory Lane	50	20	0	0	1 000 000 + 600 000 = 1 600 000	0	1 600 000

C *Economic impact survey*

To calculate the cost of flooding, multiply:
A by £20 000
B by £30 000
C by £40 000
D by £30 000.
Add these figures together to get a total cost.

D *Potential impacts of flooding in a Birmingham street*

Activities

1. Look at map B.
   a. Now think of an area near you where you could investigate the potential impact of flooding.
   b. Enter your town or village into the Environment Agency map. Then zoom into the map to identify particular streets that might flood.
2. Look at table C.
   a. Complete a similar survey for at least two streets in your local area. You may need to visit the streets to count the dwellings and businesses.
   b. Draw a suitable graph or chart to display the cost of flooding for one of these streets.
   c. Are there ways you could improve your survey and the value of your results?
3. Look at photo D. Take a photo in your area and annotate the potential effects of flooding.
4. Choose one extra activity and complete it for your local area.

Go further

Conduct fieldwork comparing the potential impacts of flooding in two contrasting places such as low- and high-income areas.

Practice question

a. Suggest an enquiry question or a hypothesis you could use to investigate the impacts of climate change. (1 mark)
b. Suggest two fieldwork activities that would help with your investigation. (2 marks)

What is this topic about?

In the UK we are surrounded by a variety of landscapes. It's great for geographers! We don't need to travel far for a change of landscape. What makes each landscape distinctive are the **geomorphic processes**, such as weathering and erosion, that shape them. These, in turn, are influenced by the underlying **geology**, or rocks, by the climate and by human activities.

Perhaps one of the most distinctive landscapes in the UK are the white cliffs of Dover. It is the first landscape you see when you arrive in England by boat from Europe. The cliffs stand guard, overlooking the English Channel where, over the centuries, Britain has often been threatened with invasion. They are made of white chalk and are slowly being weathered and eroded, disappearing at the rate of about 10 mm a year.

Your enquiry

In this chapter you will investigate the specification's two key questions.

What makes a landscape distinctive? You will:

- Identify the key elements that make a landscape.
- Distinguish between a natural and built landscape.
- Locate areas of upland and lowland landscapes in the UK.
- Recognise the extent of glaciation on the UK landscape.
- Understand the influence of geology, climate and human activity on landscapes.

What influences the landscapes of the UK? You will:

- Investigate the geomorphic processes that shape the landscape along rivers and coasts.
- Identify coastal and river landforms, and explain how they were formed.
- Study two landscapes in the UK – the Jurassic Coast and the Thames Basin.
- Examine the influence of geology and climate on the two landscapes.
- Explore the influence of human activity on landscapes, including management.

Look at the photo opposite.

- What makes the white cliffs of Dover a distinctive landscape?
- What is natural about this landscape? How do you think it was formed?
- Have people influenced the landscape at all? If so, how?
- What evidence is there that the landscape is changing?

On this spread you will consider the main landscape elements and what makes each landscape distinctive.

What's in a landscape?

A landscape is more than just scenery. Every landscape tells a story. Each landscape is created by the interaction of natural and human factors. The main elements of any landscape are:

- physical – like mountains and valleys
- water – including rivers and lakes
- living – like trees and grass
- transitory (temporary) – like weather and seasons
- human – like buildings and other land uses.

A *Flatford Mill today*

The Hay Wain

One of the most popular 'landscapes' in the UK is an old painting by John Constable called *The Hay Wain* (painting B). You have probably seen it somewhere before – it could even be hanging on a wall at home! It shows a traditional rural scene, with horses pulling a large cart, or hay wain, across a river.

The painting was based on a real landscape at Flatford Mill on the River Stour, on the county border between Suffolk and Essex. The landscape has changed very little today. The cottage is still there, though the horse and cart have gone (photo A)!

B The Hay Wain, *painted by John Constable in 1821*

Natural and built landscapes

Sometimes it is easy to recognise the natural and human factors that interact to form a landscape, but not always. For example, a mountain is obviously natural and a city was obviously built by people. But the forest growing on the mountain might have been planted by humans and the river flowing through the city was once natural (photos C–F).

C *Forests of conifers on the Cambrian Mountains in Wales*

D *The River Exe following through Exeter*

E *Mixed farming in England*

F *The Grand Union Canal in Warwickshire*

Activities

1 Look at the landscape in painting B. Identify at least one of each type of landscape element:
- **a** physical
- **b** water
- **c** living
- **d** transitory (not lasting)
- **e** human.

2 **a** Why do you think *The Hay Wain* is such a popular painting of landscape in the UK?
- **b** Compare the same landscape today in photo A. How is it the same and how has it changed?

3 Look at photos C–F.
- **a** Identify natural and human factors that interact to create each landscape.
- **b** Choose one landscape that you think best represents the UK. Justify your choice.

Go further

a Think about your local area or walk around the area to look at the landscapes. Choose a landscape that could represent your area. Find a photo of the landscape or take your own. Explain why you chose it.

b Describe the natural and human factors that interact to create this landscape.

Your enquiry

What makes a landscape distinctive?

As you work through this section, make notes about anything that will help you to answer the question. You will bring your ideas together to do an activity at the end of spread 3.5.

3.2 Upland and lowland landscapes

On this spread you will compare the distribution of upland and lowland landscapes in the UK and how they were affected by glaciation.

Relief map of the UK

The relief map of the UK can be divided into **upland** and **lowland** areas (map C). Uplands, including mountainous landscapes like the Grampians and Pennines, are found mainly in the north and west, particularly Scotland, Wales and northern England (photo A).

Lowlands, including flat and rolling landscapes like the Cotswolds and South Downs (photo B), are found mainly in England, below an imaginary line from Flamborough Head in Yorkshire to the Bristol Channel.

A *The Cuillin Hills on the Isle of Skye, Scotland*

B *The South Downs in Sussex, England*

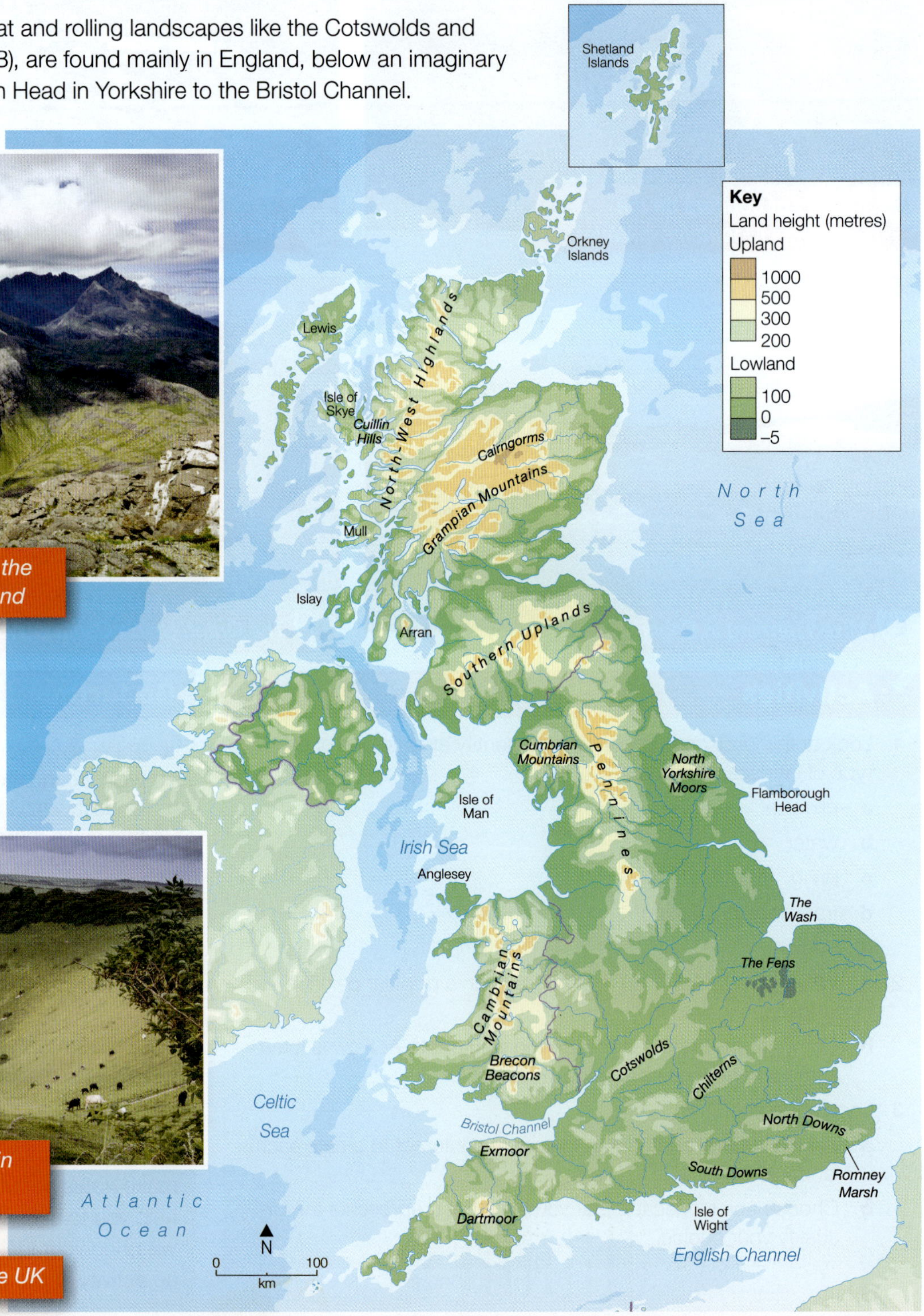

C *Relief map of the UK*

Glaciation in the UK

The UK landscape has not always looked like it does today. During the Quaternary period (see Chapter 2), much of Britain went through periods of **glaciation**, when the land was covered in ice. Over the 2.6 million years of the Quaternary period, the climate alternated between cold ice ages (glacial periods) and warm interglacial periods. The ice grew thicker and advanced over the land during the ice ages. It thinned and retreated during the interglacial periods (map D).

Glaciation made a lasting impression on the landscape. In upland Britain, the ice carved huge, **U-shaped valleys** and eroded steep mountain peaks and ridges. In non-glaciated lowland areas, **periglacial** conditions during the ice ages had an impact on the landscape (diagram E).

E *Impact of glaciation in the South Downs*

During the ice age (periglacial conditions)

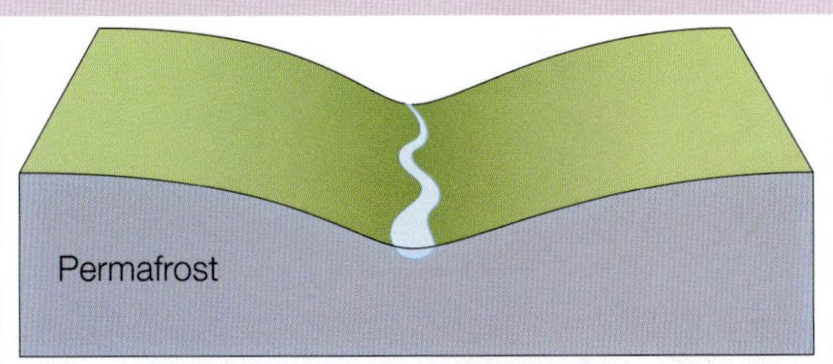

- Water below ground in the chalk remains frozen throughout the year (permafrost).
- Rivers flow over the surface of the chalk, which is normally permeable (allows water through).
- Water in the river erodes chalk to form a valley.

After the ice age (present day)

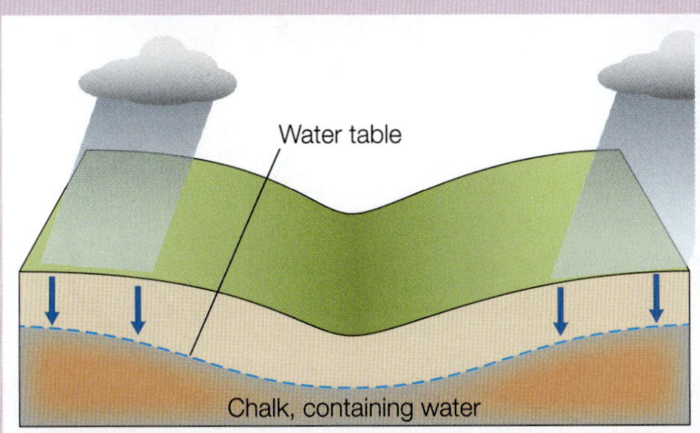

- Permafrost thaws out and the chalk becomes permeable again.
- Rainwater percolates into (moves through) the chalk and no longer flows over the surface.
- The river disappears, leaving a **dry valley** on the surface.

D *Quaternary period glaciation in Britain*

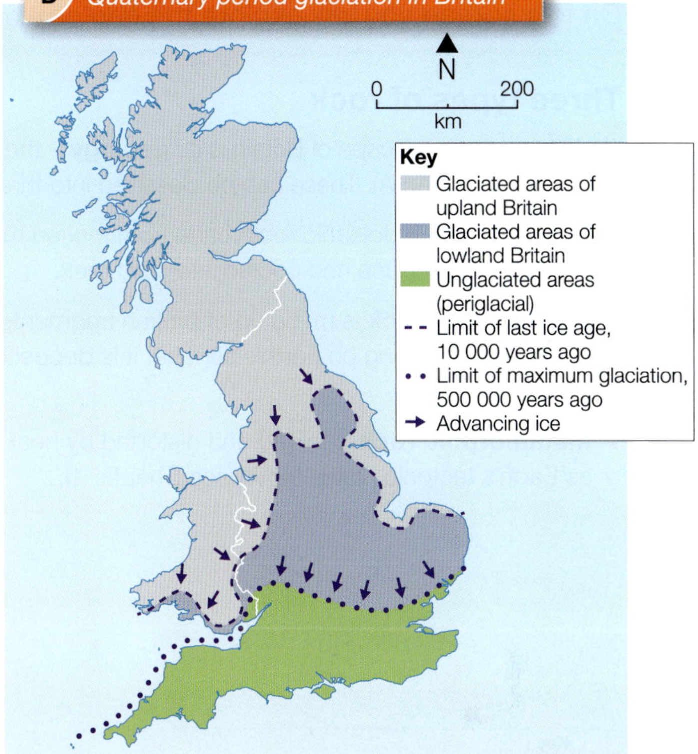

Activities

1. Look at map C.
   - **a** Name at least five mountainous upland landscapes in the UK.
   - **b** Name five lowland landscapes in the UK.
2. Look at photos A and B. Compare the two landscapes in upland and lowland UK. Mention:
   - the height and shape of the land
   - any surface water (or lack of it)
   - any vegetation cover (or lack of it).
3. Look at map D. Describe the distribution of:
   - **a** glaciated areas
   - **b** unglaciated areas.
4. Compare map D and panel E with photos A and B.
   - **a** Were the Cuillin Hills or South Downs ever covered by ice?
   - **b** Explain the impact of glaciation on the landscape in each photo.

Practice question

Describe the distribution of upland and lowland landscapes in the UK. (4 marks)

Geology and landscape

On this spread you will explore the geology of Britain and how it affects the landscape.

Three types of rock

Underlying the landscape of Britain is its **geology** – the rocks that make the land (map A). These can be classified into three types:

- **Igneous rock** is volcanic rock made from molten material brought up to Earth's surface and cooled into solid rock.
- **Sedimentary rock** is made up of broken fragments of rock worn down by weathering on Earth's surface. It is deposited in layers, often under water.
- **Metamorphic rock** is folded and distorted by heat and pressure as Earth's tectonic plates move (see Chapter 1).

In Britain most igneous and metamorphic rocks are older than the sedimentary rocks. They are found to the north and west, particularly in Scotland. Most sedimentary rocks are found in England. They become progressively younger the further south and east you go.

A *Geology of Britain*

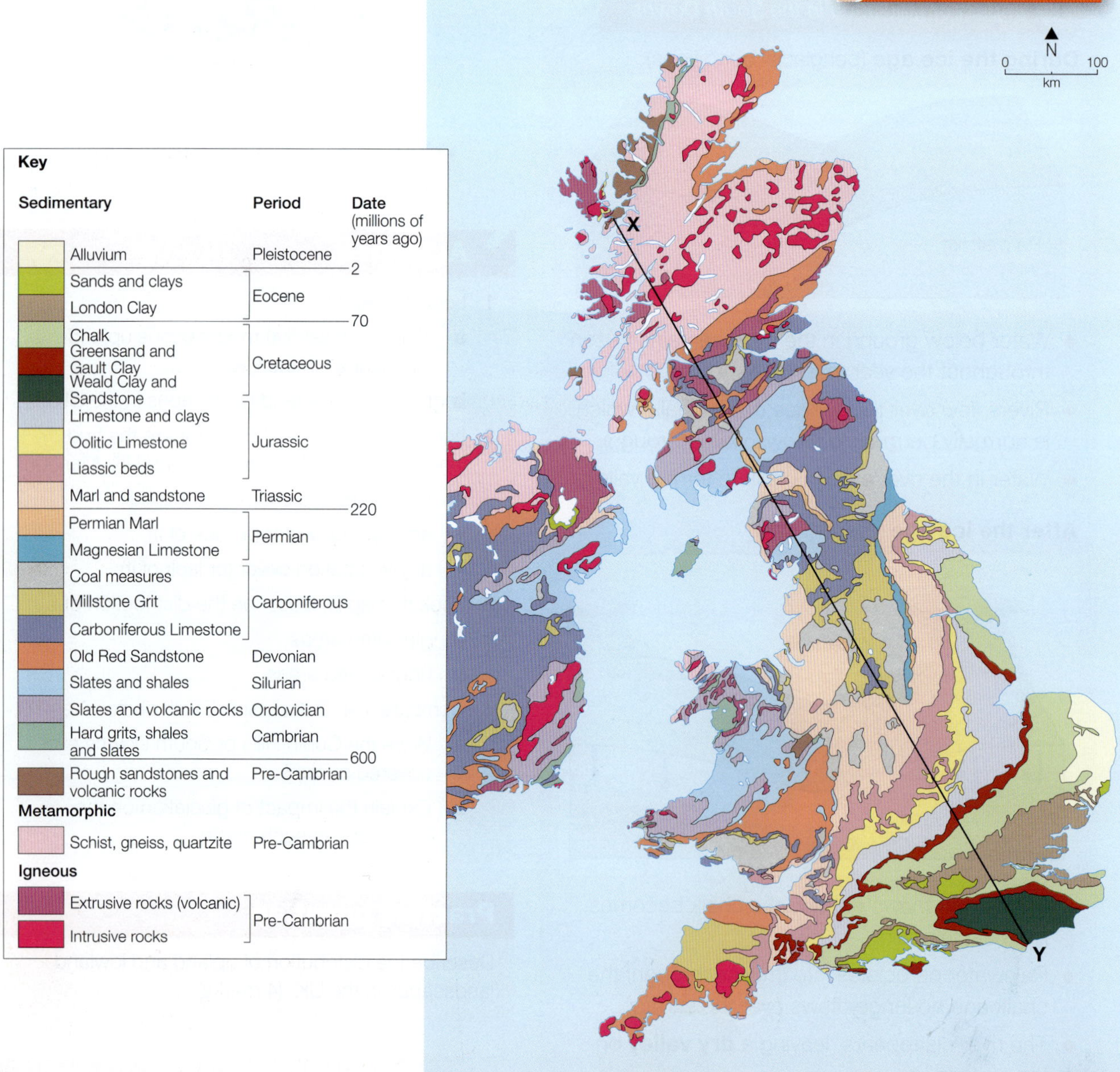

Rocks and landscape

Geology has a strong influence on landscapes. In many cases, it is possible to identify the underlying rock just by looking at the landscape. The different properties of each rock – hardness, permeability and structure – give rise to the varied landscapes of the UK.

Two rocks commonly found in different parts of the UK are granite and chalk. For example, granite is found in the Cuillin Hills on the Isle of Skye in Scotland and chalk in the South Downs of England (see spread 3.2).

B *Granite*

Formation: Granite is an igneous rock found where molten rock cools down slowly below Earth's surface. Because it cools slowly the crystals are large.

Composition: Crystals are made from quartz (colourless), feldspars (white or pink) and mica (silver or black).

Properties: Granite is a very hard rock that weathers and erodes slowly. It can be used to make kerbstones. It is also impermeable, so water does not pass through it.

Soils and farming: The soil on granite is acidic and not very fertile. Often it is also waterlogged. It is not good for growing crops.

C *Chalk*

Formation: Chalk is a type of limestone, a sedimentary rock formed in layers, or strata, on the seabed. Later, the strata were folded and pushed up to form new land above sea level.

Composition: Chalk is made of calcium carbonate from the shells of tiny sea creatures that sank to the bottom of the ocean.

Properties: Unlike other types of limestone, chalk is soft and crumbly, so it is easily weathered and eroded. It is also permeable, so water drains through it.

Soils and farming: The soil on chalk is alkali, quite fertile and well drained. It is ideal for growing grass and also suitable for crops.

Activities

1 Look at map A.

- **a** Follow the line XY from the Isle of Skye to the Sussex coast. List the rocks along the line in the order in which you would cross them.
- **b** In general, what do you notice about the age of the rocks you cross from north to south? Don't forget to refer to the key for the map.

2 Compare the rocks on map A with the relief of the UK on map C on spread 3.2. Name the underlying rock for each of the following areas:

- **a** Grampian Mountains
- **b** Cotswold Hills
- **c** Southern Uplands
- **d** South Downs
- **e** Pennines
- **f** Romney Marsh.

3 Look at photos B and C.

- **a** Write two sentences to describe the appearance of granite and chalk.
- **b** Explain the different appearance of the two rocks.

4 Compare photos B and C above with landscapes A and B on spread 3.2. Explain the difference between:

- **a** the height and shape of the land
- **b** the amount of surface water
- **c** the way the land is used by people.

Practice question

Explain how the geology of the UK gives rise to distinctive landscapes. Refer to one or more named rock types. (6 marks)

Climate and landscape

On this spread you will find out how climate varies across the UK and how it affects the landscape.

The effects of climate and weather on landscape

The UK enjoys a mild climate, without the extremes of temperature and rainfall found in some parts of the world. But there is enough variation in climate and weather to influence the landscape.

Rocks are slowly worn down by **mechanical**, **chemical** and **biological weathering**.

Mechanical weathering

This is caused by temperature changes, especially on mountains (photo A). Large temperature variation between day and night leads to **freeze-thaw weathering** (diagram B).

A *Winter in the Cuillin Hills*

B *Freeze-thaw weathering*

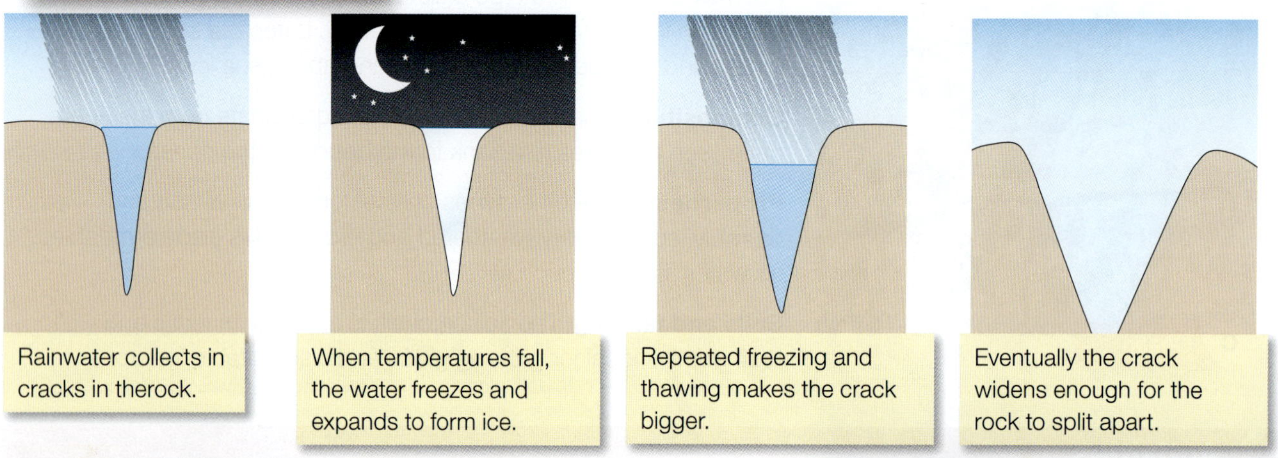

Chemical weathering

Rain dissolves carbon dioxide in the air, forming weak carbonic acid. So rainwater is naturally acidic. The action of rainwater causes chemical weathering in rock.

Some types of rock are more easily weathered by rainwater. For example, limestone and chalk, both composed of calcium carbonate, slowly dissolve in acid. In some limestone areas of the UK, the surface is weathered to form **limestone pavement** (photo C). Rainwater collects in the cracks, dissolving the rock, to leave the raised pavement.

Biological weathering

This type of weathering is caused by plants or animals. Tree roots can force their way into rocks, splitting them apart. Smaller plants, like mosses, grow on the surface of the rock, slowly making it crumble. Even animals burrow into softer rocks and people can have an effect with their trampling feet.

C *Limestone pavement at Malham in Yorkshire*

Climate and weathering

The rainfall map of the UK shows that average rainfall varies from less than 600 mm a year in the east of England to over 2400 mm a year in north-west Scotland (map D). Temperatures also vary, with the main variation between upland and lowland areas (graphs E and F).

This variation affects the UK landscape in different ways:

- Uplands are most likely to experience both high rainfall and freezing temperatures, so weathering is most likely to occur there.
- Areas with higher rainfall are likely to have more water in rivers, so more erosion occurs there (see spread 3.8).
- Higher rainfall is also likely to cause more rockfalls and landslips (see spread 3.6).

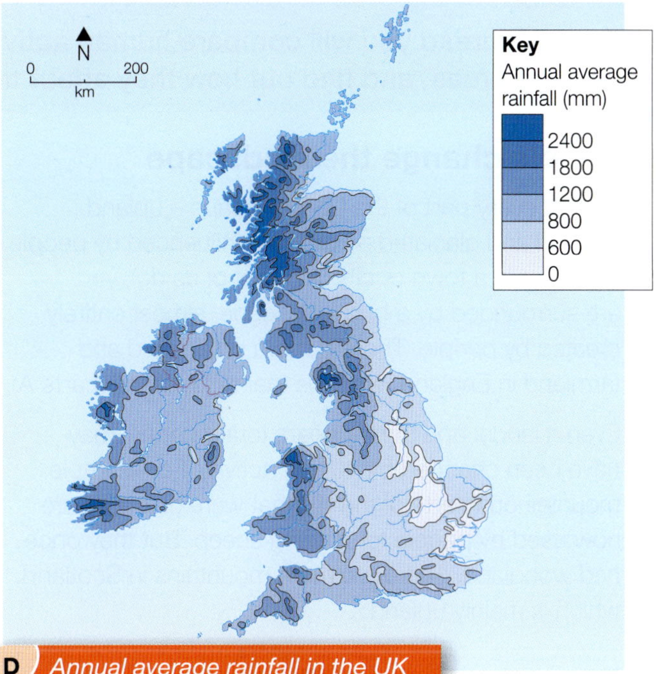

D *Annual average rainfall in the UK*

E *Climate on the Isle of Skye*

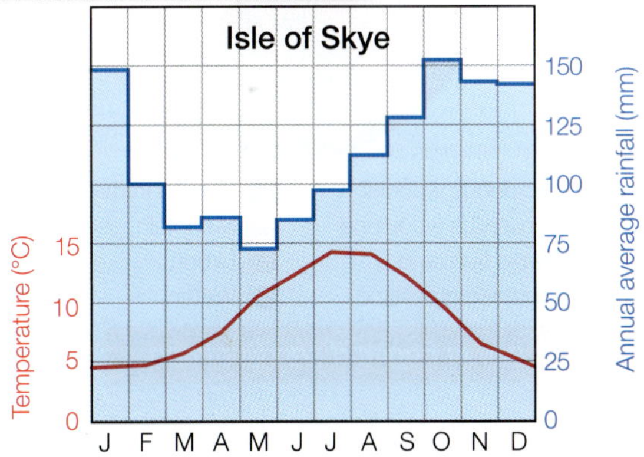

F *Climate in Sussex*

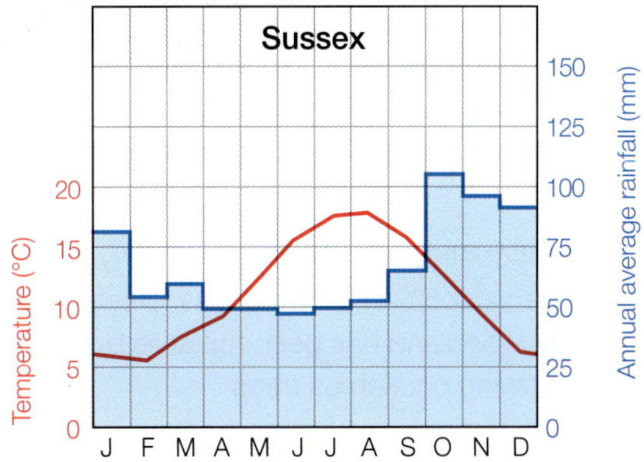

Activities

1 Suggest which type of weathering happens in photo A. Explain why.

2 Look at diagram B. Freeze-thaw weathering can affect the built landscape as well as the natural landscape. For example, it can burst pipes. Draw a sequence of sketches to show how freeze-thaw weathering could burst a pipe. This drawing will start you off:

3 Look at photo C. Draw an annotated sketch to explain how limestone pavement is formed.

4 Look at map D.

- **a** Describe the rainfall pattern in the UK. Mention the parts with most and the parts with least rain.
- **b** What effect could this pattern have on the landscape in different parts of the UK?

5 a Using graphs E and F, compare temperature and rainfall in the Isle of Skye and Sussex.

- **b** What influence could climate have on the landscape in the two areas?

Practice question

Describe how the climate of the UK gives rise to distinctive landscapes. Refer to named locations in your answer. (6 marks)

3.5 Human activity and landscape

On this spread you will compare human activities in upland and lowland areas, and find out how they affect the landscape.

People change the landscape

Almost every part of the UK landscape – upland, lowland and glaciated – has been influenced by people. If you live in a town or city (as most of us do), you are surrounded by a built landscape, almost entirely created by people. There is more urban land and farmland in England, which is mainly lowland (charts A).

Even a landscape that appears totally natural may have been changed by human activity. For example, mountainous parts of the UK that were glaciated are now used by people for grazing sheep. But they once had woodland. There are more mountains in Scotland, which is mainly upland.

Disappearing woodland

Go back a few thousand years and most of the landscape of Britain was covered by woodland. Throughout history, people have chopped down trees to make space for farmland, build towns or obtain wood for burning. By the time of the Norman conquest in 1066, only 15% of England was covered in woodland (graph B).

That amount continued to decline until early in the 20th century when just 5% of England was forested. Since then a lot of woodland has been replanted, much of it with fast-growing coniferous trees.

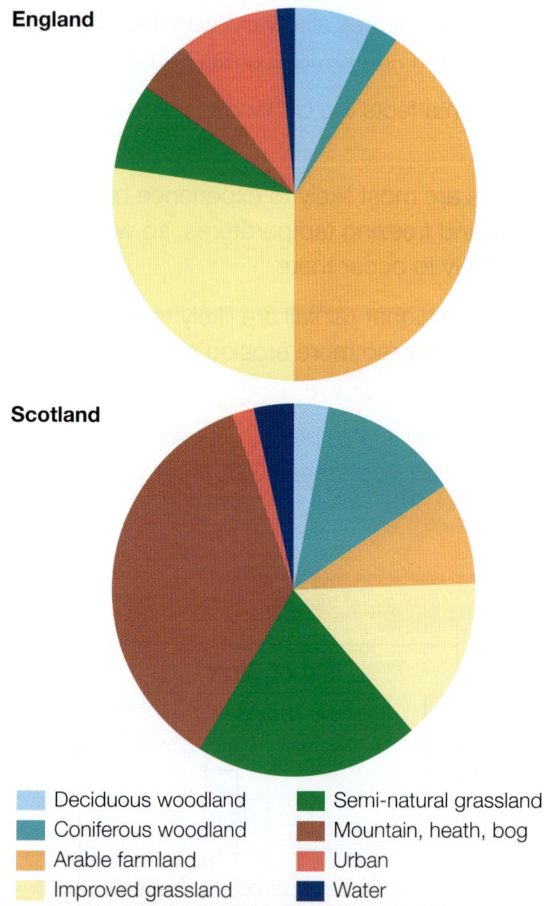

A *Land use in England and Scotland*

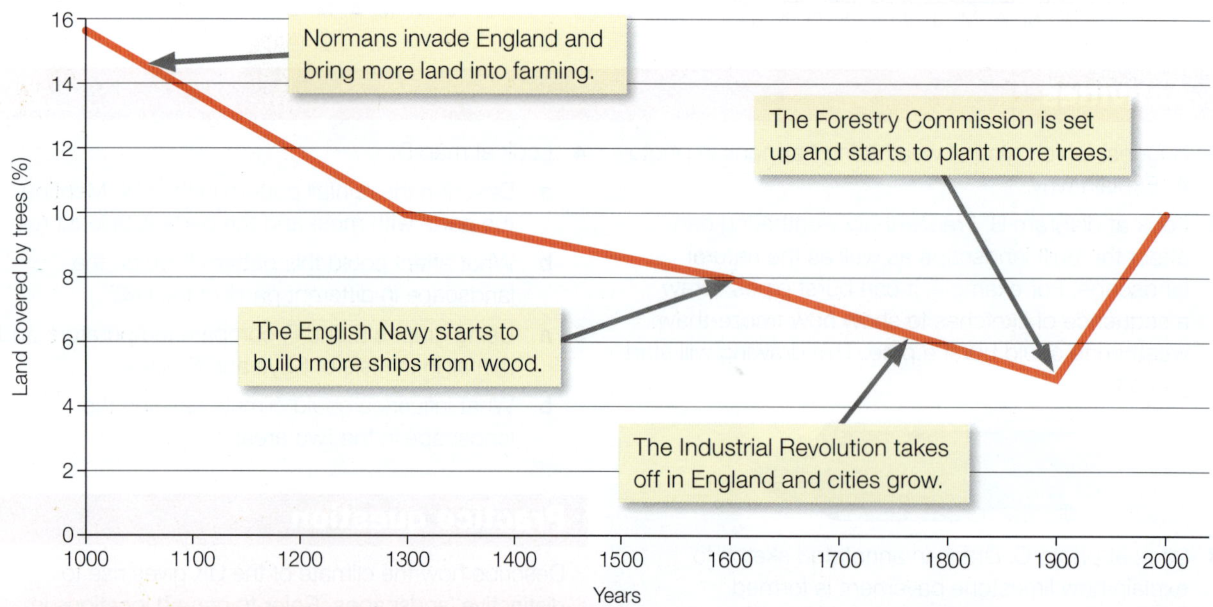

B *Changing woodland cover in England since 1066*

C *Human activities in an upland area of the UK that was once glaciated*

Activities

1 Look at graph B.
- **a** Describe the changes in woodland area in England since 1066.
- **b** Explain the changes from the events labelled on the graph.

2 Look at the pie charts in A.
- **a** Compare land uses in Scotland and England.
- **b** Explain the different land uses. (Clue: look again at map C on spread 3.2, showing the upland and lowland areas of the UK.)

3 Look at photo C. For each human activity, carry out a simple environmental impact assessment. Give a score from 0 to 4 for each impact, where 0 is no impact and 4 is a major impact.

	0	1	2	3	4
Tourism					
Visual impact					
Noise impact					
Pollution impact					
Traffic impact					
Wildlife impact (loss of natural habitats)					
Human impact (number of people)					
Wind farms					
Visual impact…					

4 Which human activities do you think are more suitable in upland areas, and which are not? Use the results of your environmental impact assessment to help you to decide. Justify your decision.

Go further

Carry out an environmental impact assessment for at least four human activities in your local area. Which activities have the greatest environmental impact?

Your enquiry

What makes a landscape distinctive?

Think of a landscape you know well or choose the landscape in one of the photos in this section.

- Draw a sketch or take a photo of the landscape. Or you could do a painting like Constable's!
- Annotate your sketch to describe the distinctive characteristics of the landscape.
- Try to explain how geology, climate and human activity influence the landscape.

Shaping landscapes

On this spread you will investigate the main agents and processes of landscape change in the UK.

Agents of change

The UK landscape is constantly changing. Some of these changes happen so gradually you won't notice them. For example, it might take thousands of years for a river to carve a valley. Other changes happen as you watch – slowly, like the sand moving up and down a beach, or suddenly, like a cliff collapsing.

There are three main agents of change forming our landscape in the UK:

- **Rivers:** They flow by gravity over the land from their source to the sea. There are rivers all around the UK, but especially in the north and west where there is more rain (photo A).
- **Glaciers:** They once covered much of the country (see spread 3.2) like giant rivers of ice. Although there are no glaciers here now, they left their mark on the UK landscape (photo B). Glaciers are still at work in other countries today.
- **The sea:** It surrounds the whole of the UK, making us an island (or group of islands to be precise). Over the centuries the sea has changed the coastline of the UK (photo C).

Geomorphic processes

Similar geomorphic processes are at work in rivers, glaciers and at the coast, shaping the landscape. However, these processes can only still be seen at work along rivers and coasts in the UK.

- **Erosion** wears away the land and removes broken fragments of rock.
- **Transport** carries away the eroded material.
- **Deposition** leaves transported material behind, building up new land.

Rivers and the sea erode, transport and deposit by means of water. Rivers get their power from flowing water. The faster the water flows, the more power it has. The sea gets its power from waves, formed by wind blowing over the surface.

Rivers and the sea both erode material by hydraulic action, abrasion, attrition and solution (diagram E). They transport material by traction, saltation, suspension and solution (diagram D).

A *The River Barle valley on Exmoor in Somerset*

B *Lingmell Beck flows through a valley formed by a glacier in the Lake District*

C *The coast of Cornwall*

D *How a river transports material (the sea transports material in a similar way)*

E *How waves erode (rivers erode in a similar way)*

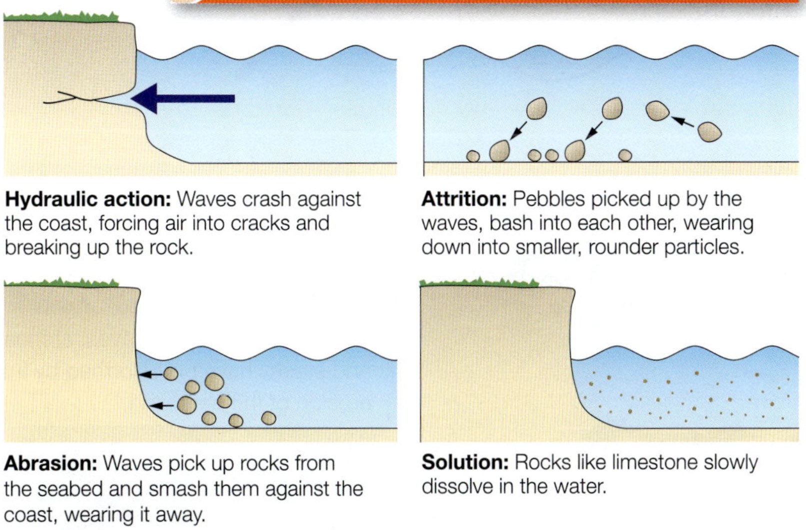

Mass movement

Together, erosion and weathering (see spread 3.4) can lead to **mass movement** of rocks. For example, **landslides**, or **land slumps**, can happen on the steep slopes of valley sides and coastal cliffs (diagram F). Weathering and erosion speed up the process.

F *Landslide on a slope*

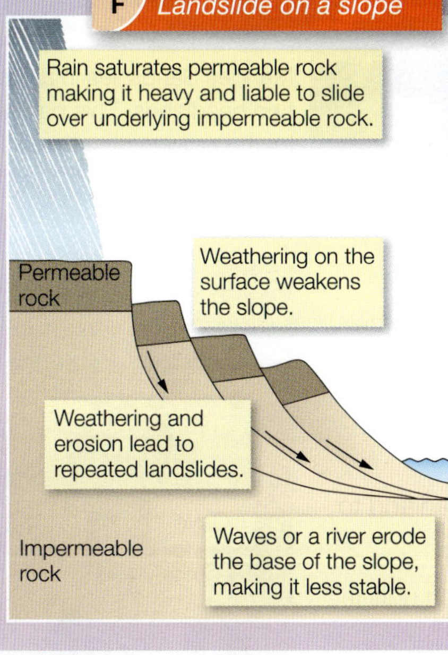

Activities

1 Look at photos A, B and C. For each photo:
- **a** Identify the main agent shaping the landscape.
- **b** Describe the landscape it has formed.

2 Think carefully about the photos.
- **a** In what ways is photo B the odd one out?
- **b** In what ways is photo C the odd one out?

3 Look at diagram D. Explain how the way a river transports material would be affected by:
- **a** the speed at which water is flowing in the river
- **b** the type of rock over which the river flows.

4 Look at diagram E. Draw similar diagrams of a river channel to show how a river erodes.

5 Look at diagram F. Explain how each of these factors contributes to a landslide:
- **a** rainfall **b** weathering
- **c** erosion.

Your enquiry

What influences the landscapes of the UK?

As you work through this section, make notes about anything that will help you to answer the question. You will bring your ideas together to do an activity at the end of spread 3.17.

Coastal landforms

On this spread you will explore the coast to find out how erosion, transport and deposition are changing the coastline.

Coastal erosion

The UK's coastline is continually being shaped by the power of waves. The resulting landforms depend mainly on the type of rock, or geology, along the coast (diagram A). **Headlands** form where hard, more resistant rocks are slow to erode. Between the headlands, **bays** are formed where soft, less resistant rocks easily erode. There are headlands and bays around the Marloes Peninsula in Wales (photo B).

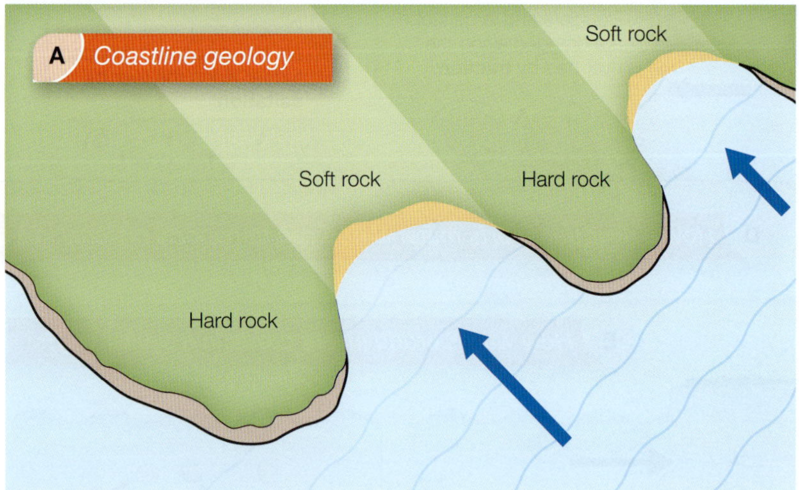

A *Coastline geology*

B *Marloes Peninsula in Pembrokeshire, Wales*

Headland erosion

Most coastal erosion happens when waves are at their most powerful. Even headlands can be slowly eroded.

- Waves attack the base of the cliff, forming a notch that is deepened into a **cave**.
- The cave wears all the way through the headland to form an **arch**.
- Weathering eats away at the roof of the arch until it collapses, leaving a **stack**.

C *Coastal landforms in Wales*

Transport and deposition

A **beach** is made of material, or sediment, deposited by the waves. The power of the waves also moves the material and changes the shape of the beach. Waves transport material up the beach in the **swash** and back down again in the **backwash**.

Waves approach the beach at an angle because of the direction of the prevailing wind. So material is transported along the coast by the process of **longshore drift** (diagram D). This process transports material northwards along the Welsh coast. At the River Dyfi estuary, material is deposited to form a **spit** across the mouth of the river (photo E).

D *Longshore drift*

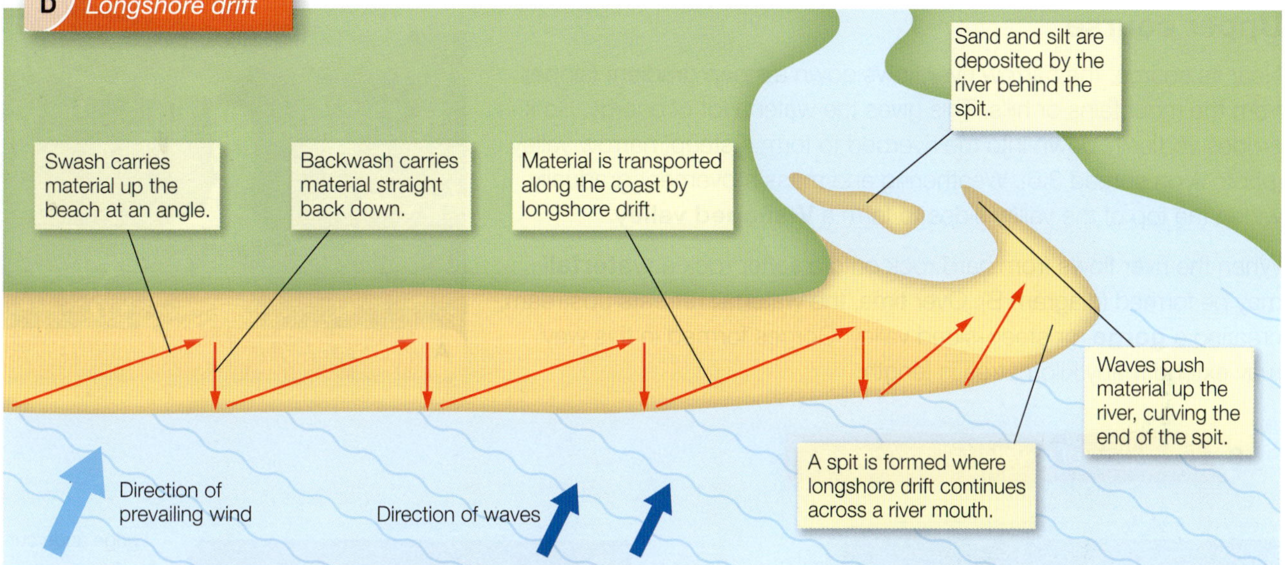

Activities

1 Look at diagram A. In your own words, explain why headlands and bays are formed.

2 Look at photo B. Draw a sketch of Marloes Peninsula. Mark and label the features on your sketch.

3 Look at photo C.

   a Draw a labelled sketch to show the landforms in the photo.

   b Draw two more sketches: one to show what the headland might have looked like in the past and the other to show what it could look like in the future. Annotate your sketches to explain the changes.

4 Look at photo E. Draw an annotated sketch of it, writing the following labels at the correct points:

- The coast to the south is eroded by waves.
- Eroded material is transported northwards by longshore drift.
- Material is deposited across the mouth of the river to form a spit.
- The river deposits sand and silt behind the spit.

5 Look at diagram D. Add arrows to your sketch for activity 4 to show the direction of:

- the prevailing wind
- longshore drift
- the river flowing to the sea.

E *The Dyfi Estuary on the Welsh coast, looking south*

Practice question

Outline the role of erosion in shaping a headland. Refer to a named example you have studied. (4 marks)

3.8 River landforms

On this spread you will travel down a river, from source to mouth, to learn about the landforms.

Rivers also shape the landscape. As it flows from its source to its mouth, a river shapes the valley in which it flows. The river itself also changes. A typical river goes through three stages, although not every river is the same (read about the River Thames on spreads 3.14–3.17).

Upper course

Near its source, the narrow river flows down a steep gradient (slope) from the mountains or hills. This gives the water a lot of energy. So it erodes vertically down into the riverbed to form a steep, narrow valley (photo A on spread 3.6). Weathering and mass movement gradually widen the top of the valley sides to form a **V-shaped valley**.

When the river flows from hard rock onto a softer rock a **waterfall** may be formed (diagram B). Over time, the waterfall retreats upstream, creating a **gorge**, or steep-sided valley. Gorges formed in this way may extend many kilometres in length.

A High Force, a waterfall on the River Tees

B Formation of a waterfall and gorge

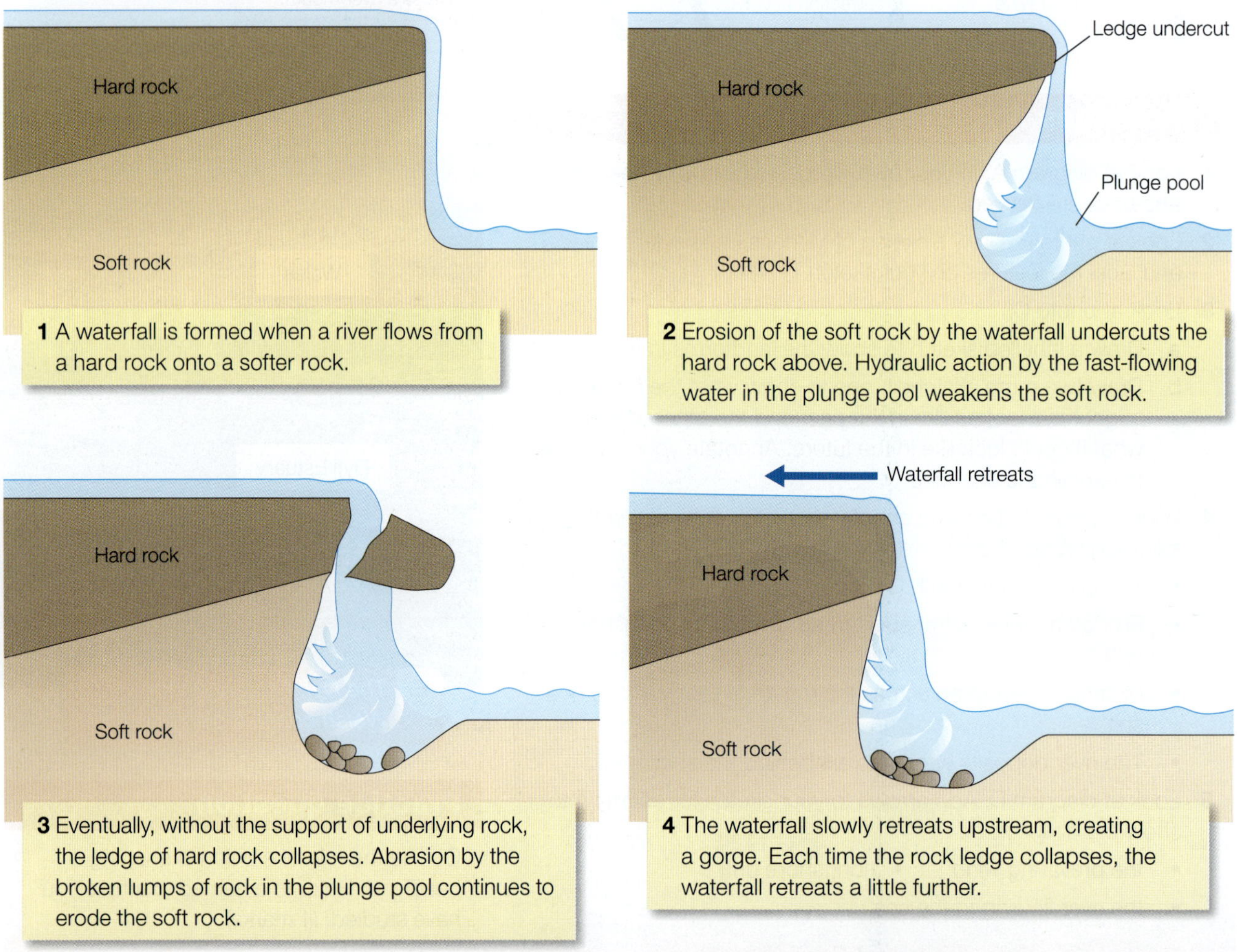

Middle course

Further from the source, the gradient gets gentler. The river widens as other tributaries flow into it and the valley gets wider too. The river begins to erode laterally (sideways) to form **meanders**, or bends (photo C). Continued erosion narrows the neck of the meander until the river breaks through, cutting off the meander to form an **ox-bow lake** (diagram D).

C *Wide meanders in the River Severn*

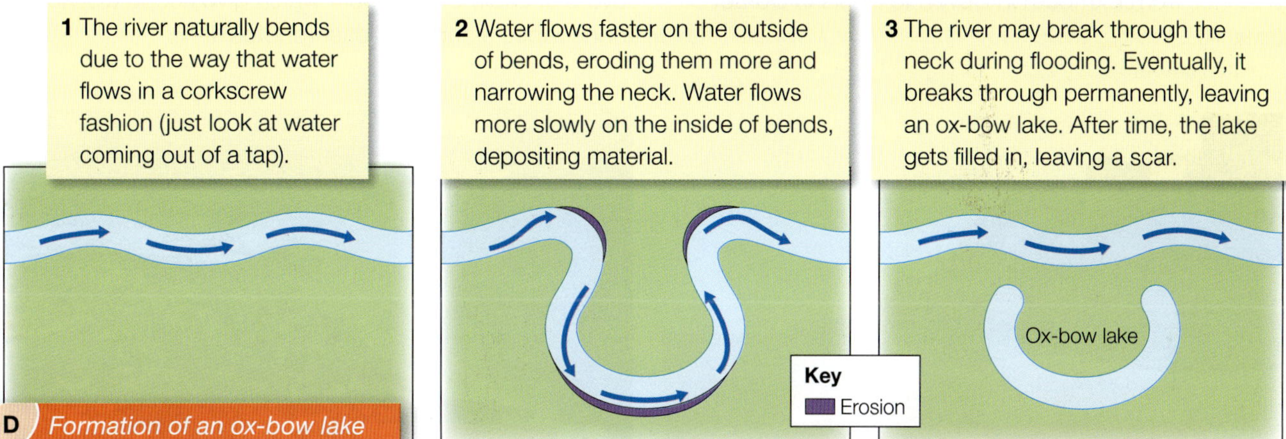

D *Formation of an ox-bow lake*

Lower course

As the river nears its mouth, it widens further and its valley becomes flatter and wider. The river may flood, depositing fine silt, or alluvium, on the valley floor to form a **floodplain**. Material deposited close to the river raises the height of the river banks to form **levees** (diagram E).

E *Formation of a floodplain and levee*

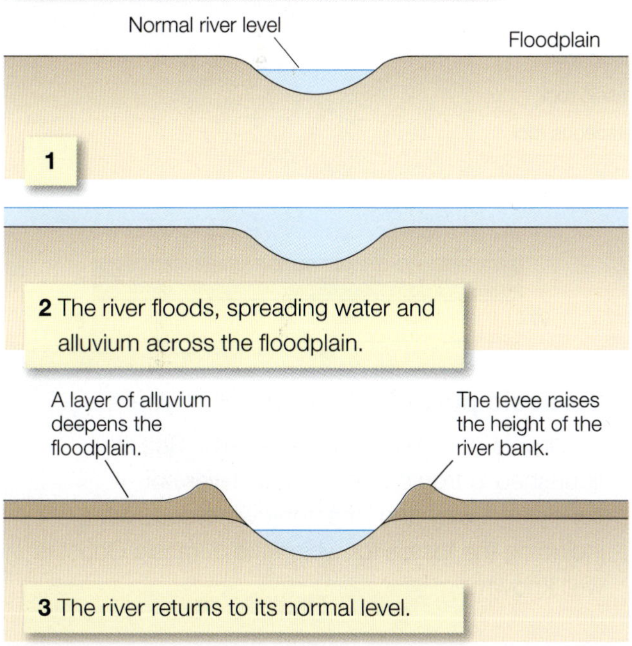

Activities

1 Look at photo A and diagram B. Draw a labelled sketch of the waterfall in the photo. Label:
- the layer of hard rock
- the layer of soft rock
- the plunge pool.

2 Think how the position of the waterfall will change over thousands of years. Draw another sketch to show how its position will change or describe what you expect to happen.

3 Look at photo C and diagram D. Draw an annotated sketch of the photo to explain how the meander could develop into an ox-bow lake.

4 Complete a large table like the one below to compare the three stages of a typical river. Describe the main features of the river and valley at each stage. List the landforms you would find.

	River	Valley	Landforms
Upper course			
Middle course			
Lower course			

Practice question

Describe the formation of one river landform you have studied. You can use diagrams to help you.
(4 marks)

The Jurassic Coast

On this spread you will travel along the Jurassic Coast in Dorset and Devon to identify landforms and the geology.

Case study

Jurassic Coast

The coast of Dorset and East Devon is known as the Jurassic Coast. It is one of the most beautiful stretches of coastline in the UK and is listed as a UNESCO World Heritage Site. The site includes 155 km of cliffs and beaches from Exmouth in Devon to Poole in Dorset (map A).

The name 'Jurassic' comes from the age of the rocks, dating from the Triassic, Jurassic and Cretaceous periods, spanning 185 million years (see map A on spread 3.3). The coastline is made up of alternating layers of sedimentary rock, which give rise to distinctive landscapes along the coast (photos 1–4).

1 *Cliffs near Sidmouth*

2 *Chesil Beach*

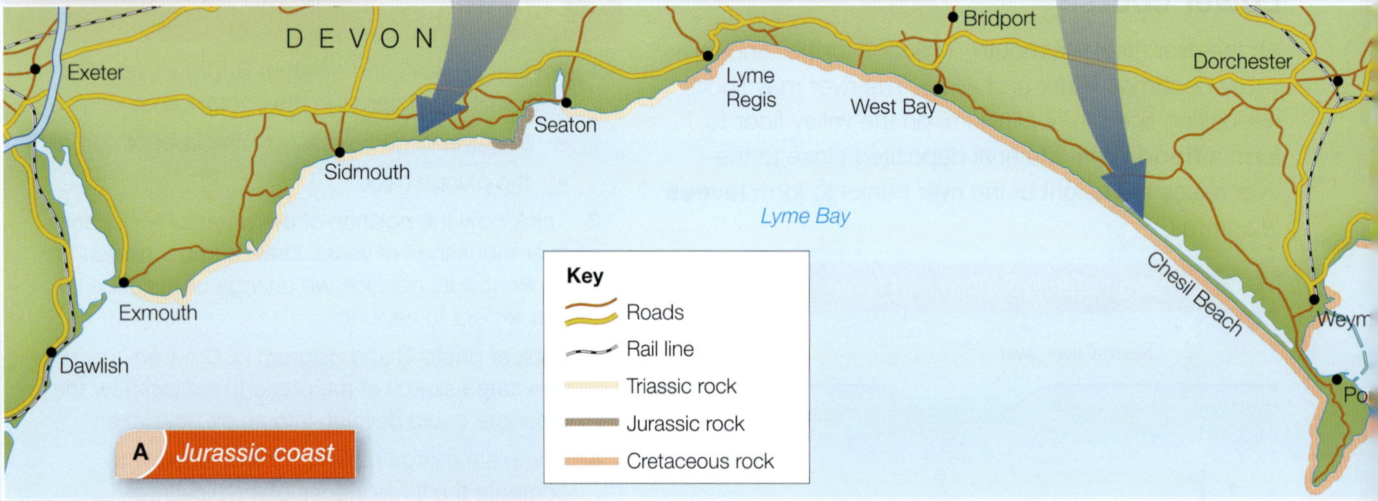

A *Jurassic coast*

B *Formation of the Jurassic Coast (i)*

Triassic period (250–200 million years ago)

Rocks were formed in desert conditions. Rivers flowing through baking desert deposited thick layers of sediment. Rocks laid down in this period include pebble beds, sandstone and mudstone.

Jurassic period (200–140 million years ago)

Sea levels rose, flooding the deserts. Sea life flourished in tropical seas. Thick layers of sandstone, clay and limestone were deposited, including the fossils for which the Jurassic Coast is well known.

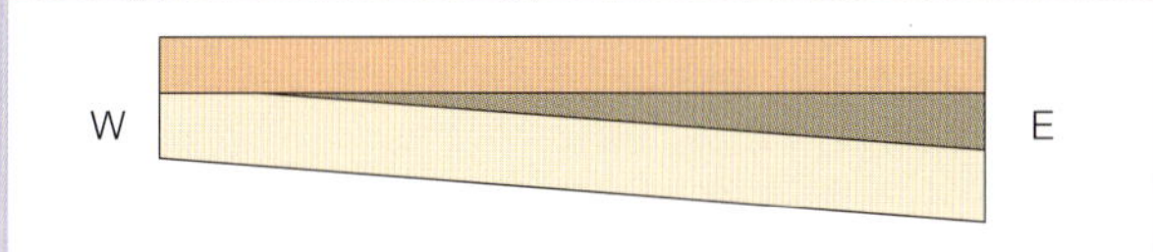

Cretaceous period (140–65 million years ago)

The sea level dropped. New rocks were formed in swamps and lagoons. Then Earth movements tilted the rocks and the layers were eroded. Later seas covered the land again, depositing more clay, sandstone and chalk.

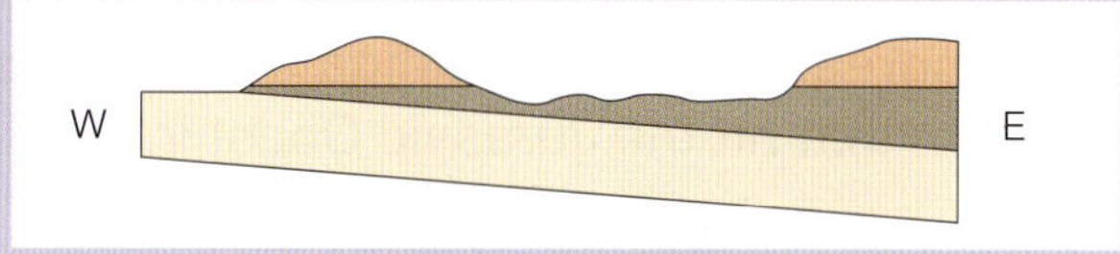

Quaternary period (2.6 million years ago to the present)

Erosion created hills and valleys in the landscape. A series of ice ages and changes in sea level took place. When the last ice age ended 10 000 years ago, sea levels rose to create the modern coastline.

C *Formation of the Jurassic Coast (ii)*

3 *Lulworth Cove*

4 *Old Harry Rocks*

Activities

1. Look at photos 1–4.
   - **a** Identify the coastal landforms in the photos. Match the landforms listed below with one or more photos:
     - cliff
     - bay
     - beach
     - headland
     - stack
     - lagoon.
   - **b** What geomorphic processes might be at work in each photo to create the landscapes you see?
2. Describe the landscape in each of photos 1–4. Write a sentence about each landscape. Mention the landforms you identified in activity 1.
3. Look at map A and diagrams B and C.
   - **a** Travelling along the coast, where do you find the oldest and youngest rocks?
   - **b** Explain why the age of the rocks changes as you travel from east to west.
4. Look at photo 3. Explain how Lulworth Cove was formed, using this simple geological map. (Clue: limestone is a hard rock, clay is soft and chalk is in between.)

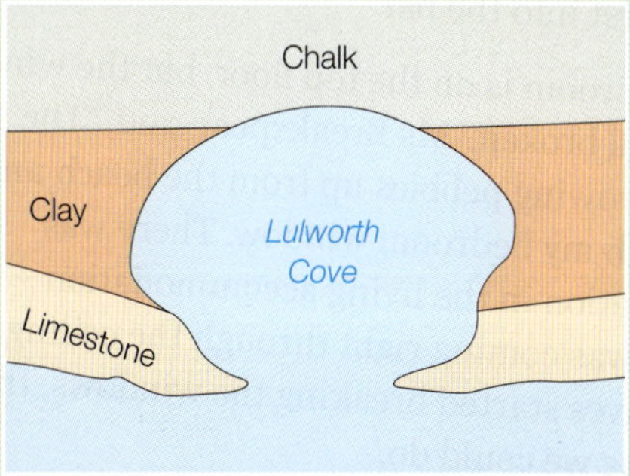

Practice question

For a named coastal landscape in the UK, explain how the landscape was formed. (4 marks)

Battered coastline

On this spread you will investigate the impacts of the storm that hit the Jurassic Coast in 2014.

Case study

Winter storm, 2014

On 5 February 2014, a huge storm battered the Devon and Dorset coast. One of the most dramatic impacts was the damage caused to the main railway line from Exeter to Plymouth that runs along the sea wall at Dawlish (photo A).

The track was left hanging in mid-air and the line had to be closed for two months for emergency repairs. It was not the first time that the railway had been damaged and it is unlikely to be the last. Dawlish is just one of many places on the Jurassic Coast that can easily be eroded, particularly in storm conditions.

Along the Jurassic Coast, other places were affected by the storm:

- Portland Bill, at the eastern end of Chesil Beach, was cut off by floodwater and buildings were damaged (extract B).
- West Bay was affected by rock falls from the cliffs, which meant the coast retreated by a few more metres (photo C).
- Lyme Regis, popular with tourists and fossil hunters, was pounded by waves but has coastal defences to protect the town.

A Damage to the railway line at Dawlish in 2014

The sea threatens 'business as usual'

Jackie Breakspear is the landlady of Cove House Inn at Portland, Dorset. She spent Wednesday morning trapped in her pub – incredibly, waves nearly 20 metres high were crashing over the top of the three-storey building. They smashed windows and burst into the bar.

'My bedroom is on the top floor, but the window was still broken,' Ms Breakspear said. 'The sea was throwing pebbles up from the beach and through my bedroom window. There was devastation in the living accommodation – the water was coming right through the ceilings. Once the waves started breaking the windows, there was nothing we could do.'

Cove House Inn is built of Portland Limestone. Although this type of stone is stronger than most building materials, pebbles hurled by the sea have left dents in it.

The pub has now reopened for business, but every window is boarded up against the next big storm. But on Britain's shrinking shoreline, how long can things really remain 'business as usual'?

B Adapted from newspaper reports, 8 February 20

Cliffs collapse

The yellow sandstone cliffs at West Bay are part of the UK's eroding coastline. Made of alternating layers of soft sandstone and harder, sandy limestone, the cliffs can easily be eroded. Although protected by a pebble beach, waves often attack the base of the cliffs, leading to rock falls. This happened again in February 2014.

The impact of climate change

The series of storms that hit the UK coastline in the winter of 2013/14 could have been a result of our changing climate (see Chapter 2).

- Warming global temperatures lead to higher sea levels as ice sheets melt and water in the sea expands. Sea level around the UK is predicted to rise by up to 600 mm by 2100.
- A warmer atmosphere makes intense storms, similar to the one that hit the south coast in 2014, more likely. They create larger, more powerful waves.
- More intense rainfall leads to greater weathering of cliffs and mass movement of rock, causing more rock falls and landslides.

Climate change is likely to have an impact around the whole UK coast (map D). The impact will be greatest on coasts with cliffs made of softer rock, which are easily eroded, and on low-lying coasts that could flood.

C *Rock falls at West Bay*

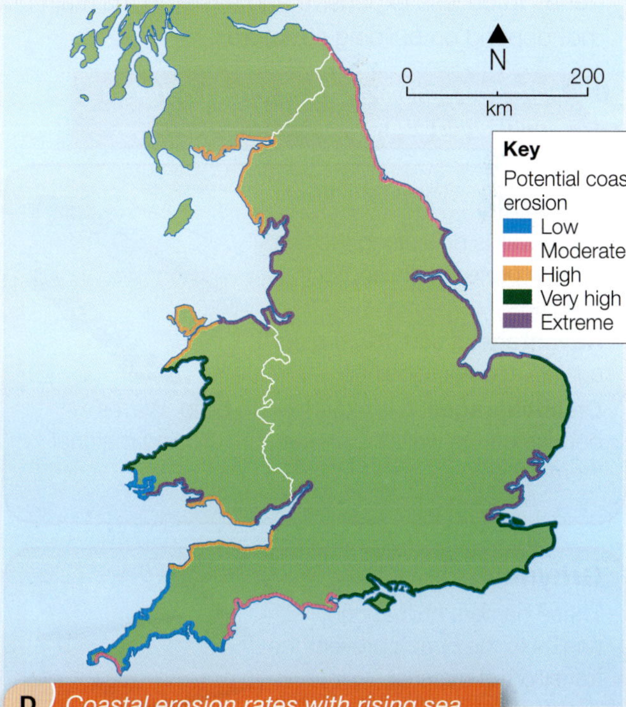

D *Coastal erosion rates with rising sea levels for England and Wales*

Activities

1 Look at photo A.
   a Locate Dawlish and the railway line on map C on spread 3.9. Describe the railway route.
   b Suggest what the problem with this route is.
2 Read extract B.
   a Locate Portland on map A on spread 3.9. Describe its location.
   b Explain why Portland can easily be affected by storms. (Clue: prevailing wind is from the south-west.)
   c If you were Ms Breakspear, would you re-open the pub or move away? Give your reasons.
3 Look at photo C. Draw a labelled sketch to explain the rock falls at West Bay. Include the following labels: cliff, soft sandstone layer, harder limestone layer, beach, rock debris.
4 Look at map D and an atlas.
   a What is the risk of coastal erosion on the Jurassic Coast?
   b Name five areas of England and Wales that face extreme coastal erosion.

Go further

Find out about any storm events that have happened since 2014 and what impact they had on the UK coast.

- When did the event occur?
- What stretches of coastline were affected?
- What damage was caused?

Practice question

For a named coastal landscape, assess the influence of weather and climate on the landscape. (6 marks)

3.11 Coastline management

On this spread you will explore ways in which the coast can be protected and devise a shoreline management plan.

Case study

Management strategies

Coastlines like the Jurassic Coast need to be managed to balance natural processes, like erosion, against the need to protect people, property and infrastructure (photo A). A combination of two types of strategy can be used (panel B):

- **Hard engineering** works against natural processes, such as erosion, to protect the coastline. It tends to be more expensive and structures continually need to be maintained or rebuilt.
- **Soft engineering** works with nature to protect the coastline. It tends to be less expensive and easier to maintain because it does not depend on building structures.

A *The sea wall and rock armour at Lyme Regis, examples of hard engineering*

B *Hard (red panels) and soft (green panels) engineering strategies*

Sea wall

A sea wall is built along the sea front to protect cliffs and/or buildings.

Advantage: It can also protect against coastal flooding.

Disadvantages: It is expensive to build. Waves bounce off the wall onto the beach, eroding material there.

Offshore reef

Concrete blocks and boulders are positioned offshore, forcing waves to break before they reach the coast.

Advantage: The beach keeps its natural appearance.

Disadvantages: The reef is difficult to maintain and does not stop erosion completely.

Groynes

These barriers built at right angles to the beach prevent the transport of material.

Advantage: This allows the beach to build up as a natural defence against erosion.

Disadvantages: Groynes can be unattractive. Preventing longshore drift can also starve other beaches along the coastline of material.

Beach nourishment

This makes the beach higher or wider by importing more sand.

Advantages: It is cheap and keeps the natural appearance of the beach.

Disadvantage: It requires constant maintenance to replace beach material that is washed away.

Rock armour

Large boulders are placed along the coastline to absorb wave energy.

Advantages: Rocks are relatively cheap and efficient.

Disadvantages: It can be unattractive and makes access to the beach dangerous.

Managed retreat

When coastal land is considered to be of low value, it is allowed to erode or flood naturally.

Advantages: This is cheap and encourages natural development of beach or salt marsh.

Disadvantage: People who lose buildings or land may need to be paid for their losses.

Shoreline management plans

Local councils in England and Wales are required to produce a **shoreline management plan (SMP)** for their stretch of coastline. The main options for an SMP are:

- do nothing
- hold the line
- retreat the line
- advance the line (table C).

The decision about which option to use varies from place to place along the coast. Councils need to consider:

- the number of people threatened by coastal erosion or flooding, and what their properties are worth
- the cost of replacing infrastructure such as roads or railways
- historic features and natural landforms, and their economic value as tourist attractions.

C *Options for coastline management*

Management option	Description
Do nothing	Allow natural processes to happen. This is most likely when land or property has a lower value than the cost of sea defences.
Hold the line	Use hard or soft engineering to maintain sea defences. This expensive option is worth it for a town where the value of property is higher than the cost of sea defences.
Retreat the line	Create a gap in the existing sea defences to allow land to flood naturally. This cheaper option allows features like sand dunes or salt marshes to develop and provides a natural barrier to further erosion.
Advance the line	Build new coastal defences further out to sea, creating new land. This most expensive option involves hard engineering. It is essential for projects like a new estuary airport.

D *Cliffs near Sidmouth*

Activities

1 Look at photo A.

  a Why has a sea wall been built at Lyme Regis? (Clue: look back at spread 3.10.)

  b Is a sea wall a good way to protect the coast here? Give reasons for your answer.

2 Look at panel B. Compare the advantages and disadvantages of hard and soft engineering strategies in a table like the one below.

	Advantages	Disadvantages
Hard engineering		
Soft engineering		

3 a Find the following places along the Jurassic Coast on map A on spread 3.9: Chesil Beach, Lulworth Cove, Dawlish, West Bay.

  b Look at table C. From what you have learned about each place, suggest what the best coastal management option would be. Give a reason for each of your choices.

4 Look at photo D. You have to devise a shoreline management plan for the coast here.

  a First consider these questions:
  - What needs to be protected? Why?
  - What would the best strategy be to protect the coast?
  - Would protection be worth the cost, do you think?
  - What is the best shoreline management option?

  b Write a short report and draw a sketch map to outline your plan, answering the questions above.

Practice question

Discuss the factors that need to be considered in managing a coastal landscape. (8 marks)

Landscape fieldwork – 1

On this spread you will plan a coastal fieldwork investigation using suitable fieldwork activities.

A day at the seaside!

There are natural landscapes to investigate everywhere in the UK, even around cities. One of the most popular locations for fieldwork is a coastal landscape. No one in the UK is more than a two-hour drive from the coast and most of us are much closer than that.

There are good reasons for choosing a coastal location for fieldwork:

- It is often possible to see geomorphic processes at work or, at least, see recent evidence that they have happened.
- There are opportunities to use a variety of fieldwork activities, such as measuring, counting, sketching and surveys.
- The coast is a fun place to be – just remember to stay safe! (photo A)

Planning a coastal fieldwork investigation

1 Decide on the topic you are going to investigate. For example, you could investigate longshore drift.

2 Think of a question you could investigate or a **hypothesis** you could test. For example, you could investigate the question, *'In which direction is longshore drift happening?'* or test the hypothesis, *'Longshore drift on this coast is from west to east.'*

3 Choose suitable fieldwork activities that will help you to investigate your question or test your hypothesis (panel B).

A *Doing fieldwork on the coast*

B *Coastal fieldwork activities*

Beach profile survey: Measure the gradient at different points along the beach. This can influence whether waves erode or deposit material.

Beach material survey: Sample the materials along the beach to find how shapes and sizes change.

Wave survey: Measure the movement of pebbles or other objects along a beach as a result of longshore drift.

Coastal management survey: Record the management strategies used along the coast on a map and take photos or draw sketches.

Cost/benefit analysis: Score the costs and benefits of different management strategies and add them to obtain an overall cost/benefit score.

C *Fieldwork equipment for measuring longshore drift. How could you use it?*

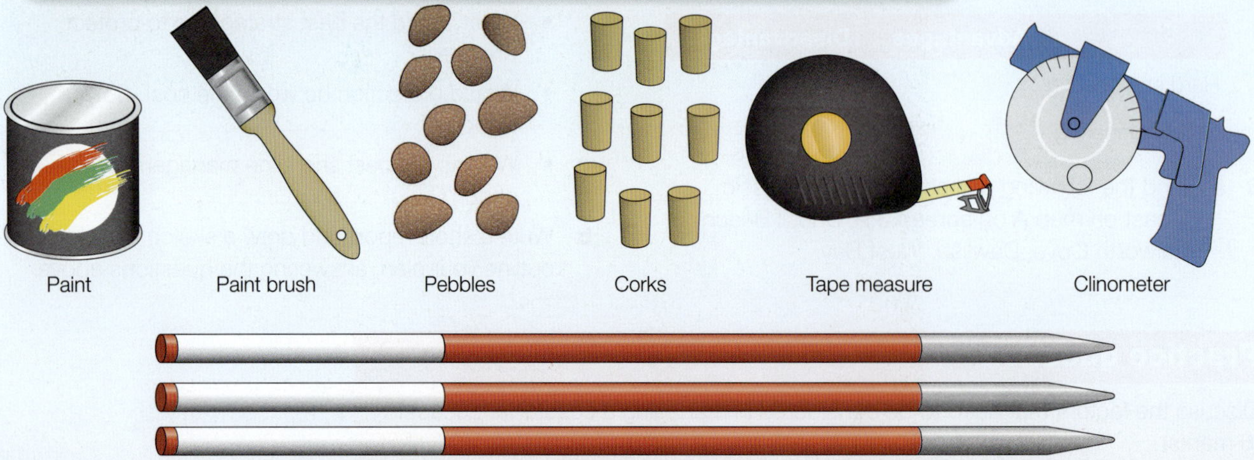

Very angular	Angular	Sub-angular	Sub-rounded	Rounded	Very rounded
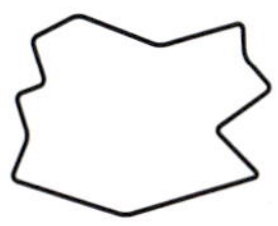	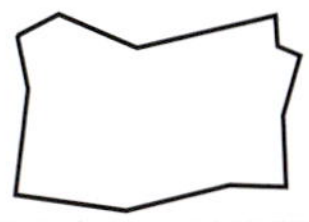	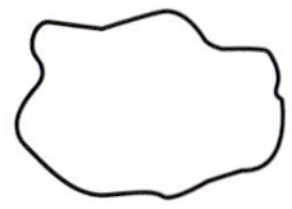		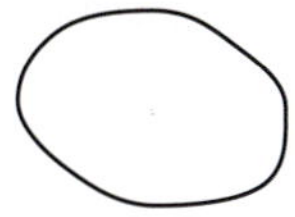	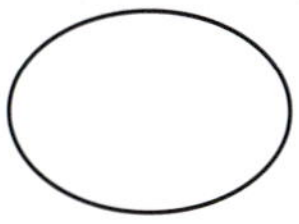
Every edge and point is sharp. There are no smooth surfaces.	One or two points and edges are smooth but most are sharp.	The shape is irregular with a few sharp points and edges.	The shape is irregular but there are no sharp points or edges.	The shape is almost smooth but with a few irregular edges.	The shape is perfectly smooth.

D *Beach material survey*

Safety first!

The most important thing to remember when doing any fieldwork is to stay safe. Coasts can be dangerous, so you need to take particular care when doing fieldwork there. Here are some precautions you should take:

- Know the times of high tide and low tide when you plan a walk along the coast.
- Don't walk too close to the edges of cliff tops.
- Don't go too close to the base of a cliff, even if you are wearing a hard hat.
- Take care on rocks, especially at low tide.
- Don't go close to breaking waves, especially on sea walls or steeply sloping beaches.

Cost	Score (0–4)	Benefit	Score (0–4)
Construction of coastal defence		Protection of buildings and land	
Maintenance of coastal defence		Protection of infrastructure	
Negative impact on landscape		Protection of tourist amenities	
Loss of tourist income		Protection of valuable landscape	
Impact further along the coast		Reduction of flood risk	
Compensation to landowners		Reduction of deaths or injuries	
Total		**Total**	

E *Coastal cost/benefit analysis*

Activities

1. Think of a question or hypothesis about coastal landscapes that you could investigate or test through fieldwork.
2. Look at panel B. Pick three fieldwork activities you could use to investigate your question or test your hypothesis. You could think of other activities too.
3. Look at the fieldwork equipment in drawing C. Suggest how you could use any, or all, of the equipment for measuring longshore drift.
4. Look at chart D.
   - **a** What happens to pebbles as they are carried along the beach by waves? (Clue: think about the types of coastal erosion in diagram E on spread 3.6.)
   - **b** Explain how the shape of the pebbles at different points along the beach could provide evidence of longshore drift.
5. Look at table E. Apply the cost/benefit analysis to a coastal landscape. You could use it to help you to decide if a sea wall should be built in front of the cliffs near Sidmouth (photo D on spread 3.11). Give a score of 0–4 for each cost and benefit. What would be the total cost and benefit of a sea wall?

Practice question

Devise a risk assessment for doing coastal fieldwork.

- **a** Identify two hazards that you might find at the coast. (2 marks)
- **b** Suggest what precautions you might take to reduce the hazard. (2 marks)
- **c** Explain how these precautions might reduce the hazard. (2 marks)

Landscape fieldwork – 2

On this spread you will plan a river fieldwork investigation with the help an Ordnance Survey map.

You might get wet!

Another popular location for fieldwork is rivers. The advantage of rivers over coasts is that they are usually closer. There are rivers in towns and cities (see spreads 3.14–3.17 on the River Thames), or in the surrounding area. In common with coasts, there are good reasons for doing fieldwork on rivers:

- You can see geomorphic processes at work.
- There are opportunities for a variety of fieldwork activities.
- Rivers are fun – so long as you stay safe! (photo A)

Planning a river fieldwork enquiry

1. Decide on the topic you are going to investigate. For example, you could investigate changes along a river.
2. Think of a question you could investigate or a hypothesis you could test. For example, you could investigate the question, *'How does a river change along its course?'* or test the hypothesis, *'The river becomes deeper and faster along its course.'*
3. Choose suitable fieldwork activities that will help you to investigate your question or test your hypothesis (panel B).

A *Students doing fieldwork on a river*

B *River fieldwork activities*

Channel survey: Measure the width of the river channel and its depth at different points across and/or along the channel.

Flow survey: Measure the velocity of the water flowing at different points across and/or along the river.

River load survey: Record the shapes and sizes of the river's load at different points across and/or along the river.

Valley slope survey: Measure the gradient of the river valley sides at different points along the river.

Water quality survey: Assess water quality at different points on the river by counting the types and numbers of plants and animals in the water.

C *Fieldwork equipment you can use on a river*

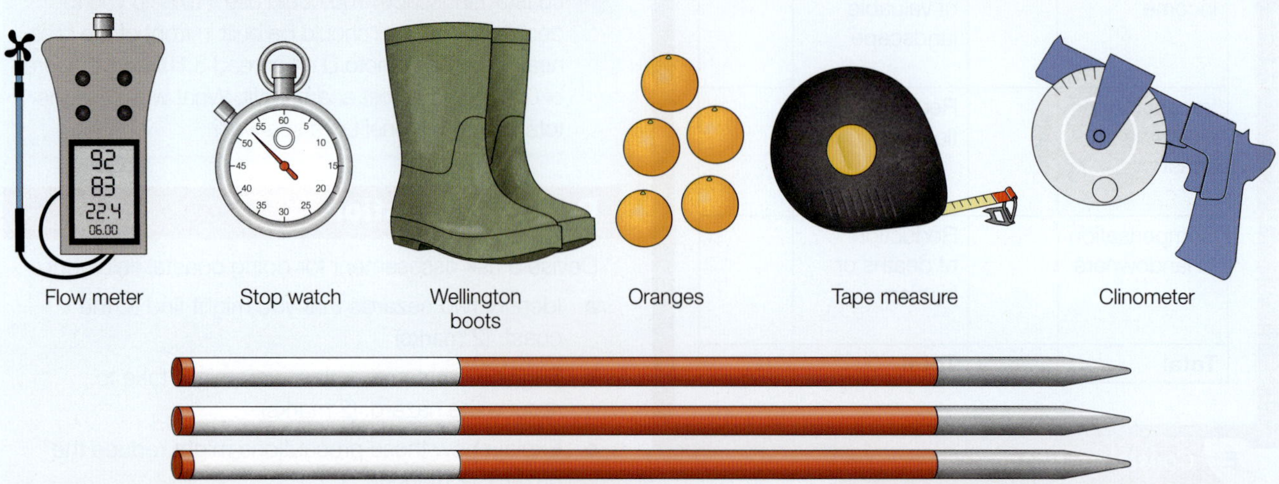

Distance across river (metres)	0	0.5	1.0	1.5	2.0	2.5	3.0	3.5	4.0
Depth of channel (mm)	0	120	260	450	600	830	920	740	0

D *Channel survey*

Choosing your river

One drawback with river fieldwork is that not all rivers are suitable. Some important factors to consider before choosing a river for fieldwork are:

- **Size of the river:** Some rivers are too wide or deep for fieldwork. Remember you will probably need to get into the river if you are taking measurements.
- **River flow:** Never get in a river when it is flowing fast – it is too dangerous even if it is quite shallow! Other rivers dry up in summer, so they don't have enough flow.
- **Land ownership:** You need to check there is public access to the river you want to investigate.

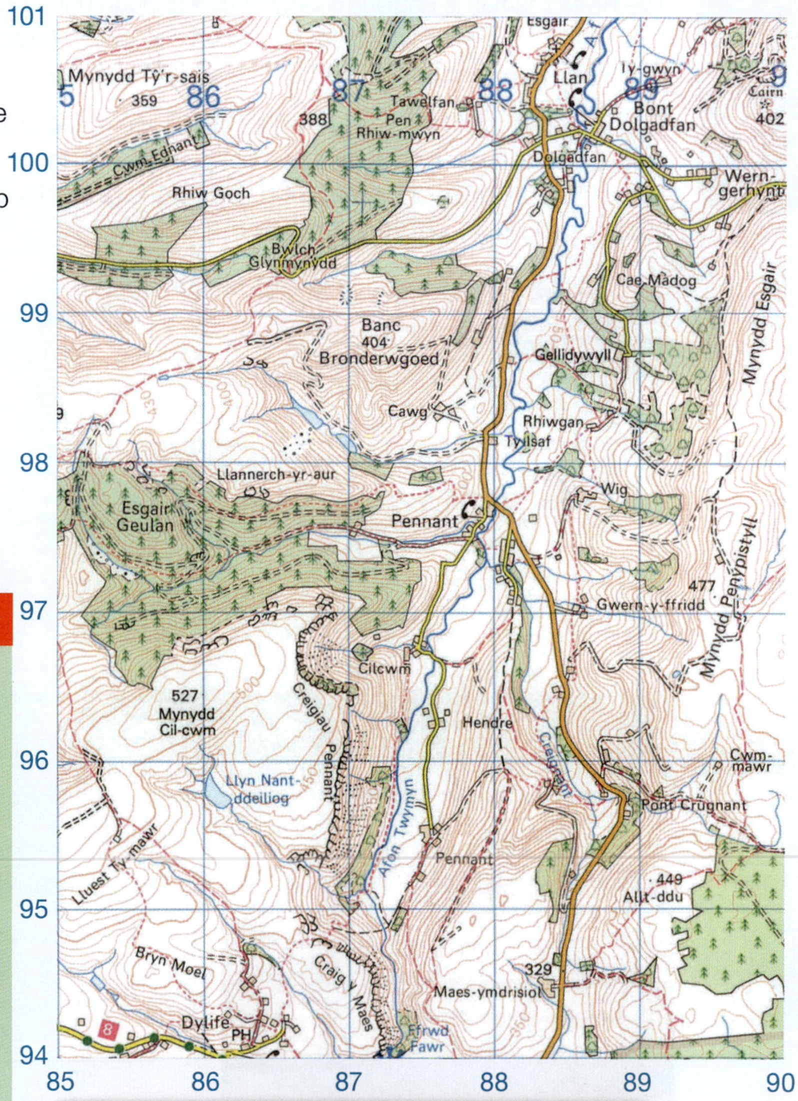

E *OS map showing Afon Twymyn, a small river in Wales; OS map, scale 1:50 000*

Activities

1. Think of a question or hypothesis about rivers that you could investigate through fieldwork.
2. Look at panel B. Select three fieldwork activities you could use to investigate your question or test your hypothesis. You could think of other activities too.
3. Look at the fieldwork equipment in drawing C. Suggest how you could use some, or all, of the equipment to investigate how a river changes along its course.
4. Look at chart D.

   Draw a cross-section to show how the channel depth varies across a river. Plot your points on a simple grid like the one below.
5. Plan a river fieldwork investigation on the river Afon Twymyn in map E.
   - **a** Would it be a suitable river to investigate your question or test your hypothesis in activity 1? If not, think of another question you could investigate.
   - **b** Choose two or more locations to carry out your fieldwork. Explain why you chose them.
   - **c** Give grid references for your locations.
   - **d** Predict the answer to the question you want to investigate or whether your hypothesis will be correct. Give reasons for your prediction.
   - **e** Suggest how an OS map could be useful when planning a fieldwork investigation on a river.

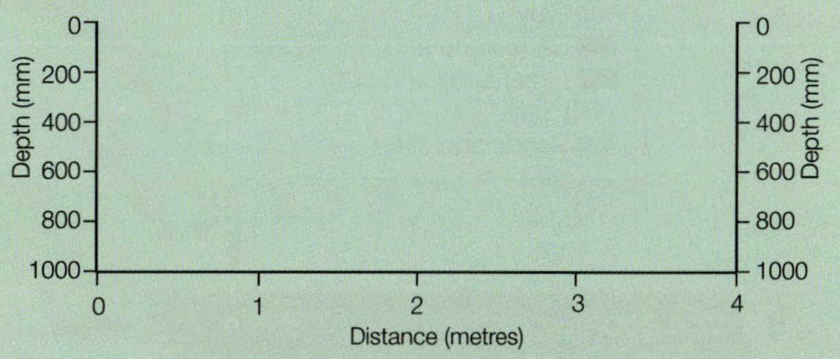

The Thames Basin

On this spread you will follow the River Thames from source to mouth, looking at landforms, geology and human activity.

Case study

Introducing the Thames Basin

The River Thames is England's longest river (though not as long as the River Severn in England and Wales). It is 346 km from source to mouth (map A). It is the most densely populated **river basin** in the UK, with over 13 million people, including London.

The Thames rises in the Cotswold Hills, which are made of Jurassic Limestone. It then flows eastwards, through a gap between the Chilterns and Berkshire Downs (photo 1), both made of permeable chalk. For the rest of its course it flows over impermeable London Clay all the way to the North Sea.

1 The Goring Gap

A The course of the River Thames

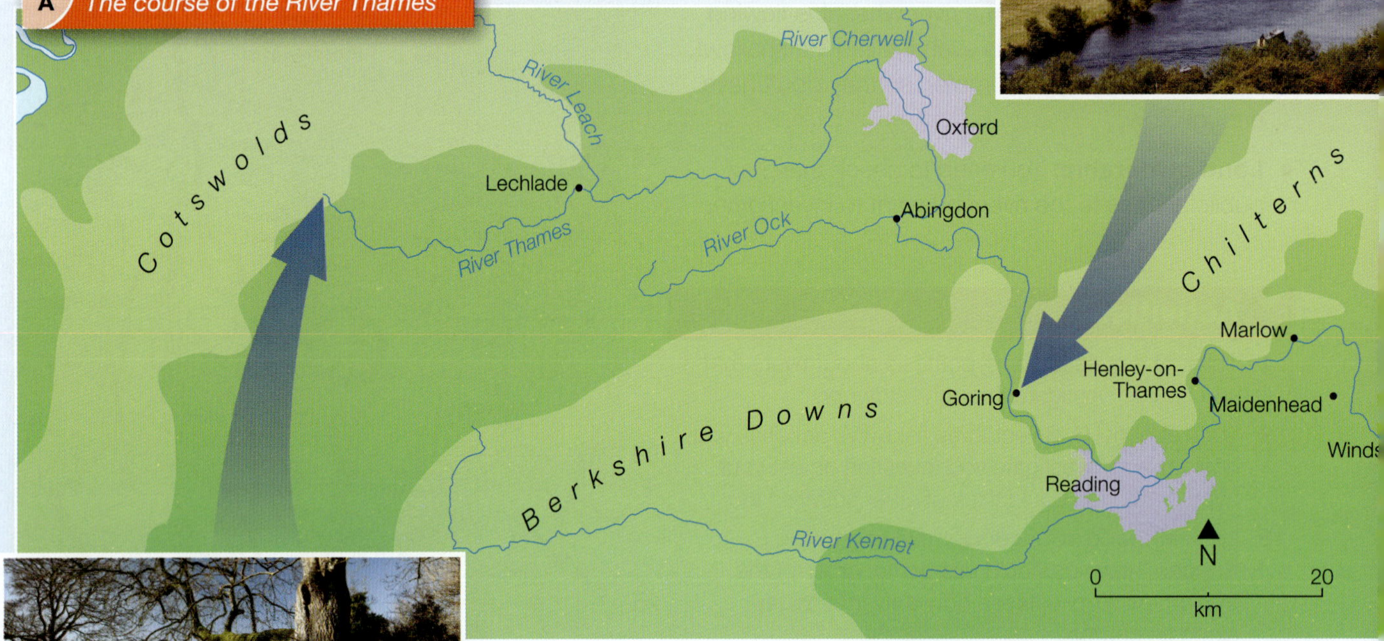

2 The source of the Thames

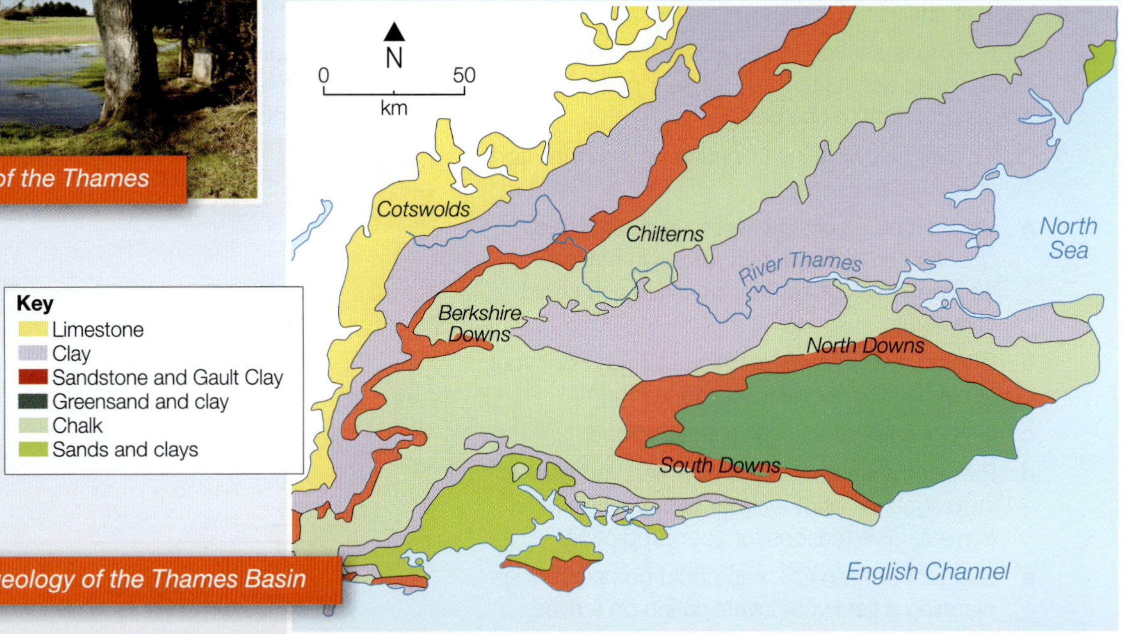

B The geology of the Thames Basin

London's water

The south-east region, through which the Thames flows, is the driest region in the country. It has an average of 690 mm of rain a year. This puts pressure on water resources in a region already struggling to meet the demand for water. A growing population makes matters worse.

London gets 40% of its water supply from **groundwater** in the chalk that lies beneath the city (cross-section C). Boreholes are drilled through the clay to obtain groundwater from the chalk **aquifer**, an underground layer of water-bearing rock.

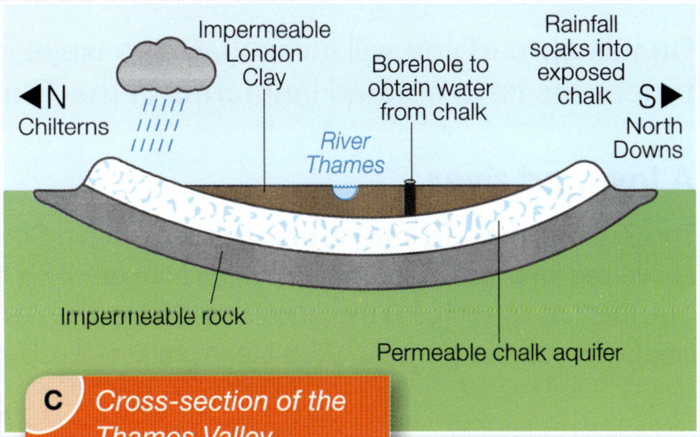

C *Cross-section of the Thames Valley*

3 *Isle of Dogs, London*

4 *Dartford Bridge over the Thames*

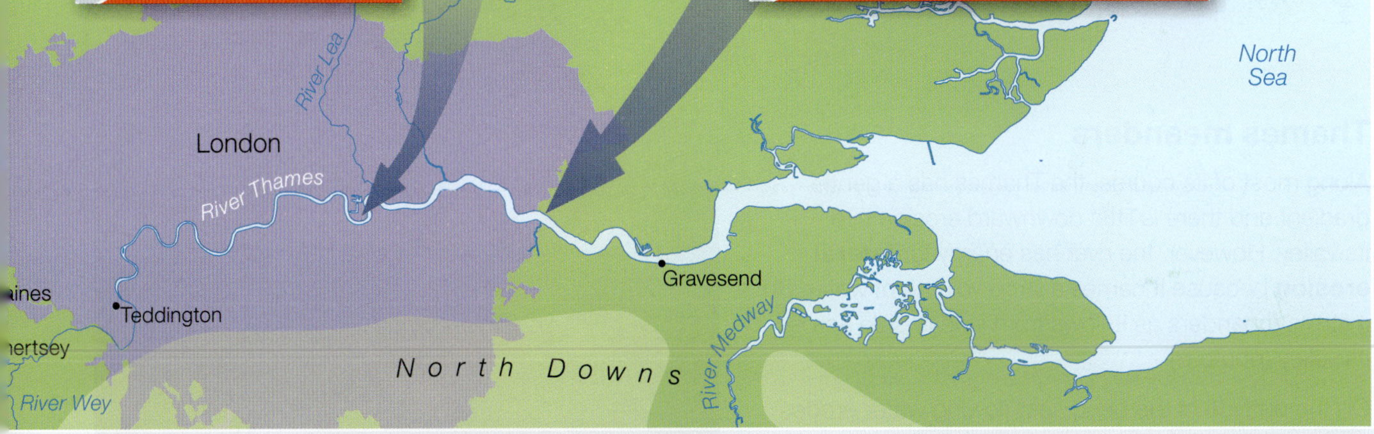

Activities

1 Look at map A.

- **a** Name five tributaries of the River Thames, in order from source to mouth.
- **b** Explain why the Thames gets wider as it flows to the sea.

2 Look at photos 1–4.

- **a** Identify examples of each of the following natural landscape features in the photos:
  - valley
  - floodplain
  - estuary
  - meander
  - source.
- **b** Identify ways in which human activity has changed the landscape in each photo.

3 Compare maps A and B.

- **a** Over which type of rock does the Thames flow along most of its course?
- **b** Why do rivers flow on this type of rock?

4 Look at cross-section C.

- **a** Explain how London obtains part of its water supply from groundwater.
- **b** With a major river flowing through the area, why is it necessary to use groundwater?

Practice question

For a named river basin in the UK, describe its location. You could draw a labelled sketch map to help you. (4 marks)

3.15 Landforms along the Thames

On this spread you will investigate the ways in which geomorphic processes have created landforms in the Thames valley.

Case study

A lowland river

The River Thames is a lowland river. Its source in the Cotswolds is only 108 metres above sea level (profile A). With less force from gravity, it lacks the energy of an upland river that has its source in the mountains. So the Thames does not have some of the landforms you find on rivers with a steep gradient, such as waterfalls or gorges.

A *Long profile of the River Thames from source to mouth*

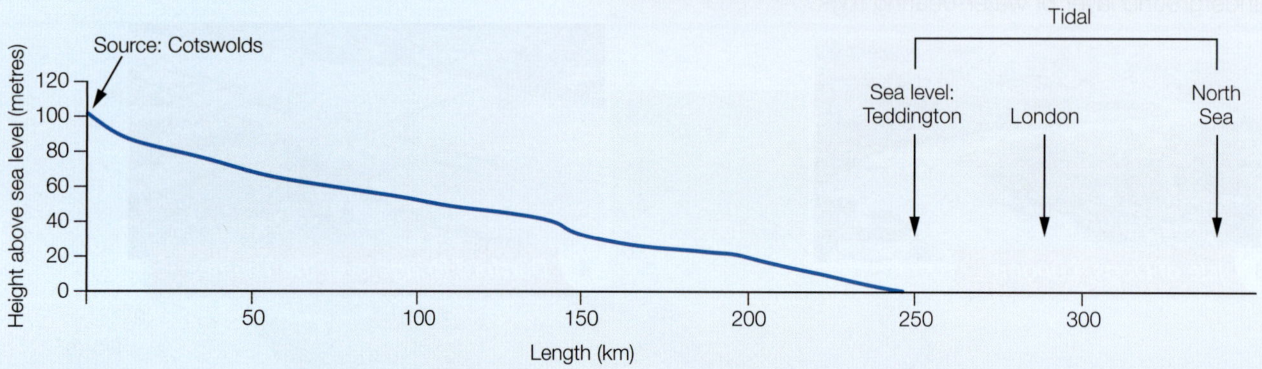

Thames meanders

Along most of its course, the Thames has a gentle gradient and there is little downward erosion into its valley. However, the river has energy for **lateral erosion** because it carries a large volume of water. So it forms meanders as it flows from side to side across its valley (photo B).

On the outside of the bend, fast-flowing water erodes the bank. On the inside of the bend, slow-flowing water deposits its **load** (diagram C). This slowly changes the course of the river. For most of its length, the Thames meanders across its floodplain, changing its course over thousands of years. (See also spread 3.8.)

B *A meander on the Thames near Oxford*

C *Meander cross-section*

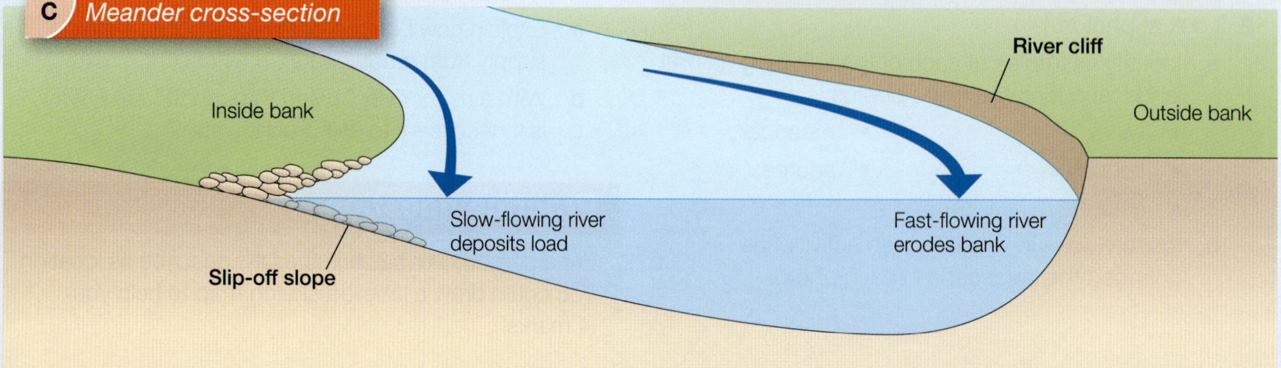

A rejuvenated river

The River Thames has flowed more or less on its present course since the ice age that began half a million years ago. Since then, temperature changes have led to the advance and retreat of ice sheets, and to sea-level changes (see Chapter 2).

Past sea-level changes affected the power of the Thames to erode its valley. When the sea level fell, the river gained energy because it had to drop further to sea level and eroded downwards into its old floodplain. This process is known as **rejuvenation**. When the sea level rose again, the river lost its energy and began to deposit more material on its floodplain (diagram D).

D *Formation of river terraces in the Thames valley*

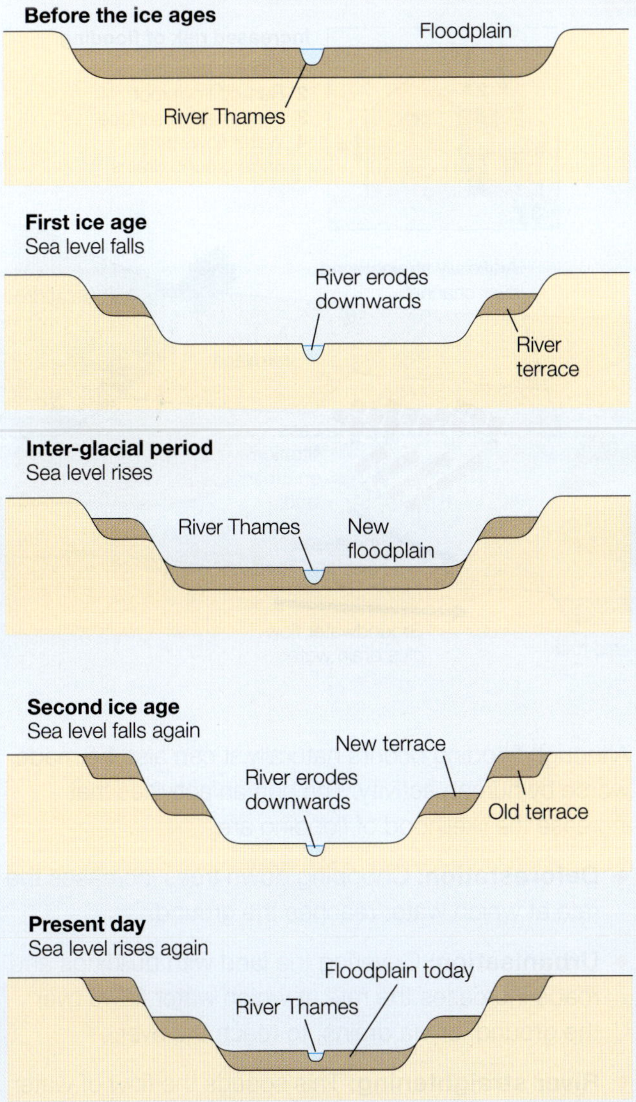

River terraces

This process was repeated during each ice age. Each time the river was rejuvenated, it formed a new floodplain at a lower level, leaving the old floodplain as a **river terrace** at a higher level.

Today, river terraces are found along the Thames Valley west and east of London. Many parts of London have developed on river terraces, above the river level and safe from flooding.

Activities

1 Look at the long profile of the River Thames on graph A.

- **a** Describe the gradient of the Thames from source to mouth.
- **b** Explain why landforms like waterfalls and gorges are not found on the Thames (see spread 3.8).

2 Look at photo B and diagram C.

- **a** Draw a labelled sketch of the meander in the photo. Label these features: meander, floodplain, river cliff, slip-off slope.
- **b** What might the shape of the meander be in thousands of years if lateral erosion continues? Draw another sketch to show what it may look like.

3 Look at the diagrams in D. Sequence the following events in the correct order:

- The river erodes downwards.
- A new floodplain is formed.
- Sea level falls.
- Sea level rises.
- The old floodplain forms a river terrace.
- The river loses power.
- The river is rejuvenated.
- The river floods and deposits material.

4 Explain how a river terrace is formed using the events you sequenced in activity 3.

Practice question

For one landform in a named river basin, explain how geomorphic processes have led to its formation. (4 marks)

3.16 Flooding on the Thames

On this spread you will find out about a flood on the River Thames and the factors that caused it.

Case study

Thames flood, 2014

February 2014 saw the worst flooding along the River Thames in 40 years. More than 5000 homes and businesses were flooded in the village of Wraysbury (photo A) and the nearby town of Staines, to the west of London.

The flood resulted in thousands of people being evacuated. Many homes were badly damaged and the estimated cost of clearing up after the flood was over £500 million. It was months before many people were able to move back into their homes.

A *Flooding in Wraysbury, 2014*

B *Factors that reduce or increase the risk of flooding*

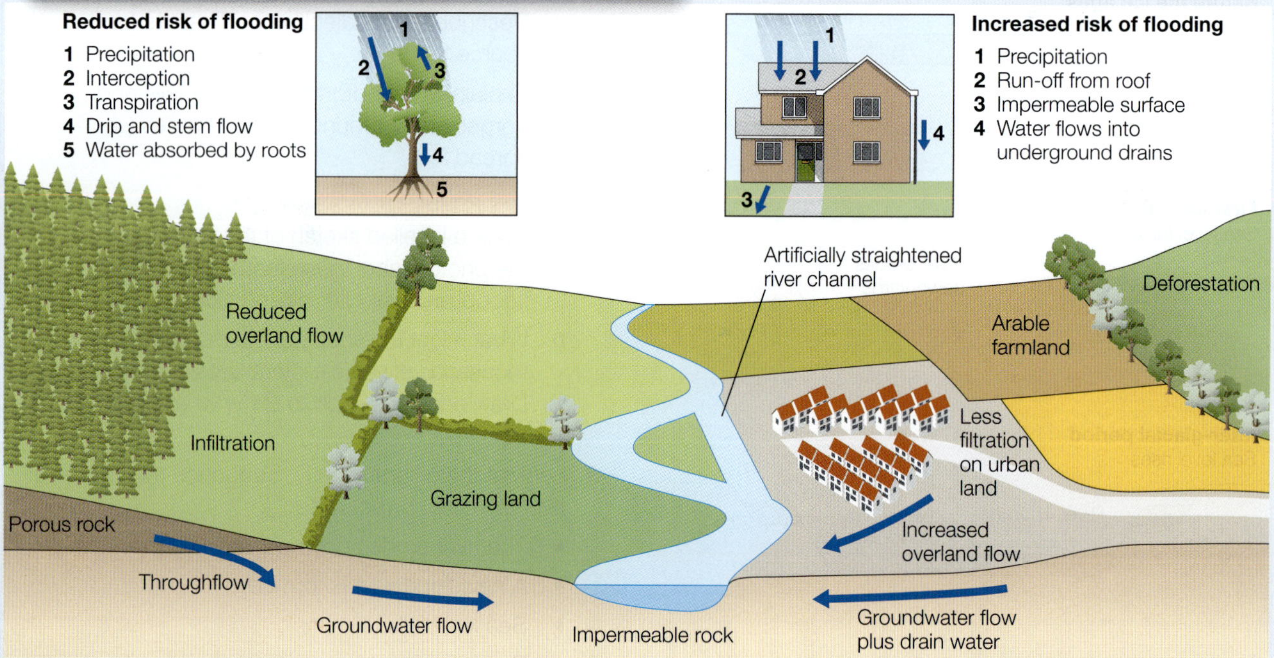

Causes of flooding

Flooding occurs when a river carries more water than its channel can hold or when water cannot infiltrate (seep) into the ground. Floods are most likely to occur when:

- the ground is impermeable rock
- the soil is saturated with water after weeks or months of rain
- the ground is frozen
- the rainfall is so intense that it cannot soak into the ground.

Although flooding occurs naturally, it can also be made worse by human activity. The human activities that increase the likelihood of flooding are:

- **Deforestation:** Chopping down trees increases the rate at which water reaches the ground.
- **Urbanisation:** Covering the land with buildings and roads increases the rate at which water flows over the ground, or via drains, to reach the river.
- **River straightening:** This speeds the flow of water in the river, reducing the risk of flooding in one area, but increasing the risk of flooding downstream.

Before the flood

The winter of 2013/14 was the wettest on record in southern England. Between 1 December 2013 and 19 February 2014, 487 mm of rain fell, over twice what is expected in a normal winter (map C). The cause was a series of Atlantic storms that lashed the UK with heavy rain throughout the winter (see spread 3.10).

On 9 February, after months of rain, the ground was saturated (wet through) and water levels in the Thames were very high. The Environment Agency, which is responsible for flood protection in the UK, issued 14 severe flood alerts along the Thames. They warned that property and lives were at risk (map D). Fortunately, when the floods came a day later, no lives were lost.

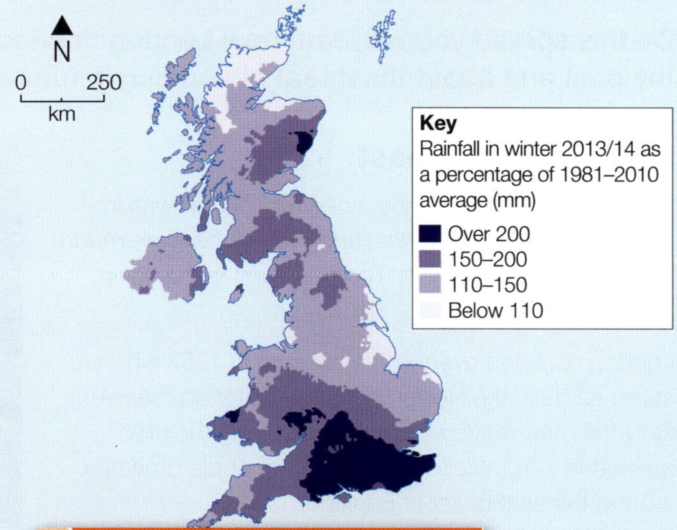

C *The UK's wettest winter, December 2013 to February 2014*

The Jubilee River

The flooding was made worse in Wraysbury by a new artificial channel on the Thames called the Jubilee River. Constructed for the Queen's Golden Jubilee in 2002, it was designed to take overflow water from the River Thames to reduce flooding around Maidenhead, Windsor and Eton. At a cost of £110 million, it is the most expensive man-made river in the UK.

The Jubilee River takes a shorter route than the Thames, so water reaches its destination faster. In 2014 it carried water quickly downstream to Wraysbury and Staines, increasing flooding there. Since the 2014 floods, there are now plans for a further **flood relief channel** to Teddington, as part of a £500 million project to reduce the flood risk on the Thames west of London.

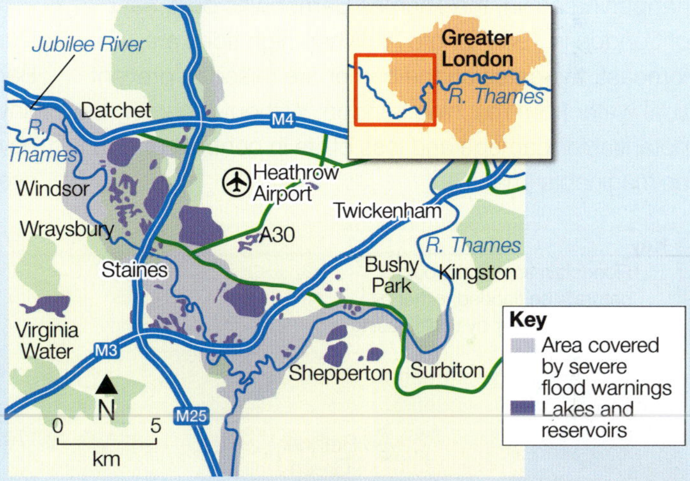

D *Area of severe flood warnings along the Thames in February 2014*

Activities

1. Look at drawing B.
   - **a** Compare what happens when rain falls on woodland with when it falls on an urban area.
   - **b** Explain why building more urban areas increases the risk of flooding.
2. Draw a table to list the natural and human causes of flooding.
3. Look at map C.
   - **a** Describe the distribution of rainfall across the UK in the winter of 2013/14.
   - **b** This map does not show the actual amount of rainfall. What is it showing and why was this important?
4. Look at map D.
   - **a** Explain how building the Jubilee River increased the flood risk in Wraysbury and Staines.
   - **b** Do you think building a further flood relief channel west of London is a good idea? Give reasons for your opinion.

Practice question

For a river basin you have studied, discuss whether human activities in the river basin help to increase or reduce the risk of flooding. (8 marks)

3.17 Flood threat to London

On this spread you will learn how London flooded in the past and about the threat of flooding in future.

Case study

Flooding in the past

The last major flood to affect central London was in 1928. Fourteen people died and thousands were made homeless when the River Thames burst its banks in January that year.

London escaped even worse flooding in 1953 when a storm surge in the North Sea drove water up the river. Only the man-made embankments along the river prevented it from flooding. Over 300 people drowned around the east coast of England.

The Thames Barrier

In response to these events, and the threat of it happening again, the Thames Barrier was built, east of London in 1984 (photo A). When high tides are forecast, the gates on the barrier are raised to prevent tidal water from reaching London. Without a barrier, the potential for damage and loss of life in central London is unimaginable (map B).

A *The Thames Barrier*

During the stormy winter of 2013/14, the Thames Barrier was closed more than 50 times, more than in any previous year. There are already proposals for a new barrier, high enough to cope with the predicted rise in sea level during the 21st century.

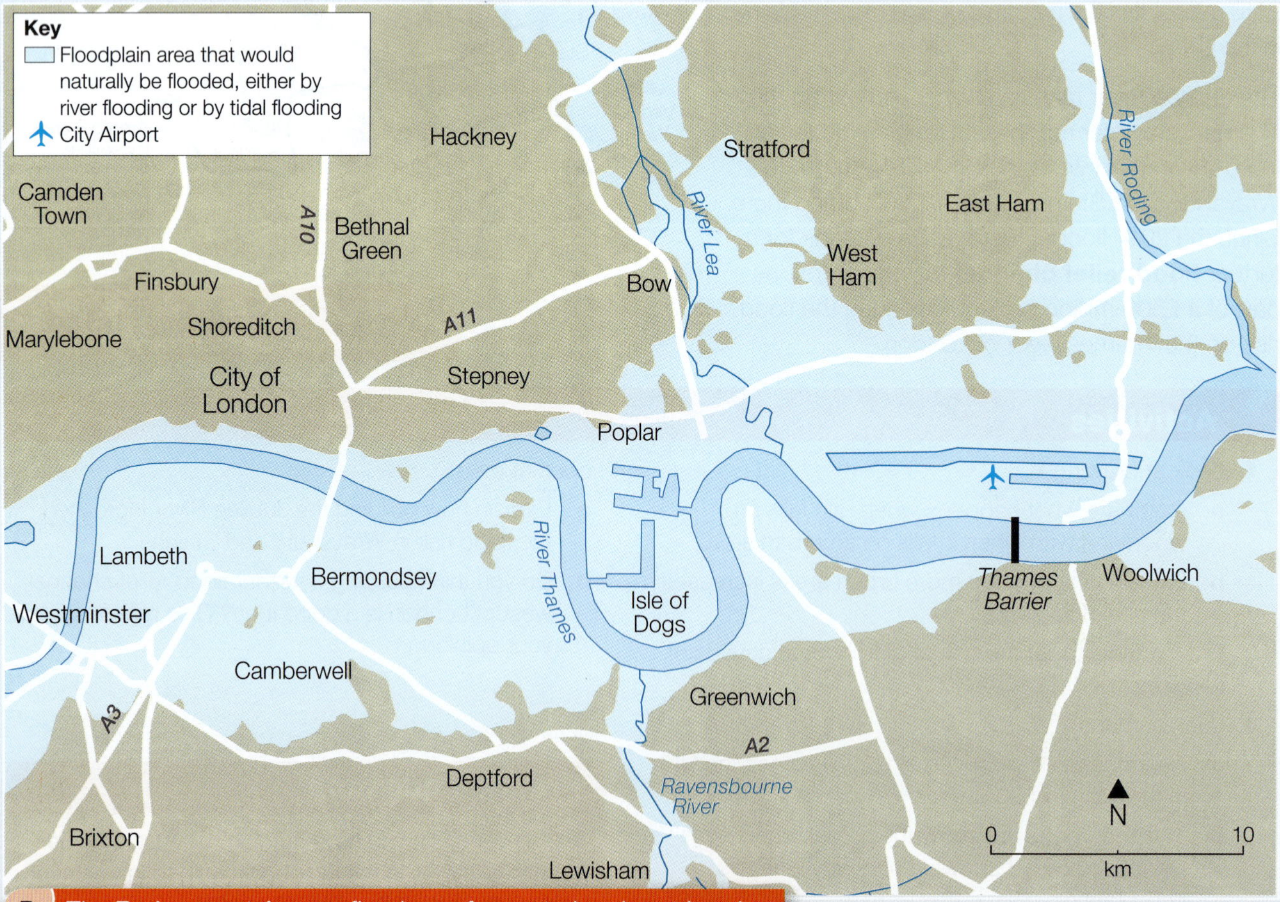

B *The Environment Agency flood map for central and east London*

Barking Riverside

The best advice these days is, wherever possible, to avoid building on floodplains. But if that advice had been available in the past, London might never have been built!

There is a severe shortage of housing in south-east England, the region in which the Thames Basin is located. As part of the solution, a major new residential development at Barking Riverside is being built to the east of London (photo C).

The site is downstream from the Thames Barrier, so it will get no flood protection. Instead, the plan is to transform the flood risk into a feature of the landscape by doing the following:

- Rather than barriers, water will be allowed to spread naturally over the floodplain and fill local creeks (inlets from the river).
- There will be green corridors within the development with water as a feature.
- Trees and parkland will help to slow the rate of infiltration and increase water absorption.
- Residential areas within the site will be on raised land to protect them from flooding.

C *Barking Riverside, a new development beside the Thames*

Activities

1 Look at photo A.
   - **a** Why was the Thames Barrier built?
   - **b** Has it been successful, do you think? Give reasons.

2 Look at map B. Which of these attractions in London is threatened by a flood?
   - Tower of London
   - Greenwich Observatory
   - St Paul's Cathedral
   - Houses of Parliament
   - Olympic Stadium
   - O2 Arena

   You will need to refer to a tourist map of London online.

3 Look at photo C.
   - **a** What views would each of these people have about the Barking Riverside development?
     - The Mayor of London, who wants more homes in London
     - A young person living in London who needs a home
     - A Wraysbury resident who experienced flooding in 2014
     - An Environment Agency flood officer responsible for flood protection
   - **b** On balance, do you think Barking Riverside is a good idea or not? Give reasons.

Go further

Find a flood map on the Environment Agency website for the area where you live.

- **a** Could your home be flooded? Could your school be flooded?
- **b** Name any other local areas at risk from flooding.
- **c** Does the risk come from river or coastal flooding?
- **d** What flood defences are there?

Your enquiry

What influences the landscapes of the UK?

Think of a landscape you know well or choose another landscape from this section (see spreads 3.6–3.16).

Consider how each of these factors has influenced your chosen landscape:

- geomorphic processes, including weathering, mass movement, erosion, transport or deposition
- human activity, including ways in which people manage the landscape.

Write a paragraph about each factor to describe how it has influenced the landscape.

4 Sustaining Ecosystems

What is this topic about?

Without ecosystems we would be nowhere. An **ecosystem** is any community of plants and animals, and the environment in which they live. A forest, a pond, a field are examples of ecosystems. Looking out of the window all you may be able to see is buildings, but our life on Earth depends on ecosystems. They are vital for food, for water, even for the air we breathe.

The world's ecosystems are under threat – from us! Human activities are in danger of destroying the life-support systems on which we depend. At a global scale, forests are being chopped down and oceans are being over-fished. Lately, people have come to realise how important ecosystems are. Now we are learning about **sustainable management** – how to manage them more sustainably by using resources more carefully so that future generations will also benefit.

Your enquiry

In this chapter you will investigate the specification's three key questions.

Why are natural ecosystems important? You will:

- Identify the links between the components of an ecosystem.
- Find out the global distribution of some important ecosystems.
- Examine the variety of climates, plants and animals in these ecosystems.

Why should tropical rainforests matter to us? You will:

- Explore the characteristics of tropical rainforest.
- Recognise the value of tropical rainforest globally.
- Understand the impacts of human activity on rainforest.
- Investigate how rainforest is managed sustainably for tourism in Costa Rica.

Is there more to polar environments than ice? You will:

- Explore the characteristics of polar regions.
- Know how the components of the Arctic ecosystem are interdependent.
- Understand the impacts of human activity in the Arctic.
- Investigate how Arctic whaling is managed sustainably.
- Devise a sustainable plan for the Arctic, based on the Antarctic Treaty.

Look at the photo opposite.

- Why do you think tropical rainforest is being destroyed?
- Trees are lost, but what are the other impacts of destroying rainforest?
- How could this affect other parts of the planet?
- Is this something you need to be concerned about? Why, or why not?

Ecosystems

On this spread you will consider how an ecosystem works as an interdependent system with stores and flows.

Components of an ecosystem

As you know, an ecosystem is a community of plants and animals, together with the environment in which they live. There are living and non-living components in every ecosystem, all interdependent (photo A). Animals depend on plants and each other for food, while plants may depend on animals for pollination and to scatter their seeds.

The non-living parts of the ecosystem, like climate and soils, are also linked. They provide nutrients, water, warmth and shelter for the living parts of the ecosystem. Plants and animals also affect the non-living parts. They add gases to the atmosphere and nutrients to the soil. When they die, they decompose, or decay, into the soil.

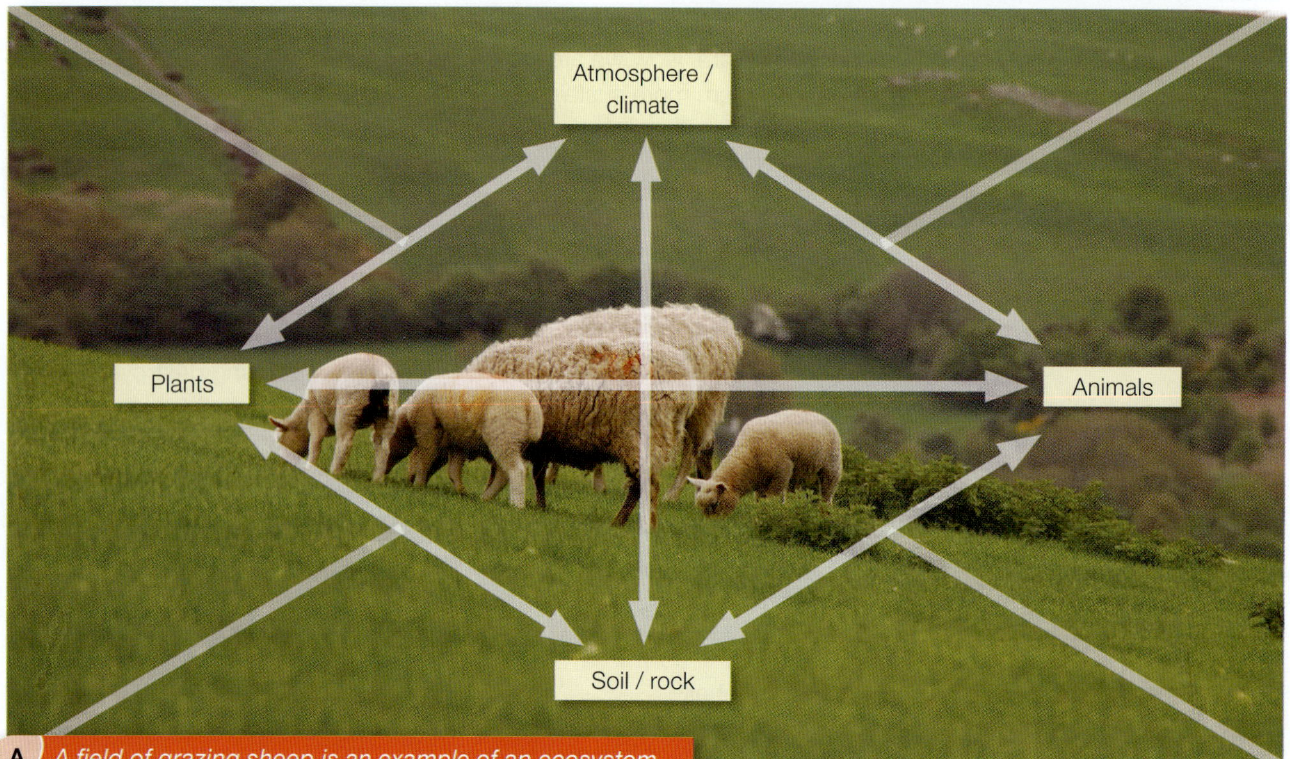

A *A field of grazing sheep is an example of an ecosystem*

The nutrient cycle

Within every ecosystem there are stores and flows of nutrients, water and energy. For example, within the **nutrient cycle**, nutrients are stored in the soil, **biomass** (the mass, or weight, of living material) and **litter** (decomposing leaves on the ground) (diagram B). The nutrients flow from one store to another in a cycle. The cycle also has inputs and outputs.

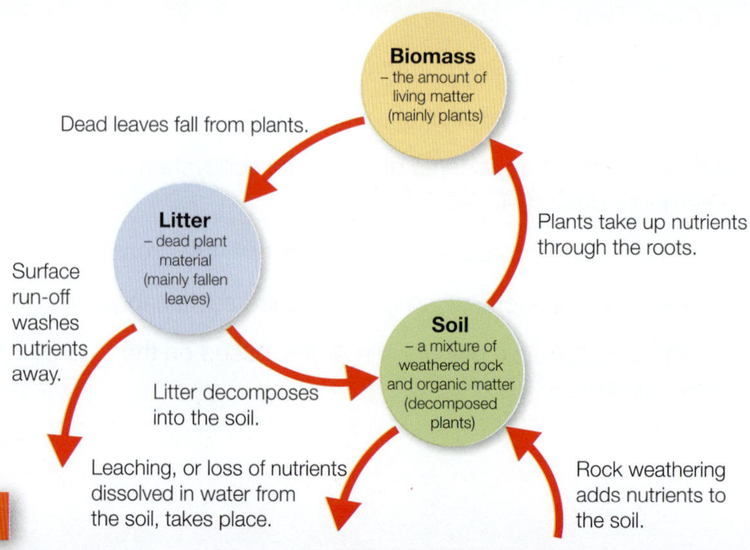

B *The nutrient cycle*

Food chains

The energy in an ecosystem comes from the Sun. Plants and animals need energy to grow and to function. Plants are known as **producers** because they harness the Sun's energy through the process of **photosynthesis** in their leaves. Animals are known as **consumers** because they eat the plants (and each other!). Together, producers and consumers form a **food chain** in the ecosystem (diagram C).

Energy is transferred up the food chain from one level to the next – from producer to primary consumer, primary to secondary consumer, and so on. Around 90% of energy is lost at each level of the food chain, so the energy that reaches the top of the food chain is a small percentage of the energy at the bottom. Energy is lost through:

- heat
- functions such as breathing and movement
- excretion
- wastage (uneaten food).

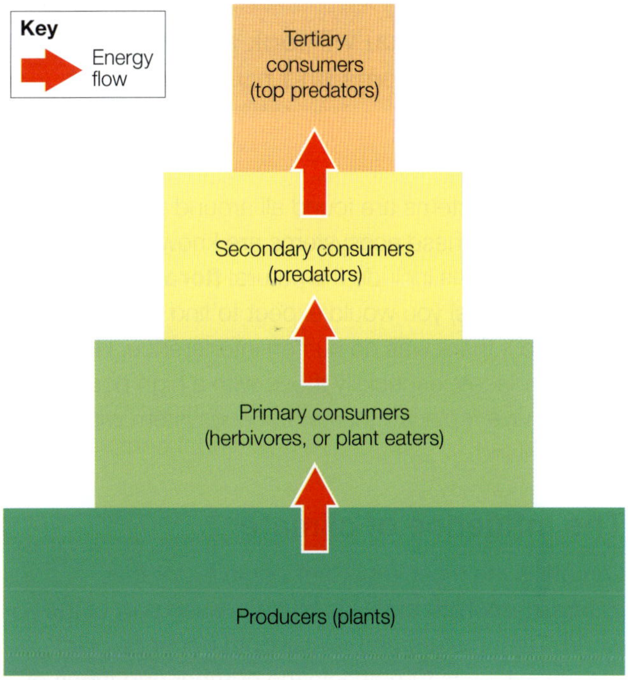

C *The food chain in an ecosystem*

Activities

1. Look at photo A. Write six sentences to describe how each component in the ecosystem is linked to the others. For example: *Animals breathe in oxygen from the atmosphere and breathe out carbon dioxide.*
2. Photo A is not a natural ecosystem. What influence do people have on this ecosystem? For example, how did the sheep get there?
3. Look at diagram B.
   - **a** Name the three stores in the nutrient cycle.
   - **b** Describe three ways in which nutrients flow around the cycle.
   - **c** Describe three inputs or outputs in/out of the cycle.
4. Think about the way water is cycled in an ecosystem. Draw a diagram similar to diagram B to show the water cycle. Start like this to show the three main stores. Then add arrows to show flows, inputs and outputs. Label the arrows to describe what happens.

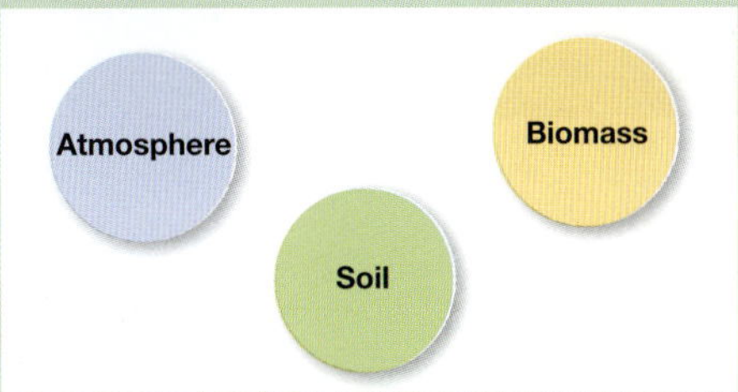

5. Look at diagram C.
   - **a** Where do producers get their energy from?
   - **b** Explain how energy is lost, moving up the food chain.

Your enquiry

Why are natural ecosystems important?

As you work through this section, make notes about anything that will help you to answer the question. You will bring your ideas together to do an activity at the end of spread 4.3.

Global ecosystems

On this spread you will look at the global distribution of some ecosystems and try to explain it.

Biomes

Natural ecosystems are found all around the world. At a large scale, these ecosystems are known as **biomes** (map B). Biomes include the natural **flora** (plants) and **fauna** (animals) you would expect to find in each part of the world if there was no human interference. However, in many places, especially those with a high population, people have replaced the natural ecosystem with cities and farmland.

The influence of climate

Climate is the most important single factor that determines which biome is found in each part of the world. Temperature and rainfall both influence plant growth, which in turn affects the number and types of animals. The distribution of biomes reflects the global climate pattern (see spread 1.1).

The most productive biomes – which have the greatest biomass – grow in climates that are hot and wet (graph A). The natural ecosystem or biome in the UK is **temperate forest** (see spread 3.5). It grows in a temperate climate, which has summers and winters. Trees in a temperate forest are mainly **deciduous**, growing during summer when temperatures are warm and shedding their leaves in winter.

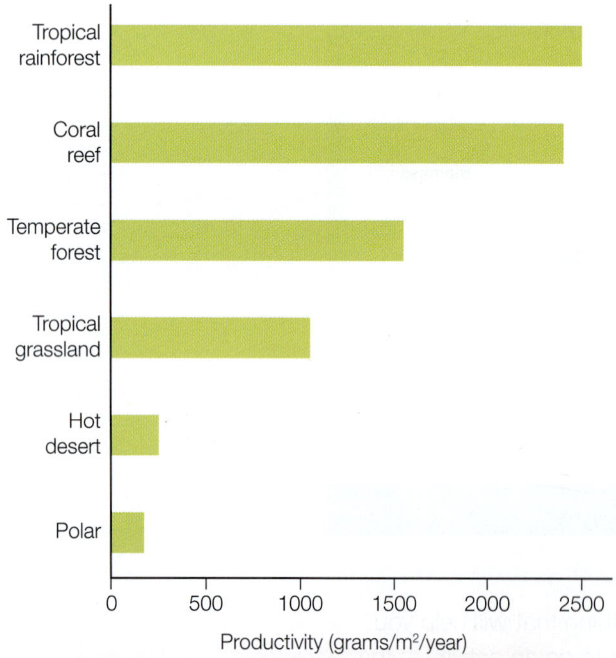

A Biomes productivity measured by biomass

B Some important world biomes

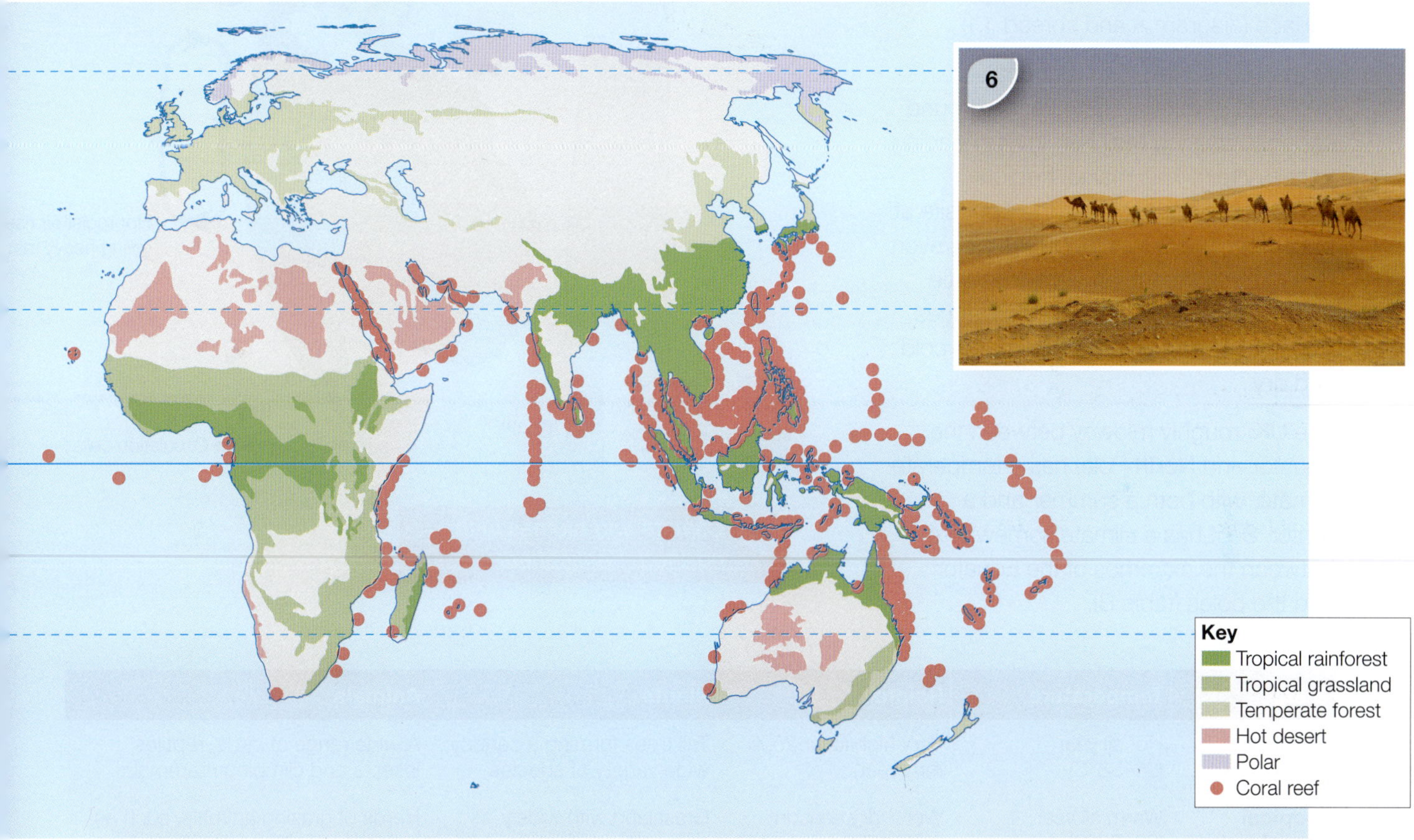

Activities

1. Match each of photos 1–6 with one of the biomes on map B.
2. Write six sentences to describe the global distribution of biomes. Mention the latitude (distance from the Equator) for each biome in your description. For example, *Tropical rainforest is found north and south of the Equator, between the Tropic of Cancer and the Tropic of Capricorn.*
3. Look at graph A.
   - **a** Which biomes are:
     - most productive?
     - least productive?
   - **b** How do you explain this from their climate and location on the globe?

Practice question

Outline the global distribution of one named biome, giving examples of locations where it is found. (4 marks)

4.3 More about biomes

On this spread you will go a bit deeper into biomes to explore the ways that plants and animals adapt to climate.

Global climate patterns

The global distribution of biomes closely mirrors the world's climate zones. So, for example, **tropical rainforest** is found in areas with a tropical climate, while **tundra** (table B) and ice are found in areas with a polar climate. The difference between tropical and polar climates is explained by their location on Earth and the angle of the Sun's rays hitting Earth's surface (diagram A and spread 1.1).

At the Equator the Sun is directly overhead at midday. It heats the ground strongly, causing moist air to rise, leading to heavy rain. This makes tropical climates hot and wet. It is the opposite at the poles. The Sun is low in the sky, even during the summer, and provides little heat. At the poles, sinking air produces little rain. This makes polar climates cold and dry.

The UK, roughly midway between the Equator and North Pole, has a temperate climate, with both a summer and a winter. So it has a climate somewhere between the extremes of the Equator and the poles (table B).

A *Climate and biomes in the northern hemisphere*

Biome	Temperature	Rainfall	Plants	Animals
Tropical rainforest	Hot all year (25–30°C)	Very high (over 2500 mm/year)	Tall trees forming a canopy; wide variety of species	A wide range of birds, reptiles, insects and climbing mammals
Tropical grassland	Warm all year (20–30°C)	Wet + dry season (500–1500 mm/year)	Grassland with widely spaced trees	Herds of grazing animals, such as zebra, and hunters, like lions
Hot desert	Hot by day (over 30°C), cold at night	Very low (below 300mm/year)	Lack of plants and few species; adapted to drought	Animals adapted to living in dry conditions, e.g. camels
Temperate forest	Warm summers + mild winters (5–20°C)	Variable rainfall (500–1500 mm/year)	Mainly deciduous trees; a variety of species	Animals adapted to changing seasons, including some that hibernate
Tundra (in polar regions)	Cold winters + cool summers (below 10°C)	Low rainfall (below 500 mm/year)	Small plants grow close to the ground and only in summer	Mainly migrating animals that live there only in summer
Coral reef	Tropical and sub-tropical oceans (16–30°C)	(Not applicable)	A wide range of seaweed, plankton and algae grows on reefs	Coral is a jellyfish-like animal and a third of the world's fish live in reefs

B *Biomes, climate and plants*

C Plants growing on the tundra in the polar region

How plants adapt

The plants in each biome are adapted to the climate of the region they live in. For example, in a polar climate, tundra plants:

- have a short growing season because it is too cold to grow for the rest of the year
- grow slowly because soil lacks nutrients in a cold climate where rocks are weathered slowly
- grow close to the ground because it is too cold and windy for trees to grow
- have small leaves and lose little moisture because there is low rainfall and only snow in winter.

D Tropical rainforest

Activities

1 Look at diagram A. Explain why it is hot and wet at the Equator and cold and dry at the North Pole.

2 Study table B. What connection do you notice between climate (temperature and rainfall) and the type of plants that grow?

3 Match each of the following animals with one of the biomes in table B. Suggest one way in which each animal is suited to living in that climate. You may have to do some extra research about these animals, but try using your general knowledge first.

  a Camel
  b Hedgehog
  c Reindeer
  d Antelope
  e Orangutan (monkey)

4 Look at photos C and D.

  a Write a short paragraph to compare the plants in tundra and in tropical rainforest. Mention their height, biomass, shape and leaf size.

  b Suggest how the plants in tropical rainforest are adapted to a tropical climate. (Clue: many of the features of a tropical climate are opposite to those of a polar climate.)

Your enquiry

Why are natural ecosystems important?

The ecosystems in this section can be damaged or destroyed by human activities. You have to make the case for better **conservation** (protection) of one of them.

a Choose one ecosystem and do some extra research on it. If you choose tropical rainforests or polar regions, you can use the later sections of this chapter for research. Otherwise, do your research online. Try to find out what makes this ecosystem:
  - unique (plants or animals are found only here)
  - essential (for local people and the planet)
  - irreplaceable. (What would happen if it disappeared?)

b Write a leaflet to persuade people that this ecosystem is important and needs to be conserved.

4.4 Tropical rainforests

On this spread you will investigate the characteristics of tropical rainforest and how the parts are interdependent.

The importance of rainforest

Tropical rainforests cover only about 5% of Earth's surface but their importance far outweighs the area they cover. They are home to about 50% of the world's **biodiversity**, or variety of plant and animal species.

Rainforests are found in parts of Asia, Africa, Australia and South America. They are spread across a region north and south of the Equator between the Tropics of Cancer and Capricorn. All areas of rainforest share similar characteristics (drawing B), including their climate (graph A).

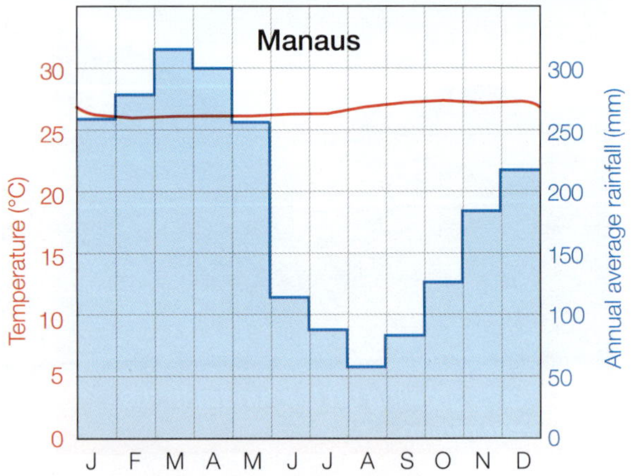

A *Climate in Manaus, in the Amazon rainforest in Brazil*

B *Characteristics of tropical rainforest*

Sloths spend much of their time hanging from branches. They are camouflaged in the forest.

Lianas and vines climb tall trees to reach sunlight at the top of the canopy.

Buttress roots support the weight of tall trees. They take up nutrients from the shallow top layer of soil.

Drip tips on leaves allow heavy rain to run off without damaging the leaf.

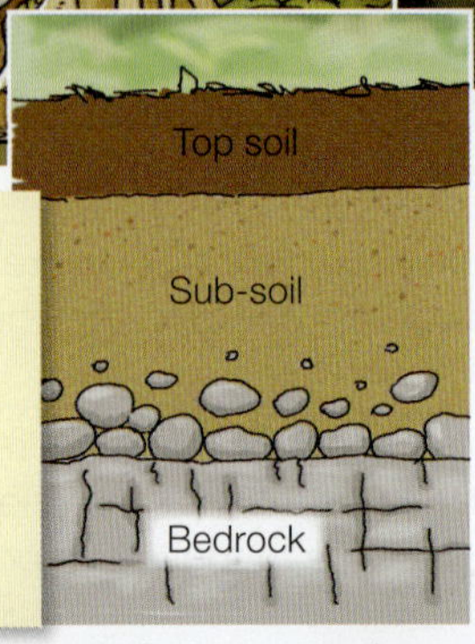

Soil profile

- The very thin litter layer rapidly decomposes in heat.
- Shallow topsoil is a mixture of decomposed organic matter and minerals.
- The sub-soil is deep due to weathering of rocks below.
- Underlying rock weathers quickly at high temperatures to form sub-soil.
- The soil is low in nutrients as they are quickly taken up by trees.

Nutrient cycle

Like other ecosystems, nutrients are cycled within the tropical rainforest ecosystem. In rainforests most of the nutrients are stored in the biomass (diagram C), mainly in large trees. The size of the circles in diagram C represents the amount of nutrients stored and the width of the arrows shows the rate at which they flow.

When trees in rainforests are chopped down the main nutrient store is lost. Without the continuous flow of nutrients, the soil quickly loses its natural fertility.

Water cycle

Rainforests also play an important role in the **water cycle** in tropical regions. The canopy intercepts the intense rain that falls throughout the year. Most of the rainfall returns to the atmosphere through evaporation and transpiration (loss of water vapour). This leads to cloud formation and the next rainstorm.

Some water reaches the soil and is drawn up, through the roots, into the trees. If the trees are chopped down, the soil is quickly eroded by heavy rain. The soil washes into rivers that are then prone to flooding. In the long term, chopping down forest means less water being cycled and tropical regions become drier.

C *The nutrient cycle in tropical rainforest*

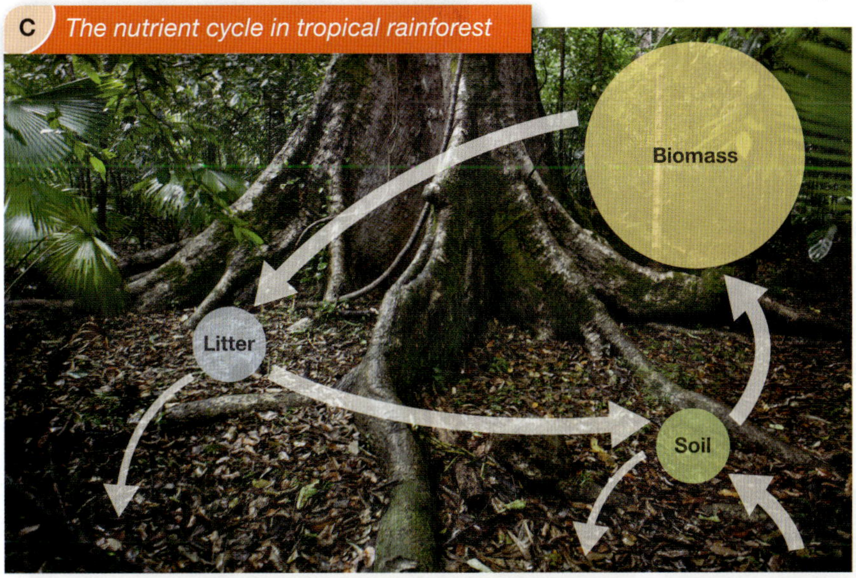

Activities

1. Look at graph A. Describe the main characteristics of a tropical climate. Write one sentence each to describe the annual pattern of:
   - **a** temperature
   - **b** rainfall.
2. Study the drawings in diagram B. Identify at least four ways in which plants or animals are adapted to living in a tropical rainforest.
3. Look at the soil profile in diagram B.
   - **a** Suggest at least three ways in which rainforest soil might differ from soil in a polar region.
   - **b** Explain why you think these differences would occur.
4. Look at diagram C and the explanatory text above.
   - **a** Compare the size of the stores and flows in tropical rainforest with a model ecosystem (diagram B on spread 4.1). What differences do you notice?
   - **b** Explain the differences from what you know about tropical rainforest.
5. Explain what impact chopping down tropical rainforest has on:
   - **a** the nutrient cycle
   - **b** the water cycle.

Your enquiry

Why should tropical rainforests matter to us?

As you work through this section, make notes about anything that will help you to answer the question. You will bring your ideas together to do an activity at the end of spread 4.8.

4.5 The value of rainforests

On this spread you will find out how people depend on rainforests and you will consider the goods and services rainforests provide.

Rainforest inhabitants

Around the world, about 60 million people now live in areas with tropical rainforest. Many of them are **indigenous people** who, over thousands of years, have developed sustainable ways of life in rainforests (photo A). The forest provides them with all they need to live.

- Food is obtained by hunting, gathering and fishing in the forest.
- Homes and boats are made from forest timber.
- Natural medicines from forest plants are used to cure diseases.
- Everyday items, such as clothes, rope, fuel, oil and glue, are made from forest plants.

A *Indigenous people in the Amazon rainforest, South America*

Shifting cultivation

Some rainforest inhabitants have developed a sustainable way of farming in the forest, known as **shifting cultivation** (panel B). One of the problems faced by people trying to cultivate, or farm, in rainforest soil is its lack of fertility. Remember, most of the nutrients in the rainforest ecosystem are stored in the trees. The soil itself has few nutrients and trees grow because of the rapid cycling of nutrients. With the trees gone, the cycle is broken.

B *Shifting cultivation*

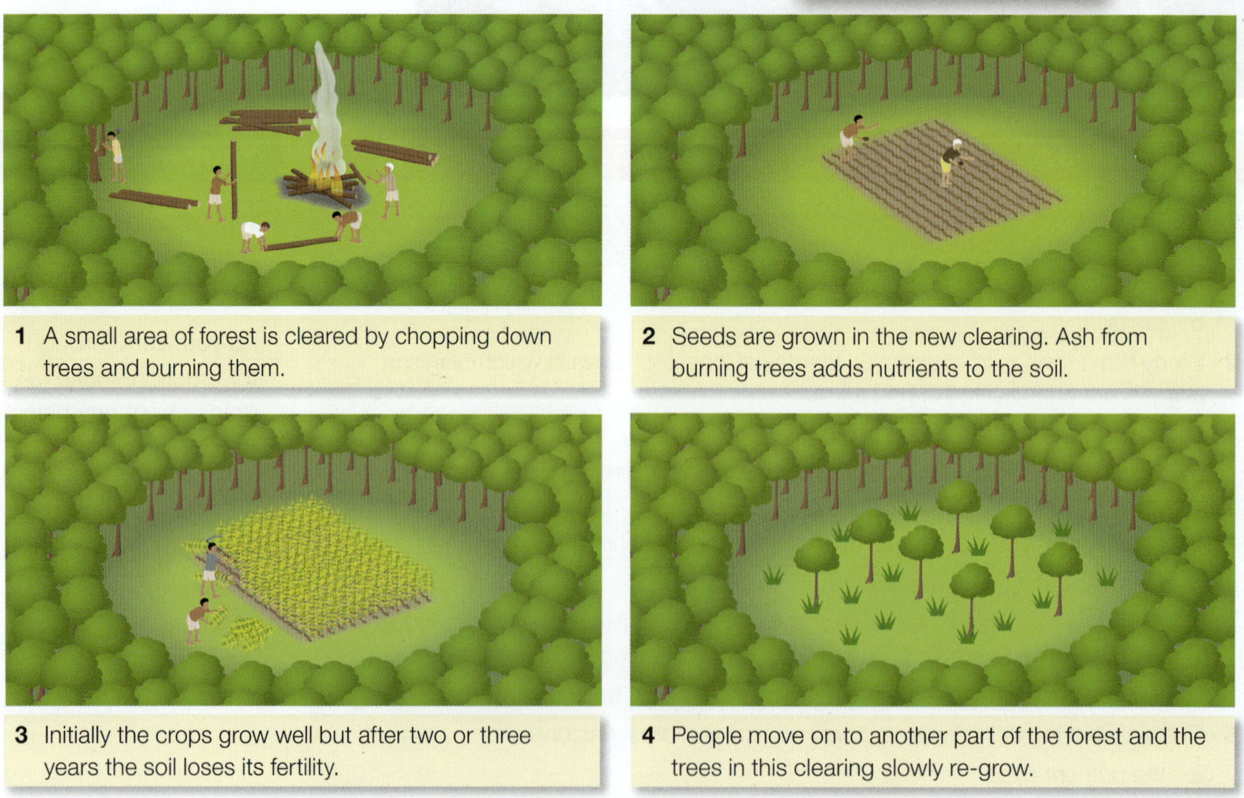

Rainforest goods and services

Tropical rainforests provide goods and services for local people and the rest of the world. Goods are the products people obtain from the forest and services are the jobs forests do for them. The value of rainforest goods and services can be put under six headings:

Climate

Rainforests help to cool the climate locally and globally. They act as **carbon sinks**, removing and storing about 15% of our carbon dioxide emissions every year, which would otherwise contribute to global warming (see spread 2.5). Huge volumes of water evaporate from them, which helps to cool Earth's surface, and create clouds that reflect the Sun's heat back into outer space.

Water

Rainforests act like huge filters to purify water before people drink it in the cities in tropical regions. The forest also acts as a sponge, controlling the flow of water to prevent floods and droughts in these regions.

Food

Many foods used by indigenous people in rainforests have found their way into diets around the world. These include bananas, pineapples, avocados, cocoa, coffee, and various nuts and spices.

Health

Many medicines come from rainforest plants. An estimated 25% of modern medicines are made from rainforest ingredients, including the drug Vincristine, which has dramatically increased survival rates from childhood cancers.

Energy

Traditionally, indigenous people used wood from the forest as a fuel for cooking and heating. Today, large dams are built on rivers in rainforests in order to generate hydro-electric power. Two-thirds of Brazil's energy is generated by hydro-electric power.

Raw materials

Indigenous people depend on rainforest materials for survival. Some of these materials are also commonly used around the world, including timber, rubber, oils and fibres (photo C).

C *Products made from rainforest materials*

Activities

1. Look at photo A. Explain why these people's lives in the Amazon rainforest are sustainable.
2. Look at panel B.
   - **a** Draw a simple flow diagram like this to describe shifting cultivation.
   - **b** Explain why this type of farming is sustainable.

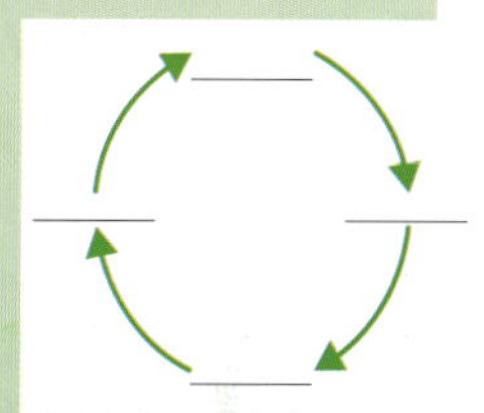

3. Study all the information on this spread.
   - **a** Make a list of all the ways in which people depend on rainforests.
   - **b** Classify all the ideas in your list into two groups – goods and services.

Go further

Find out more about one product from tropical rainforests that you use.

- Where does it come from?
- How is it produced?
- How does this affect the forest?
- Is this sustainable or not?

Practice question

Explain how people living in tropical rainforest are interdependent with the natural ecosystem. (6 marks)

4.6 Human impacts on rainforests

On this spread you will examine a range of human activities in tropical rainforests and carry out an environmental impact assessment.

Rainforest exploitation

Most of the world's tropical rainforests are found in **low-income developing countries (LIDCs)** and **emerging and developing countries (EDCs)** in South America, Africa and Asia. The Amazon rainforest in South America is the largest area of tropical rainforest in the world (map A). Straddling six countries, it covers an area half the size of Europe.

The priority for many LIDCs and EDCs is economic development. Rainforests are seen as a resource to **exploit** for short-term economic benefit, either by felling trees for timber (panel B) or by using the land on which trees grew (panels C–E). This exploitation of timber and land has led to large-scale **deforestation**.

A The Amazon rainforest

Rainforests' value as 'natural capital'

Tropical rainforests have value as 'natural capital' – a resource that, if used wisely, can provide an income for many years. However, once chopped down, it is gone forever. Like real capital (money), it can be spent all in one go or invested to last a lifetime.

Governments have begun to realise the value of rainforests for now and for the future. The rate of deforestation in the Amazon has slowed down (graph F), but the pressure to exploit the forest continues.

B Logging

Logging involves felling trees for timber. Indigenous people have done this on a small scale for thousands of years. Now, with the building of roads into rainforests, logging is done on an industrial scale. The timber is often exported to the USA or Europe for furniture and construction. Tropical hardwood trees, like teak, are very popular.

C Open-cast mining

Mineral extraction takes place in rainforests, especially for gold, copper and diamonds. The forest and soil are removed from large areas for open-cast mining. Chemicals used in the mining process pollute the rivers. Oil and gas are also found deep below ground. Drilling for them can lead to air and water pollution. Building roads or pipelines leads to further destruction of rainforests.

D *Cattle ranching*

Agriculture in rainforests first involves clearing trees in order to use the land. Poor farmers clear small areas of forest to feed their families. When the soil loses fertility, they move on to clear a new patch of forest, rather like shifting cultivators. In the Amazon, most farming is carried out by big companies that clear vast areas of land to grow soya beans or graze cattle. Brazil is one of the world's largest producers of beef and soya.

E *Tourism*

Tourism is an alternative way to use rainforests without chopping the trees down. Some countries have created protected areas of forest where deforestation is banned. Indigenous people can be involved in protecting the forest and continue to lead their traditional way of life. Foreign tourists like to visit rainforests and this can help the local economy.

Activities

1 Look at map A.
   - a Name the countries that share the Amazon rainforest.
   - b Estimate the total area of the Amazon forest. Use the scale on the map to help you.

2 Look at graph F.
   - a Describe how the rate of deforestation has changed in the Amazon rainforest.
   - b Suggest why this has happened.

3 Study panels B–E. Then carry out an environmental impact assessment for four human activities in rainforests.
   - a Make a copy of the table opposite. For each human activity in the first column, give a score from 0 to 5 for its impact on the trees, soil, water and atmosphere. (0 is no impact and 5 is a major impact.)
   - b For each activity, add the scores together in the sixth column to work out an environmental impact score from 0 to 20.
   - c Rank the four activities according to their environmental impact.

4 Write a report for the Brazilian government to recommend how tropical rainforest should be used. You should consider the environmental impact and the possible economic benefit for each activity.

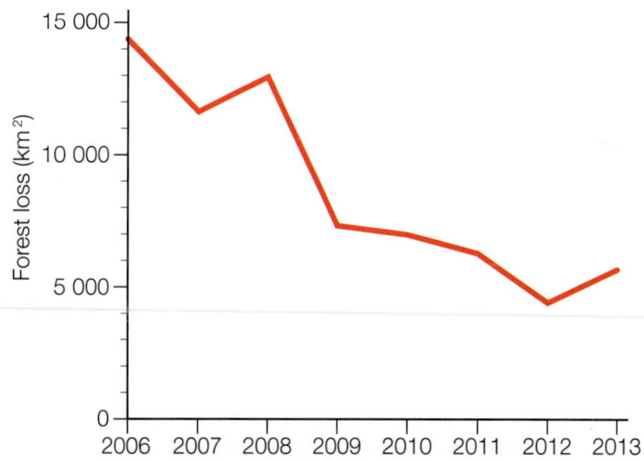

F *Amazon deforestation, 2007–2013*

	Trees	Soil	Water	Atmosphere	Environmental impact score
Logging					
Mining					
Farming					
Tourism					

Practice question

Describe and explain the impact of one human activity on tropical rainforest. (4 marks)

Costa Rica – sustainable rainforest management

On this spread you will learn what Costa Rica does to protect its rainforest and manage it more sustainably.

Case study

Ecosystems in Costa Rica

Costa Rica is a small country in Central America. The whole country is more than a hundred times smaller than the Amazon rainforest, yet it is home to 6% of the world's biodiversity and it attracts over 2 million tourists a year. Many of these tourists come to experience its tropical rainforests (photo A).

A *Rainforest and coastline, Costa Rica*

Although small, Costa Rica boasts a wide variety of ecosystems. It has:

- tropical rainforests
- cloud forests (rainforests at higher altitude)
- dry forests (in areas with lower rainfall)
- mangrove swamps (on the coast)
- beaches (an ecosystem as well as a tourist attraction)
- coral reefs (offshore)
- volcanoes (not an ecosystem, but they influence ecosystems around them).

Protecting forests

In the 1960s Costa Rica began to experience rapid deforestation, mainly due to cattle ranching and agricultural development. Between 1970 and 1990, deforestation reached one of the highest rates in the world, with 32 000 hectares of forest cut down each year.

The government recognised the problem and established protected areas of forest throughout the country. They passed laws to stop deforestation. Importantly, they gave landowners a financial incentive to protect forest by paying them to keep it rather than turn it into farmland. Today, there are 28 **national parks** and **nature reserves** in Costa Rica, and 24% of the country's land area is protected (map B).

B *National parks in Costa Rica*

Sustainable management of rainforest

Deforestation in Costa Rica reached its peak in the 1970s (graph C). Since then, the rate of deforestation has declined to almost zero. In addition to the creation of protected areas, there are other strategies for the sustainable management of forests.

- **Agroforestry:** Trees and crops are grown together. Some crops benefit from the shelter of the tree canopy. Trees also prevent soil erosion and crops benefit from the nutrients in the leaf litter.
- **Selective logging:** Trees are felled only when they reach a certain height. Young trees are allowed to mature to ensure that the height of the canopy is maintained.
- **Afforestation:** Trees are planted to replace the original primary forest that has been lost. However, this secondary forest lacks the biodiversity of primary forest as it takes a long time for species to re-establish.
- **Monitoring:** Satellite technology and photos are used to ensure that no illegal activities are taking place and that landowners follow guidelines for sustainability.

Activities

1. Look at photo A. What can you say about the physical geography of Costa Rica from this photo? Mention at least three things.
2. Look at map B.
   - **a** Describe the distribution of national parks in Costa Rica, particularly in relation to land over 1000 metres and the coastline.
   - **b** Suggest why some of the national parks extend into the sea. (Clue: think about the range of ecosystems in Costa Rica.)
3. Look at graph C.
   - **a** Describe how the rate of deforestation in Costa Rica has changed.
   - **b** Explain why this happened.
4. Suggest what the benefits of rainforest protection would be for these people:
   - **a** a forest landowner
   - **b** a resident of San José, capital of Costa Rica
   - **c** an American tourist in Costa Rica
   - **d** you.

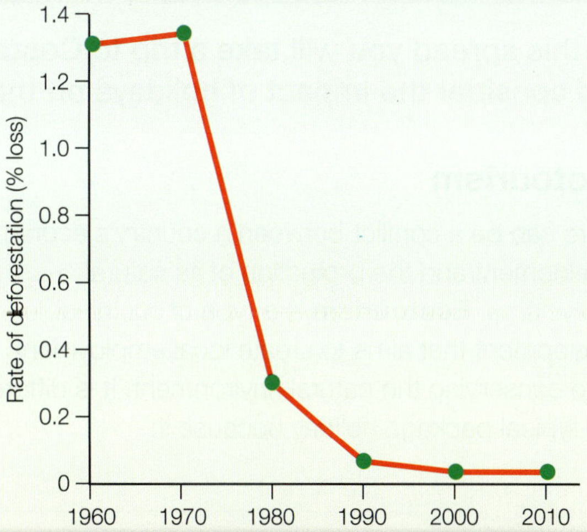

C *Rates of deforestation in Costa Rica since 1960*

A wildlife corridor

Costa Rica, together with neighbouring countries, is trying to create a continuous wildlife corridor along the length of Central America from Mexico to Panama. It is called the Mesoamerican Biological Corridor and it involves all the countries in the region. The aim is to reconnect the fragments of rainforest to provide routes for forest animals to migrate, rather than be trapped in isolated patches of forest.

Go further

Find out more about national parks in the UK.

- How many are there? Draw an outline map of the UK with the national parks marked and labelled.
- Where are they located?
- What are their main aims?
- What natural ecosystems are found there?

Practice question

Using a named example of an area of tropical rainforest you have studied [Costa Rica], evaluate how successful one sustainable management strategy has been. (6 marks)

Ecotourism in Costa Rica

On this spread you will take a trip to Costa Rica as a tourist and consider the impact of holidays on the environment.

Ecotourism

There can be a conflict between a country's economic development and the protection of its natural ecosystems. **Ecotourism** is a type of sustainable development that aims to create local employment while conserving the natural environment. It is different to a typical package holiday because it:

- does not damage the natural environment
- is on a relatively small scale
- consults with the local community about planned developments
- ensures that infrastructure improvements benefit local people as well as tourists.

Samasati Nature Retreat

Samasati Nature Retreat is a prime example of Costa Rica's wilderness and exuberance. Your experience begins by driving up a gravel road lined with huge trees and exotic flowers, through the coconut groves and wild beaches of the Caribbean coast. You will find yourself in one of the most magnificent natural settings on Earth – the tropical rainforest.

At the top of the hill you can enjoy a superb view of the sparkling ocean below. There, a unique and secluded traditional Caribbean world awaits you.

A One example of ecotourism is the Samasati Nature Retreat in Costa Rica

B Accommodation at Samasati is designed to minimise the impact of tourism

Ecotourist activities

Samasati offers a range of activities for tourists that also are intended to have minimal impact on the environment. These include:

- bird watching
- canopy exploration (photo C)
- horseback riding
- sea and river kayaking
- walking and hiking
- white-water rafting
- yoga and meditation.

Ecotourism may not appeal to everyone and, with limited numbers, it can be a bit exclusive. However, we don't need to go to Costa Rica, or even visit a rainforest, to be ecotourists. We can enjoy sustainable holidays much closer to home. We just need to ask the right questions before we go on holiday (table D).

C *Exploring the rainforest canopy on a zip-wire*

	Questions to ask about your holiday	The impact of a typical package holiday
	Where will the money for my holiday go?	Most of the money goes to holiday companies in the UK.
	What impact will my holiday have on the environment?	The environment is changed by tourism development.
	How will the number of tourists affect the resort?	Large numbers of tourists change the character of resorts.
	What resources does my holiday use?	Luxury holidays put pressure on resources like water.
	How does my holiday affect natural ecosystems and wildlife?	Natural ecosystems are destroyed and wildlife is lost.

D *How sustainable is your holiday?*

Activities

1. Look at brochure A. Would a holiday at Samasati appeal to you? Why, or why not?
2. Look at photo B. Explain how each factor around the photo would help to make a holiday at Samasati more sustainable.
3. Look at photo C.
   - **a** Why are the activities at Samasati sustainable?
   - **b** Can you think of any unsustainable activities people do on holiday? Give examples and reasons why they are unsustainable.
4. Look at table D.
   - **a** Answer the questions for a holiday at Samasati.
   - **b** What further information do you need about Samasati to judge whether holidays are really sustainable or not?

Your enquiry

Why should tropical rainforests matter to us?

One of the aims of Samasati is to give tourists a close encounter with tropical rainforest so they learn about its importance. You are going to plan a holiday at Samasati for people to gain a better understanding of rainforest. This could include some of the activities already available at Samasati or other activities you can think of.

Answer the questions below and then plan an itinerary for a five-day trip to Samasati.

- What should tourists see or experience in the forest?
- What activities would they have to do in order to see or experience them?
- What could they learn from the holiday?

4.9 Polar regions

On this spread you will go from pole to pole to identify the characteristics of polar regions.

At opposite ends of Earth, around the North and South poles, are the two polar regions – the Arctic and Antarctic (photo A). There are many similarities between them, but there are also differences.

A *Ice in Antarctica*

Similar climates

Both the Arctic and Antarctic experience cold, dry climates with all-year ice cover on land and sea. The coldest temperature ever recorded was –89°C at the Vostok Research Station in Antarctica. This was a bit extreme, even by the standards of polar regions!

The reasons for such extreme climate are:

- Winters are long and dark with little sunlight. In mid-winter there is up to 24 hours of darkness (diagram B).
- The angle of the Sun is low in the sky, even in summer. When the Sun's rays hit the ground they spread over a wide area, so give less intense heat (diagram A on spread 4.3).
- Much of the surface is covered in ice, which reflects the Sun's energy back into space. This is the **albedo effect**.
- Sinking air at the poles brings high pressure and dry, desert-like conditions. With low temperatures, little water evaporates and few clouds are formed.

B *Summer and winter at the poles*

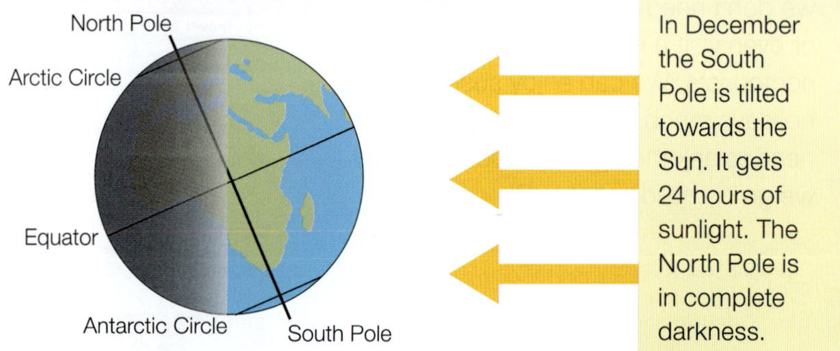

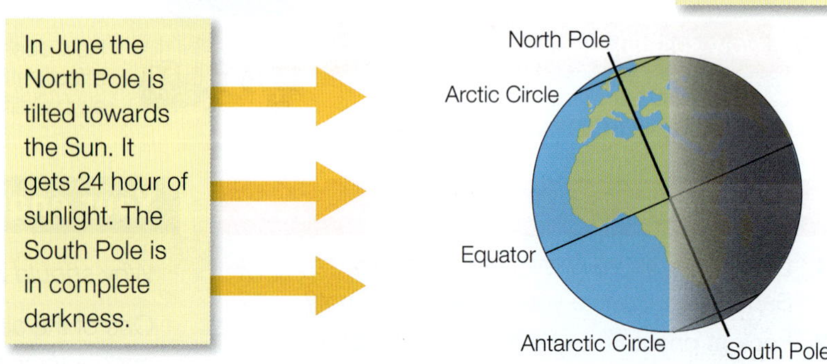

Differences between the Arctic and Antarctic

The Arctic is mainly ocean, almost completely surrounded by land. In winter the ocean is covered by **sea ice**, but in summer temperatures rise (graph C) and some of the ice melts. With climate change, the area of ocean that is ice-free each summer is increasing (see spreads 2.2 and 4.13).

The Antarctic is a continent surrounded by ocean. Most of the land suffers extreme cold in winter and temperatures never rise above 0°C. Only around the edge of the continent does the ocean moderate the climate and temperatures rise above freezing.

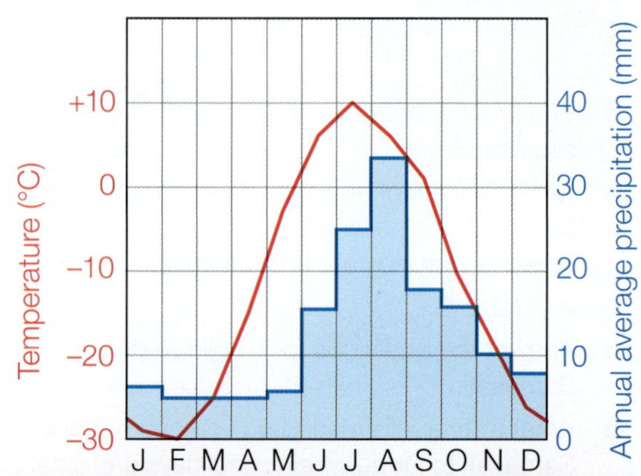

C *Climate in Pruhoe Bay, Alaska, in the Arctic*

Variations within polar regions

Ice sheets cover most of Antarctica and much of Greenland (in the Arctic). In the ocean, around the edges of the land, **ice shelves** and **icebergs** are formed (drawing D).

Further from the poles, the climate becomes less extreme and temperatures rise above freezing in summer. In the Arctic tundra, snow covers the land during the winter but melts in summer to allow a short growing season for plants. This helps animals, like the caribou, to survive (photo E). 98% of Antarctica is ice so very little vegetation grows. In the oceans, other animals, like seals and whales, thrive in both the Arctic and Antarctic.

D *Landscape features of the Arctic and Antarctic*

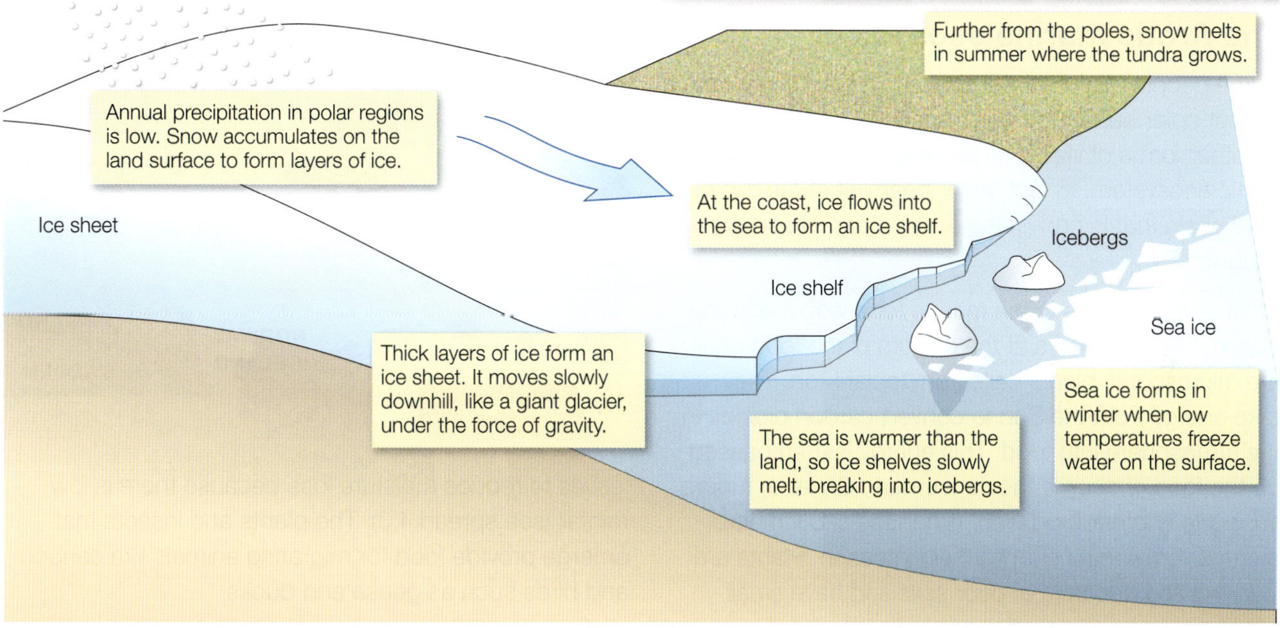

Activities

1 How typical is photo A of polar regions? Explain.

2 Look at diagram B.
  - a Explain why polar regions are cold.
  - b Explain why the coldest time of year differs between the Arctic and Antarctic.
  - c Suggest why the middle of the Antarctic is colder than the middle of the Arctic.

3 Look at graph C.
  - a Describe the annual climate pattern for Prudhoe Bay. Mention changes in temperature and precipitation.
  - b Explain the following features of the climate:
    - June is warmer than December.
    - Annual precipitation is low.

4 Look at drawing D. In your own words, describe each of these features:

   a ice sheet   b ice shelf   c iceberg   d sea ice.

5 Look at photo E. Caribou are adapted to life in the Arctic tundra. Explain how each of these adaptations help them to survive:

   a thick fur coats   b they migrate south in winter
   c a diet of low-growing plants.

E *Caribou live in the Arctic tundra*

Your enquiry

Is there more to polar environments than ice?

As you work through this section, make notes about anything that will help you to answer the question. You will bring your ideas together to do an activity at the end of spread 4.14.

Life in the Arctic

On this spread you will explore the Arctic to find out how different parts of the ecosystem are interdependent.

Arctic location

The Arctic is the region north of latitude 60°N around the North Pole. It gives its name to the Arctic Ocean, which occupies most of the region. In winter the ocean is covered by sea ice (map A).

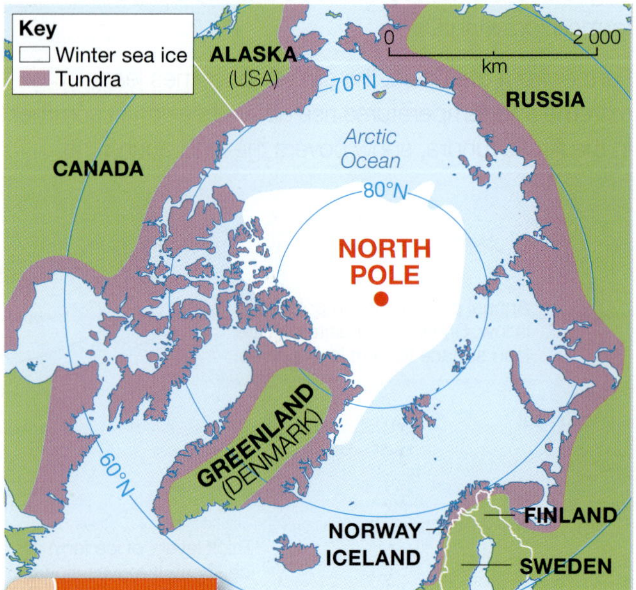

A *The Arctic*

The Arctic marine ecosystem

Some people might think that nothing lives in the Arctic except polar bears, but polar bears must depend on other forms of life. They provide evidence of the Arctic ecosystem. In fact, polar bears sit at the top of a food chain that is part of the Arctic marine **food web** (diagram B).

Most of the species in the marine food web live in the ocean. The producers at the bottom of the food chain are microscopic plants called phytoplankton. They take energy from the Sun to convert carbon dioxide in the atmosphere into food. This energy is then passed up the food chain, via the consumers. On land, tundra supports another food web. When the snow melts, summer provides a short growing season. Plants are low-growing because of poor soils and have small leaves to reduce moisture loss because there is low rainfall (see spread 4.3). The plants and insects that emerge provide food for migrating animals like caribou and birds such as geese and ducks.

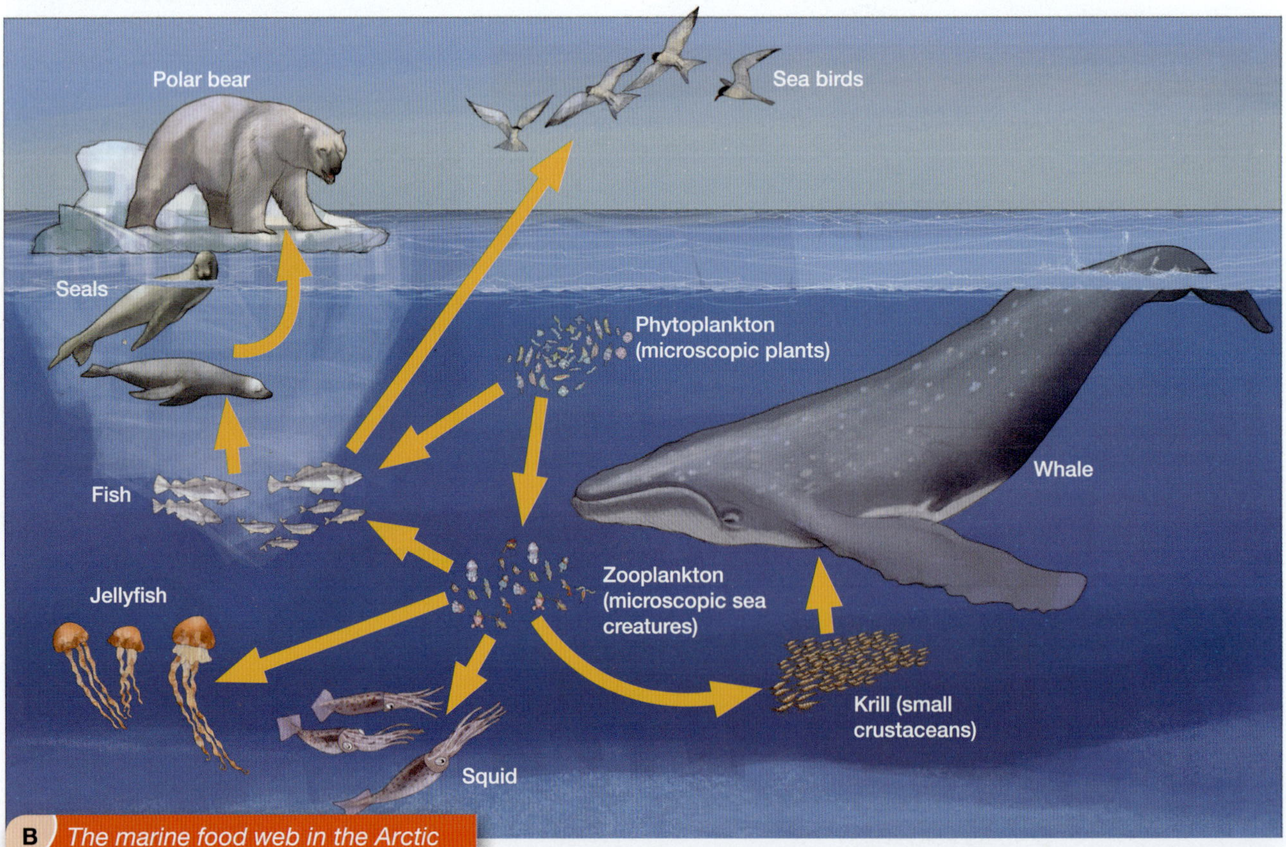

B *The marine food web in the Arctic*

Arctic soils

In those parts of the Arctic where there is any soil at all, it is generally thin and of poor quality. Low temperatures mean that the underlying rock weathers slowly and few nutrients enter the soil.

Permafrost is another feature of Arctic soils. This is the part of the ground that remains permanently frozen all year. Even in areas of the Arctic where air temperatures rise above 0°C in summer, parts of the ground can remain frozen. The upper layer of soil that thaws in summer is the **active layer**. Further from the pole, as temperatures rise, this active layer becomes deeper (diagram C).

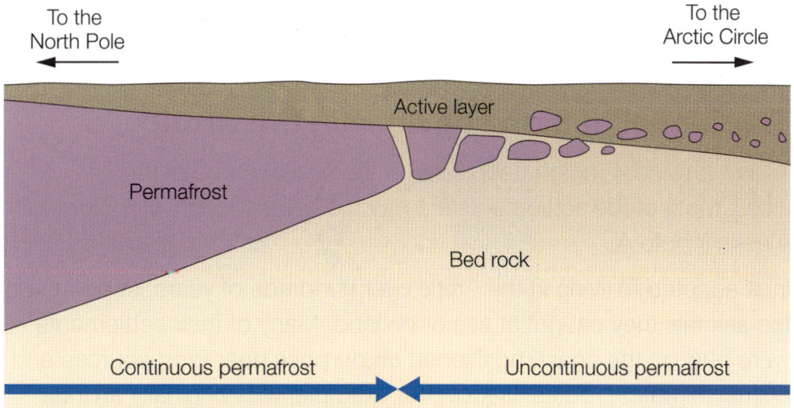

C *Permafrost in the Arctic*

Arctic homes

Living in the Arctic presents many challenges for people. Traditionally, **Inuit** – the indigenous people of the Arctic – lived in tents when they were hunting in summer and built igloos in winter.

Now people live in modern buildings, adapted to Arctic conditions (photo D). Homes are built on stilts a metre or so above the ground. This is to prevent heat from the building melting the permafrost. If that happened, the ground would become soft, the foundations would collapse and the building would sink into the ground (photo E).

D *Modern homes in the Arctic*

E *The effect of melting permafrost*

Activities

1 Look at map A. Name the countries that surround the Arctic Ocean.

2 Look at diagram B.

a Name one example of each of these types of creature in the Arctic marine food web:
- producer
- primary consumer
- secondary consumer
- tertiary consumer.

b Draw a diagram of the species in a food chain with the polar bear at the top.

c Suggest why people might get the idea that nothing lives in the Arctic (apart from polar bears).

3 Identify the ways in which water is interdependent with the following. (Clue: remember ice is also water.)

a Plants and animals b Soils c Human activity

4 Look at diagram C.

a Describe what the terms 'permafrost' and 'active layer' mean.

b Explain why they get deeper or thinner moving further from the North Pole.

5 Look at photo D. Try to explain each of these features of the buildings:

a Buildings are on stilts. b Roofs slope steeply.

c Windows are small.

Practice question

Describe the ways in which climate and soils are interdependent in the Arctic or Antarctic. (4 marks)

Human impacts in the Arctic

On this spread you will investigate the impact of human activities in the Arctic.

Traditional human activities in the Arctic

Today, around 4 million people live in the Arctic, some of them in cities. Most of the settlements are much smaller though, like Barrow in Alaska (photo A).

Inuit adapted to living in the Arctic over hundreds of years, surviving on the animals they caught at sea or on land. Many of their settlements were built on the coast, positioned on purpose near food sources and animal migration routes. Barrow, for example, was originally an Inuit settlement. It is located on the northern tip of Alaska, where whales migrate from the Arctic to the Pacific.

A *Barrow, Alaska. Notice the sea ice in the Arctic Ocean*

Seals and whales

In the 18th century, European explorers discovered the large numbers of seals and whales in the Arctic. For the next 100 years or so, boats came north to the Arctic to hunt these animals, which were valuable for their oil. By the 20th century many species were driven nearly to extinction.

There are now international agreements to protect seals and whales, limiting how many can be killed (see spread 4.12). Hunting for seals and whales in the 21st century is on a much smaller scale.

Modern human activities

The traditional Inuit way of life was sustainable. It was in harmony with the natural ecosystem and did not over-exploit it.

The Arctic ecosystem is particularly fragile and sensitive to the changes that human activities bring (map B). The most important activities these days are:

- fishing
- mineral exploitation
- drilling for oil and gas
- shipping routes
- tourism.

However, even more important than these activities, is the likely impact of climate change on the Arctic ecosystem. As global temperatures rise, the area of summer sea ice in the Arctic is shrinking (see spread 2.2). As the ice disappears and the region becomes more accessible, it opens up the possibility of even more human activities in the Arctic.

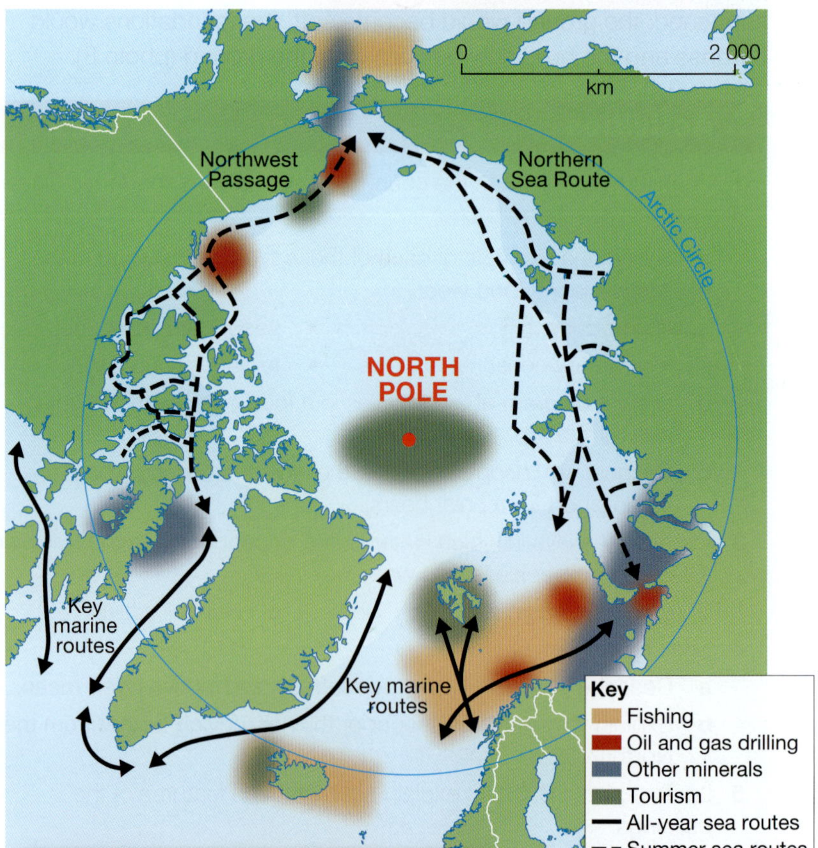

B *Present-day activities in the Arctic*

Oil and gas exploration

The Arctic has become the world's new frontier in the search for oil and gas. In 2009 new surveys revealed that the seabed beneath the Arctic Ocean holds 30% of the world's remaining untapped gas resources and 13% of the untapped oil.

Together with melting ice, this discovery brought large companies, such as Shell and ExxonMobil, to search for oil and gas (photo C). It is ironic that climate change, in part due to burning fossil fuels like oil and gas, should make it easier to drill for even more oil and gas!

C *An oil rig in the Arctic*

Threats to the ecosystem

Most of the new oil and gas reserves in the Arctic lie offshore, around the coastline. Drilling for oil poses many threats to the fragile ecosystem:

- Oil spills in the Arctic could be impossible to clean up. There is no proven method for cleaning up an oil spill in icy water.
- Response to an oil spill will be slow. There are few ports around the Arctic and it could take days or weeks for a boat to reach the spill.
- Recovery from an oil spill will be slow. In low temperatures, limited sunlight and a short growing season, it takes the ecosystem longer to recover.
- Species are at risk from human activities. Drilling for oil can disturb fish and sea animals, which then migrate to other parts of the ocean. Even worse, animals may be injured or die.

Environmental organisations like Greenpeace and WWF (World Wide Fund for Nature) campaign for a ban on drilling and other activities in the Arctic. In 2015 Shell gave up its controversial drilling operations in the Arctic Ocean for the foreseeable future. The decision was made as a result of the high cost of drilling in the Arctic and because of opposition from environmental organisations.

Activities

1 Look at photo A.
   a Describe the location of Barrow from the photo.
   b Explain why this was a good location for an Inuit settlement.
2 Look at map B.
   a Describe the distribution of modern human activities in the Arctic. Write a sentence about the location of each activity.
   b Suggest reasons for the location of each activity. For example, *Fishing is located in areas where there are lots of fish, close to countries that have a fishing industry.*
3 Look at photo C. Explain the connection between the oil industry and global climate change.
   a Why does climate change make it easier to drill for oil and gas in the Arctic?
   b How could drilling for oil and gas contribute to further climate change?

Go further

Find out more about recent changes in the Arctic.

- What is the most recent extent of summer sea ice?
- Which oil companies are still drilling in the Arctic?
- Have any new agreements been reached to limit the areas where drilling is allowed?

Practice question

Outline the impact of one human activity on the Arctic or Antarctic ecosystem. (6 marks)

Sustainable management of whaling

On this spread you will find out how whaling has changed and investigate one small-scale example of its sustainable management in the Arctic.

Case study

Sustainable whaling

People have been hunting whales in the Arctic for centuries. Inuit hunted whales in small boats using simple harpoons and ropes. They caught whales, one at a time, for their oil, meat and bones. This type of **whaling** was sustainable because the Inuit population was small and caught only what the people needed.

Whaling in the 20th century

By contrast, whaling in the 20th century grew to an industrial scale (photo A). Rather than hunt for single whales, factory ships were introduced, harvesting whales in large numbers. The ships were used to catch, transport and process the whales, which were used in the production of margarine, chemicals, cosmetics and animal feed. The number of whales killed reached an all-time high in the 20th century but had fallen dramatically by the end of the century.

- By the late 1930s, 50 000 whales were being killed every year and whale populations began to decline. Clearly, this was unsustainable.

A *Whales caught by a whaling ship*

- In 1946 the International Whaling Commission (IWC) was set up to help conserve whales and to control the whaling industry. It still sets the quotas (numbers) of whales each country is allowed to catch.
- In 1986 the IWC introduced a ban on commercial whaling. But loopholes in the ban allowed some countries to carry on. Japan continued to catch whales for what they called **scientific research**.
- Since 1986, 30 000 more whales have been killed, despite the ban. Most of the whales are caught by Japan and Norway. However, the IWC does allow a small number of whales to be caught by indigenous communities in countries like Canada and Greenland (graph B).

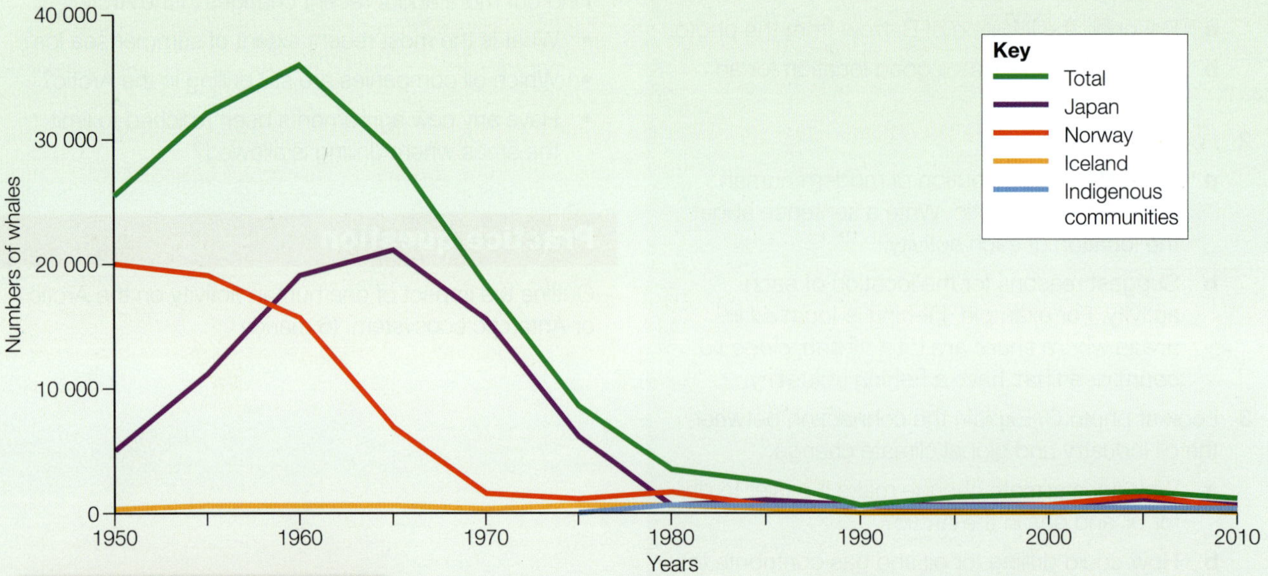

B *Whales caught since 1950*

A marine wildlife sanctuary

In 2008, Clyde River on the coast of Baffin Island in northern Canada became the country's first Marine Wildlife Area (map C). The area extends 12 nautical miles from the coast and covers 3360 km^2. It is now a sanctuary for the conservation of bowhead whales, one of the 17 whale species found in the Arctic.

Up to 2000 bowhead whales stop over annually on their migration around the coast of northern Canada. The sanctuary includes two deep offshore troughs that are rich in zooplankton on which the whales feed. As well as bowhead whales, polar bears, narwhals, seals, fish, geese and ducks benefit from the sanctuary.

A community of about 1000 Inuit live in Clyde River. They supported the establishment of the sanctuary. So you may be shocked that in 2014 the community successfully hunted and killed a bowhead whale! You may be even more surprised that they were congratulated by the environmental organisation, Greenpeace (extract D). However, whale-hunting on a small scale like this can be part of a sustainable management strategy for whale populations.

Greenpeace supports Inuit whale hunting

A dramatic 90-minute battle involving explosions and a flurry of harpoons resulted in the death of a 18-metre bowhead whale on 3 August this year. So ended the first whale hunt in 100 years by Inuit hunters from Clyde River.

Despite losing one of their six boats in the hunt, the locals were overjoyed with their catch. While they were still processing the whale's huge carcass, congratulations came from the most surprising source – Greenpeace.

This environmental group made its name with the slogan 'save the whales', but Greenpeace's new strategy is to 'make amends with indigenous peoples' and make a stand with them against oil drilling in the Arctic. So Greenpeace honoured the people of Clyde River for killing a mammal still considered to be an endangered species in parts of the Arctic.

D *Adapted from news reports, 2014*

C *The location of Clyde River, on Baffin Island, northern Canada*

Activities

1. Look at photo A. What is your opinion about whaling? Should it be banned or not?
2. Look at graph B.
   - **a** Describe how the total number of whales caught has changed since 1950.
   - **b** How has the number of whales caught by indigenous communities changed? Why?
3. Look at map C.
   - **a** Describe the annual range of bowhead whales around the coast of Canada.
   - **b** Explain why Clyde River might be a good location for a wildlife sanctuary.
4. Read extract D.
   - **a** Explain how Inuit whale hunting can be part of the sustainable management of whaling.
   - **b** Do you agree with Greenpeace's decision to support Inuit whale hunting? Give reasons for your viewpoint.

Practice question

Examine how successful one small-scale example of sustainable management in either the Arctic or Antarctic has been. (8 marks)

4.13 Sustainable management of the Arctic

On this spread you will find out how the Arctic is protected by global action at the moment and what ideas there are for the future.

Case study

A fragile environment

The Arctic is a **fragile environment**. That means it can easily be damaged by human activities such as resource exploitation in the Arctic or even by air pollution from activities in other parts of the world. Perhaps the greatest threat to the Arctic environment is from climate change due to global warming. Sea ice in the Arctic Ocean is melting. Each summer the minimum area of ice in the ocean shrinks a little further (see spread 2.2). By the middle of the 21st century, the Arctic Ocean could be like any other ocean in the summer – totally ice-free (photo A).

A *What hope is there for polar bears, which spend much of their lives on sea ice?*

Arctic wilderness

The land and ocean of the Arctic make up the largest remaining area of **wilderness** – a wild, natural area in which few people live – in the northern hemisphere. Only the Antarctic, in the southern hemisphere, is larger. There are eight countries in the Arctic and the two continents, North America and Asia, almost completely enclose the Arctic Ocean.

The future of the Arctic matters, not just to these countries but to all of us. It plays an important role in regulating global climate by reflecting more heat back into space. The Arctic also supports many unique marine and land species, such as polar bears, narwhals and reindeer, and is a breeding place for migratory birds and mammals, such as whales.

However, unlike the Antarctic, very little of the Arctic is protected. On land, some areas of ice and tundra have been given protected status as wilderness by their national governments (map B). Each country has its own laws, protecting their land and also their marine territory, up to 200 nautical miles from the coast.

B *Wilderness areas in the Arctic*

Protecting the Arctic

One way to protect the Arctic and give it a more sustainable future would be with an international treaty, or agreement. The **Arctic Council** was formed in 1996 by governments of the eight countries, together with six indigenous communities living in the Arctic. It is one of the first regional government bodies in the world devoted to environmental protection. However, up to now, there is no international treaty and the Arctic Council has no legal powers to prevent member countries carrying out harmful activities.

There are already some international treaties that could help protect the Arctic. The **Paris Agreement**, signed by 195 countries in 2015, is the first legally binding global climate agreement. It sets out a global action plan to help the world to avoid dangerous climate change by limiting global warming to less than 2°C. It might help to slow the rate at which Arctic sea ice is melting. Other international treaties that could help protect the Arctic include:

- United Nations Convention on the Law of the Sea
- Convention on Biological Diversity
- Long Range Treaty on Atmospheric Pollution.

An Arctic Sanctuary?

The environmental organisation Greenpeace has made a proposal for an Arctic Sanctuary, based on the much larger Antarctic Treaty (see spread 4.14). It wants to establish a marine nature reserve where any human activity harmful to wildlife would be banned.

The sanctuary would cover the central Arctic Ocean, beyond the 200-nautical-mile limit from the coasts of Arctic countries. This area, usually described as the 'high seas', is shared by the international community. No one country has any rights over it. The total area covered by the sanctuary would be 2.8 million km^2, much smaller than the whole Arctic Ocean, which is 14 million km^2.

Greenpeace's proposal for the Arctic Sanctuary includes:

- no fishing
- no exploration or drilling for oil or gas
- no mining on the seabed
- no military activity
- strict environmental controls on shipping.

C *The Arctic Sanctuary proposed by Greenpeace*

Activities

1. Look at photo A. Why might this become a rare sight in the Arctic in future?
2. Look at map B.
   - **a** Estimate the percentage of wilderness in the Arctic that is protected.
   - **b** Write a paragraph to argue the case for protecting a larger area of land and ocean in the Arctic.
3. Read about the Arctic Council. What do you think is:
   - **a** a benefit of the Arctic Council?
   - **b** a weakness of the Arctic Council?
4. Look at map C. How might the people below respond to the proposal for an Arctic Sanctuary? In each case, give a reason.
   - **a** An Inuit living in Greenland
   - **b** President of Russia
   - **c** Chief Executive of Shell (the oil company)
5. **a** Do you agree with the ideas in Greenpeace's proposal for an Arctic Sanctuary? For each idea, give a reason why you do or do not agree.
   - **b** What other ideas would you include in an Arctic Treaty?
   - **c** Make a list of all the ideas you would include in an Arctic Treaty.

Practice question

Describe what role global actions play in the sustainable management of a polar environment you have studied [the Arctic]. (6 marks)

4.14 Towards an Arctic Treaty

On this spread you will devise a new Arctic Treaty as a way to manage the Arctic more sustainably.

Case study

The need for an Arctic Treaty

It has been suggested that, for the Arctic to have a sustainable future, it needs its own treaty similar to the Antarctic Treaty. Despite the work of the Arctic Council to protect the environment, the Arctic still faces many threats:

- global warming is melting sea ice in the Arctic Ocean
- resource exploitation, like drilling for oil, damages the environment
- pollution from other parts of the world affects Arctic ecosystems
- populations of unique Arctic species, such as the polar bear, are declining.

The Antarctic Treaty

The Antarctic Treaty, signed in 1961 by 46 countries, has become one of the most successful international agreements of all time. In 1961, the environment in the Antarctic was also under threat. In 1991, all member countries declared their commitment to maintaining and strengthening the Treaty. In 1998, they agreed a new Environmental Protocol.

As a result, the Antarctic environment is protected and the only human activities permitted are scientific research (photo A) and a limited amount of tourism. Scientific research in Antarctica has made a big contribution to our understanding of the Earth and climate change (see Chapter 2). Research there in the 1980s discovered the ozone hole in the atmosphere (photo B). Ozone is a gas found high in the atmosphere that protects us from the sun's radiation. Pollution from human activities was destroying it.

A *Rothera Research Station in Antarctica, run by the British Antarctic Survey; 14 other countries also have research stations there*

The Antarctic Treaty

- Prohibits military bases and weapons testing.
- Requires co-operation between countries to share the results of scientific research.
- Prohibits nuclear explosions and nuclear waste disposal.
- Does not recognise any country's claims on territory.

The Environmental Protocol

- A ban on mining for 50 years, or until all countries agree to end it.
- A ban on killing or interfering with wildlife (no dogs can be taken to Antarctica).
- All waste must be removed (dumping or burning rubbish is banned).
- There are controls on fishing in the Southern Ocean around Antarctica.
- All activities must be assessed for their likely impact on the environment before they are allowed.

B *The ozone hole, discovered by scientists working in Antarctica, is shown in purple*

The Arctic and Antarctic compared

The similarities and differences between the Arctic and Antarctic (see spread 4.9) could be important when thinking about a new Arctic Treaty. They could help us to decide to what extent the Arctic Treaty should be similar or different to the Antarctic Treaty.

Similarities

- Both the Arctic and Antarctica are polar regions, close to the poles at opposite ends of Earth.
- They have extreme climates, cold all year round with long dark winters and 24 hours of daylight.
- They have both land and marine ecosystems, with plants and animals adapted to the climate (though polar bears only live in the Arctic and penguins only live in the Antarctic).
- The ecosystems are fragile and easily damaged by changes in the environment.
- They are both challenging environments for people to live in.

Differences

- The Arctic comprises eight countries and an ocean surrounded by two continents. Antarctica is a single continent surrounded by ocean.
- Indigenous inhabitants live in the Arctic. The only people living in the Antarctic are scientists working at the international research stations.
- Arctic land consists of areas of ice and tundra. Antarctica is covered by a massive ice sheet, up to 5 km thick, with only 2% of the land being ice-free.
- Countries in the Arctic each control their own land and also part of the seabed. No country owns land in Antarctica.
- There has already been a lot of resource exploitation and military activity in the Arctic, whereas none is allowed in Antarctica.

C *The South Pole, surrounded by the flags of some of the countries that signed the Antarctic Treaty*

Activities

1 Look at photo A.
  - a Why is Antarctica a good place to do scientific research?
  - b Could scientists do similar research in the Arctic? Explain your answer.

2 Read the Antarctic Treaty and Environmental Protocol.
  - a How successful do you think they have been in sustainably managing the Antarctic? Give reasons.
  - b To what extent is the Antarctic Treaty a good model for sustainable management in the Arctic? Give arguments for and against.

3 Look at photo C.
  - a Would it be possible to mark the location of the North Pole in a similar way? Explain your answer.
  - b Forty-six countries agreed the Antarctic Treaty. Why might it be more difficult for countries to agree an Arctic Treaty?

Your enquiry

Is there more to polar environments than ice?

Devise a new treaty for the Arctic modelled on the Antarctic Treaty. It will not be exactly the same because the Arctic is different from the Antarctic. First, think about these questions:

- Why are both the Arctic and the Antarctic worth protecting?
- What are the differences between the Arctic and the Antarctic?
- How would this make a difference to the treaty you write?
- Which parts of the treaty might be the same?

Now write an Arctic Treaty of not more than 200 words. You could use bullet points.

Urban Futures

What is this topic about?

Cities – you either love them or hate them! Never before has the landscape of our planet looked more urban. Cities today are growing faster than at any time in human history. Over half the world's population now lives in a city. In the UK, over 80% of us live in an urban area.

'Urbanisation' is the word we use in geography to describe the growth in the proportion of people living in cities. It's a word that keeps appearing in this chapter. You are going to find out why urbanisation happens and what the challenges and opportunities for cities are.

Your enquiry

In this chapter you will investigate the specification's two key questions.

Why do more than half the world's population live in urban areas?
You will:

- Compare urbanisation rates around the world.
- Define the terms 'megacity' and 'world city'.
- Describe the global spread of megacities and world cities.
- Understand why people live in cities and why they keep growing.
- Investigate the consequences – good and bad – of urbanisation.

What are the challenges and opportunities for cities today?
You will:

- Study two cities – Birmingham in the UK and Istanbul in Turkey.
- Consider the importance of those cities to their region, their country and the wider world.
- Understand patterns of migration to the cities.
- Explore ways of life in the cities.
- Investigate the challenges of living in the cities.
- Describe one initiative in each city to make it more sustainable.

Look at the photo opposite.

- What do you love about cities and what do you hate?
- What advantages do cities have for people who live there?
- Are cities too crowded or could more people fit in?
- Where do you think this city is? What clues can you find?

An urban world

On this spread you will compare urban growth rates in different parts of the world.

Urbanisation

In 2010 the world's population reached a new landmark. For the first time in history, half of us were living in cities.

180 000 people are added to the urban population of the world every day. With urbanisation, the proportion of people in cities grows each year. By 2030, 60% of us are expected to live in cities and by 2050 it will be 70%.

Urbanisation happens at different rates in each country (map A). The most urbanised countries are mainly the **advanced countries (ACs)** of Europe, North America and Oceania. There, the rate of urbanisation has slowed down because most people already live in cities.

The most rapid urbanisation is happening in **emerging and developing countries (EDCs)** in Asia and **low-income developing countries (LIDCs)** in Africa where the majority of people still live in the countryside.

Worldwide, the rate of urbanisation is slowing down as more people are living in cities (graph B).

A *Urban population by country*

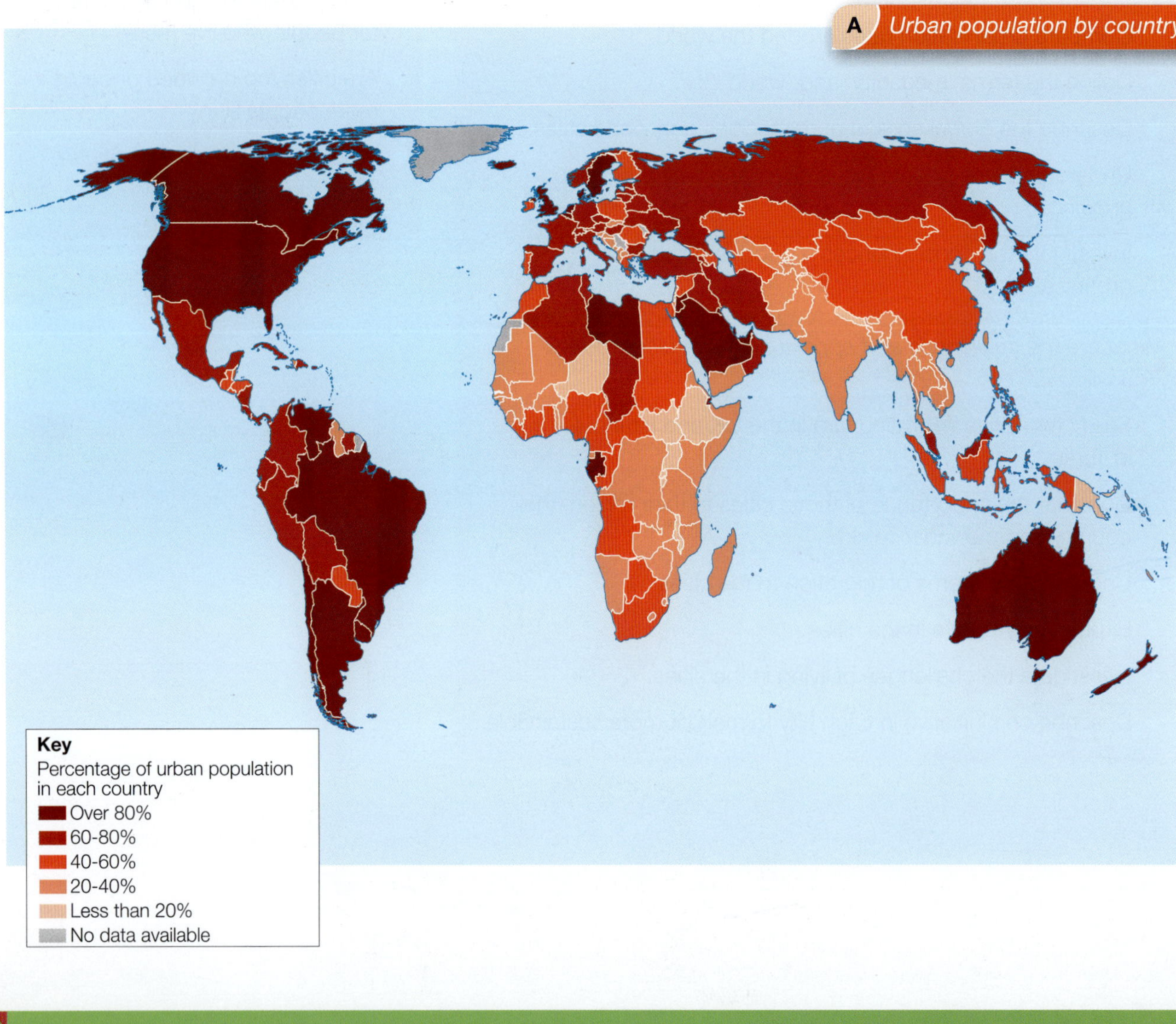

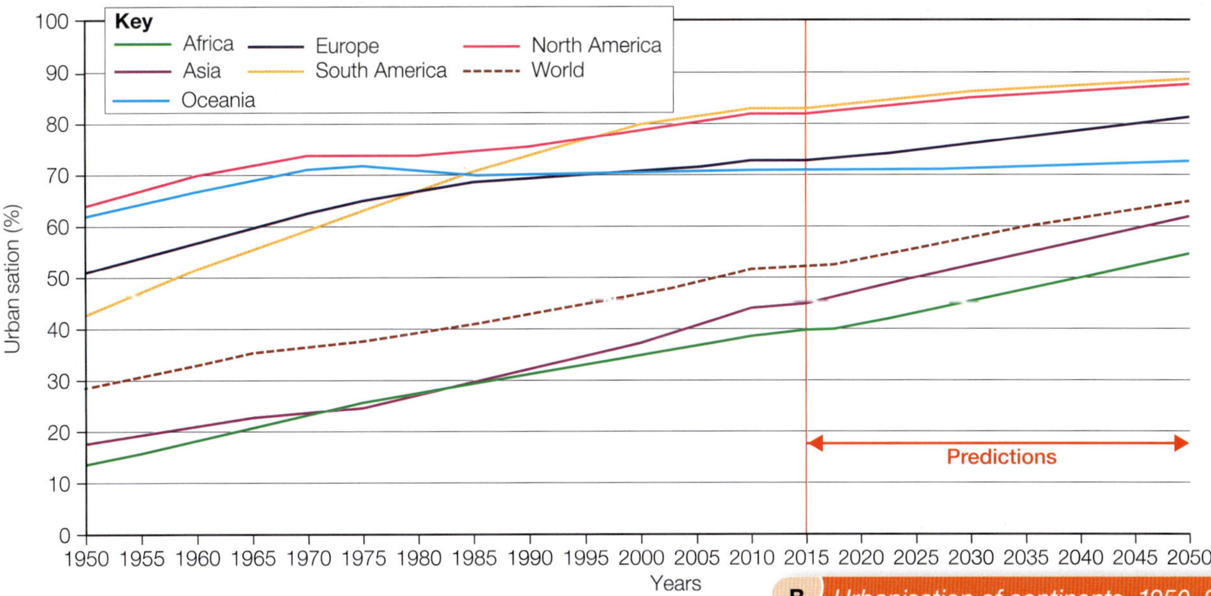

B *Urbanisation of continents, 1950–2050*

C *London – a city in the UK where over 80% of people live in cities*

D *Shanghai – a city in China where over 50% of people live in cities*

Activities

1 Look at map A. Name three countries where the urban population is:
a over 80% **b** 60–80% **c** 40–60% **d** 20–40% **e** less than 20%.
First try on your own. Then check your answers in an atlas.

2 Describe the overall pattern on map A.
- **a** In which continents are the countries more urbanised?
- **b** In which continents are the countries less urbanised?
- **c** Name any countries in each continent that are exceptions to the overall pattern.

3 Look at graph B. Write two or three sentences to describe the changing urban population for each continent. Give figures from the graph. For example, *The urban population in Europe has gone up, but the rate of urbanisation is slowing down. In 1950 just over 50% of the population was urbanised. By 2015 it was over 70% and by 2050 it will be over 80%.*

4 Look at photos C and D.
- **a** Compare the buildings in the two photos. What do they show about the age of the two cities and how they are urbanising now?
- **b** Explain the differences from what you know about the rate of urbanisation in the UK and China.

Your enquiry

Why do more than half the world's population live in urban areas?

As you work through this section, make notes about anything that will help you to answer the question. You will bring your ideas together to do an activity at the end of spread 5.7.

5.2 Super-sized cities

On this spread you will get to know what and where megacities and world cities are.

Megacities

As the world's urban population grows, so do the number of super-sized cities. **Megacities** are cities with a population of over 10 million people. London is a large city, with a population of over 8 million, but even so it does not qualify as a megacity.

Most of the world's megacities are in Asia (map A). Not only is Asia home to over half the world's population, it is also urbanising rapidly.

Factors causing some cities to grow into megacities

- **Natural population increase:** Cities tend to have better healthcare, so death rates fall and life expectancy increases when people live in cities.
- **Migration:** People migrate to cities from the countryside, or even from other countries, to find jobs or a better life.
- **Coastal location:** Many megacities have grown from port cities because a coastal location is an advantage for trade.
- **Economic development:** More trade means more business, so these cities grow into centres for production, which creates more jobs and attracts more people.

A *Megacities, 2015*

B *Megacities, 1950*

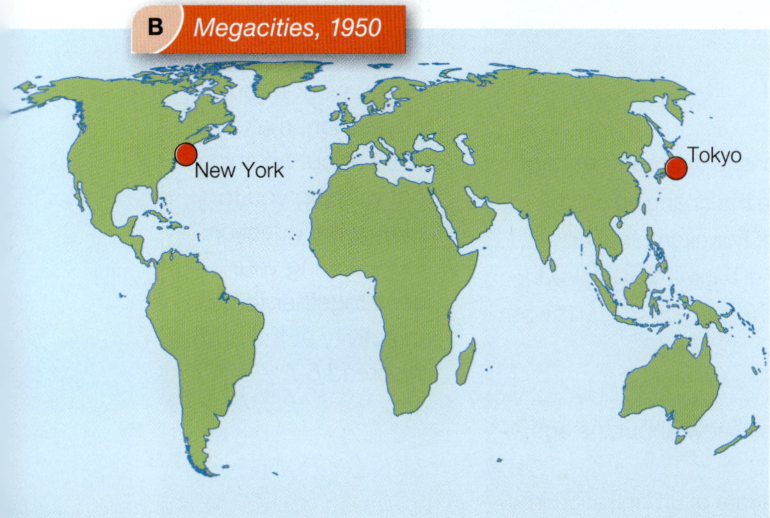

C *Megacities, 1975*

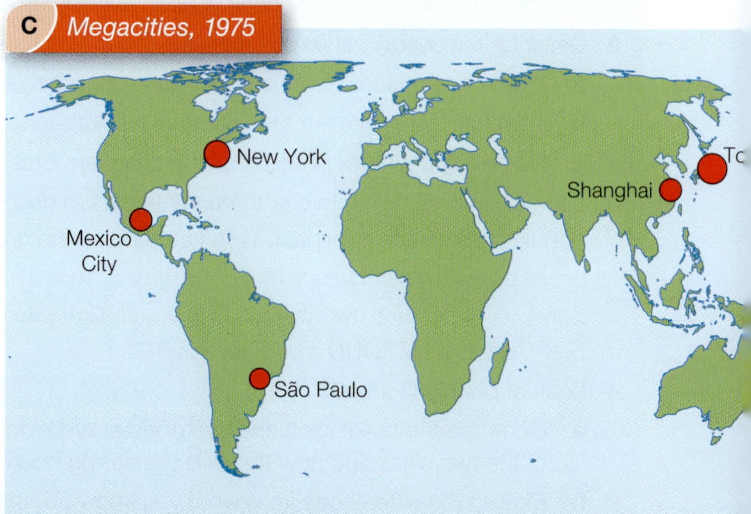

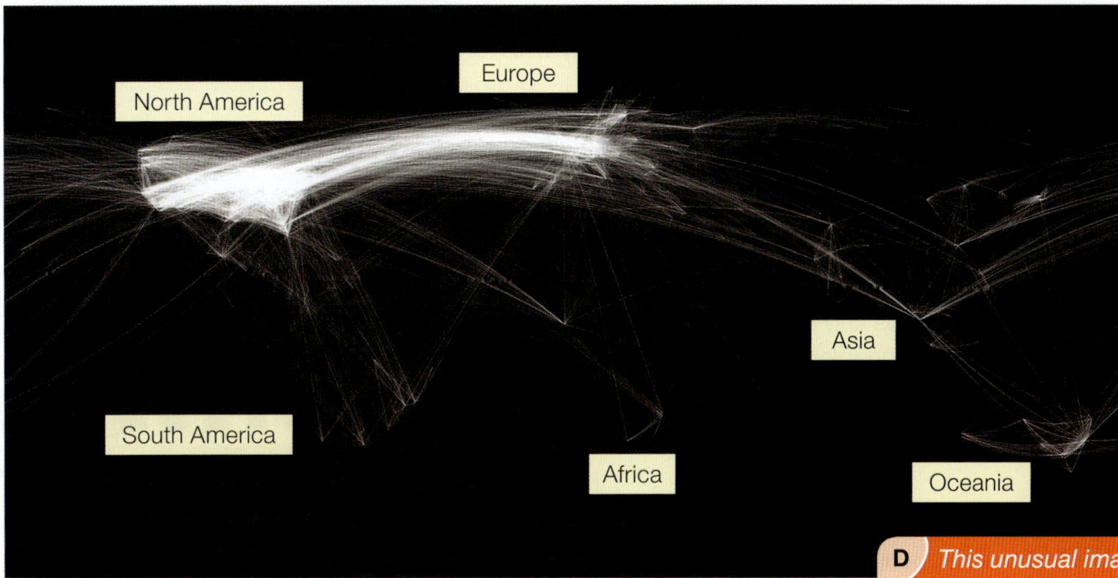

D *This unusual image was created to represent internet connections between the top world cities*

World cities

The most important cities in the global economy are known as **world cities**. Some of them are also megacities, but they don't have to be. For example, London is a world city but not a megacity.

World cities are hubs for international trade. Many international banks and the headquarters of large **trans-national companies (TNCs)** are located in world cities. They are at the centre of global communication networks (map D).

The top two world cities, London and New York, rival each other as centres of world trade. They are known as **alpha++ cities**. They are followed by another group known as **alpha+ cities** (table E).

E *Alpha++ and alpha+ cities*

Alpha++	Alpha+
London	Chicago
New York	Dubai
	Hong Kong
	Paris
	Shanghai
	Singapore
	Sydney
	Tokyo

Activities

1 Give a definition in your own words for these terms:
 a megacity
 b world city.

2 Look at map A. Rank the ten largest megacities in the world in a table. Name the country and continent in which each city is found.

Rank	Megacity	Country	Continent
1	Tokyo	Japan	Asia

3 Compare maps A, B and C. How did the following factors change between 1950 and 1975, and between 1975 and 2015?
 a The number of megacities
 b The distribution of megacities.
 Write two sentences for each pair of maps.

4 Look at image D.
 a Describe the pattern of global communication connections between world cities.
 b Explain the pattern

Go further

Choose one megacity or one world city. Find out more about the city. Suggest why it has developed into either a megacity or a world city.

Practice question

Describe the difference between a megacity and a world city, giving a named example of each. (4 marks)

How cities began and grew

On this spread you will investigate why cities grew in the first place and how their functions are changing.

The first cities

People have been living in cities for thousands of years. The first cities grew in river valleys like those of the Nile in Egypt and the Tigris and Euphrates in the Middle East. Their fertile floodplains and regular water supply helped farmers to grow crops and rear animals. The surplus food they produced allowed some people to stop farming and move into settlements. There, they learned new skills and traded with each other. Eventually, these settlements grew into cities.

A The Colosseum was a sporting arena in Rome almost 2000 years ago

BC

4000 The first cities appear in Egypt, the Middle East, India and China.

2000 Cities at the centre of the Greek empire and, later, the Roman empire grow larger.

AD

300 Rome is the first city to reach 1 million people. Later the population falls.

600 Cities decline in Europe during the Dark Ages.

1300 Cities in South America grow at the heart of the Aztec and Inca empires.

1450 The population in European cities declines further because of the Black Death.

1750 British cities grow during the Industrial Revolution. Cities in Europe follow.

1800 London becomes the first modern city with 1 million people.

1950 New York becomes the first megacity of 10 million people.

2010 Over half the world's population now live in cities – most of them in Asia.

B Timeline of urban growth

C Birmingham during the Industrial Revolution in the 19th century

Urban growth

The big growth in cities came with the Industrial Revolution in the 19th century. In Britain, inventions like the steam engine led to the development of large-scale manufacturing. Cities like Manchester and Birmingham grew as factories were built. Migration brought people to the cities from the countryside to work in the factories. There was similar growth of cities across Europe and North America.

D *The London skyline today*

Urban functions

Cities have many functions and these can change over time. The first cities were a place for people to get together to trade and exchange ideas. Since then, they have developed many other functions:

- **Market:** a place where goods and services can be bought and sold.
- **Employment:** work in manufacturing or services.
- **Administration:** organisation of life in the city and the area around it.
- **Defence:** protection of the city and its population from attack.
- **Entertainment:** such as sports, theatre, shopping and restaurants.
- **Religion:** large places of worship or a destination for pilgrims.
- **Transport hub:** destination for public transport routes and major road intersection.
- **Residential:** places for people to live in houses, apartments – even palaces!
- **Culture:** new ideas can develop, including art, technology and consumption.

Activities

1 Explain how the following activities played an important role in the development of cities:
  a farming
  b industry.
2 Look at images A, C and D. Identify all the functions for which you can see evidence in each city. List the functions and give evidence from the images.
3 Suggest how the functions of London have changed over time. Give evidence from photo D to support your ideas.

Go further

Think about the city you live in or one you know. What are the main functions of the city today? Give at least three functions and evidence of buildings to support your ideas.

Practice question

a For one city you have studied, explain why the city first developed. (2 marks)
b For the same city, explain why it grew. (2 marks)

The urban explosion

On this spread you will find out why the most rapid urbanisation is now happening in Africa.

Urbanisation in Africa

Africa is the world's least urbanised continent, but also the one that is urbanising most rapidly. Of course, those two facts are connected. More people living in the countryside means a bigger supply of people to move to the city. The majority of people in Africa still live in rural areas and work as farmers (photo A).

About 40% of Africa's 1.1 billion population lives in cities (map B) and the number rises every day. By 2030 the proportion will have risen to about 50% and by 2050 it will be 60%.

An important factor in this growth is **rural–urban migration** – the movement of people from the countryside to cities. There are many **push factors** that force or encourage people to move from the countryside (panel C). At the same time, they are drawn by **pull factors** that attract them to the city (drawing D).

A *The majority of the population in Africa are still farming in rural areas*

B *Growth of African cities*

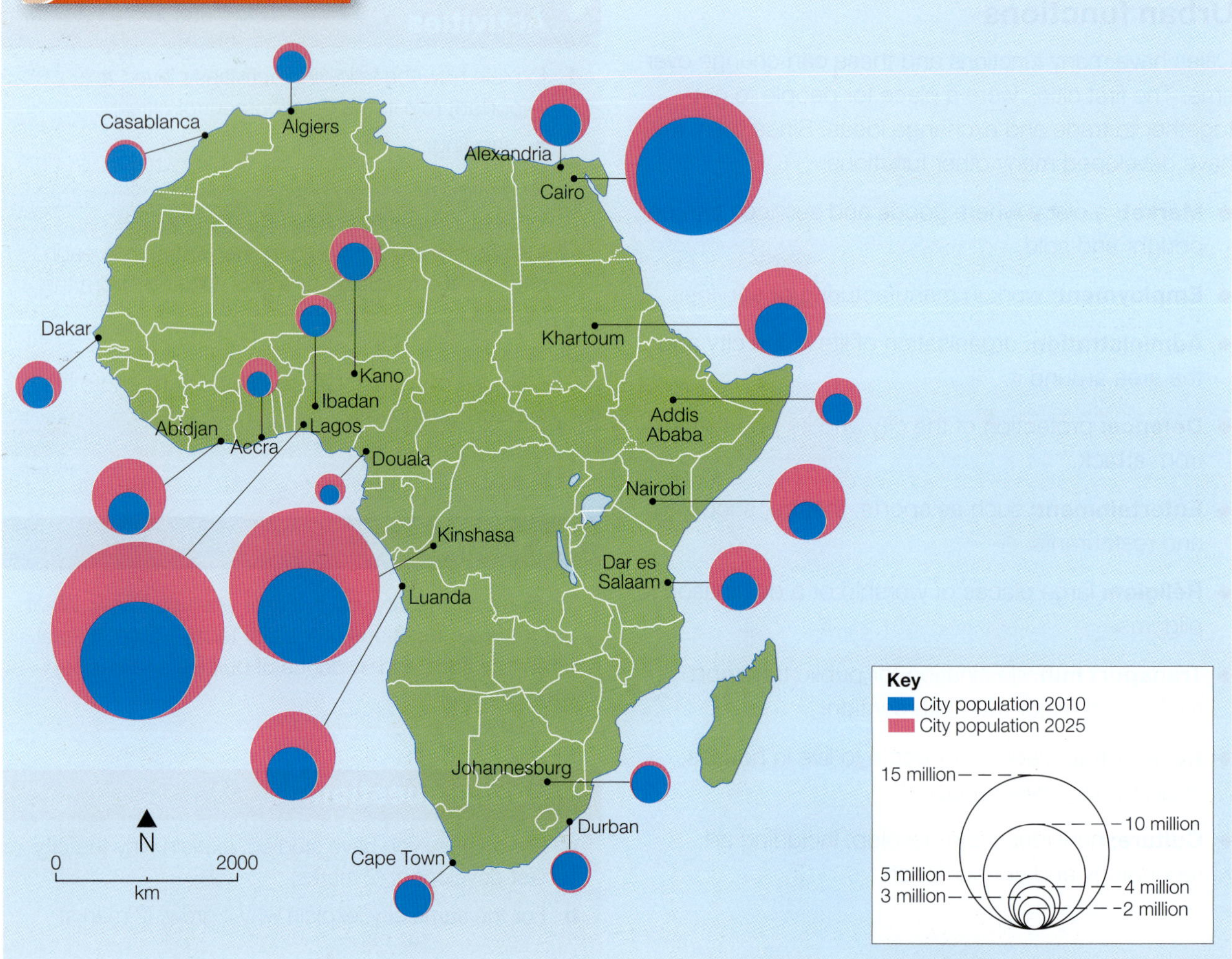

The growth of Lagos

Lagos, in Nigeria, is now the largest city in Africa (photo E). In 1960 the city still had less than a million people, but this had grown to about 15 million in 2015.

The main driver of Lagos' growth has been rural–urban migration. But another reason is **internal growth**, as a result of a higher birth rate than death rate. This is due to the city's youthful population, since most **migrants** to the city are young.

C *Push factors for rural–urban migration*

Many people want to move away from the countryside because of:

- few job opportunities
- lack of access to services such as healthcare and education
- declining income due to the falling price of agricultural products
- overpopulation, leading to farms being broken up into smaller plots
- poor transport networks
- civil war and conflict
- natural disasters such as floods or drought
- climate change, making the growing season less predictable
- food shortage and famine.

D *Push and pull factors for rural–urban migration*

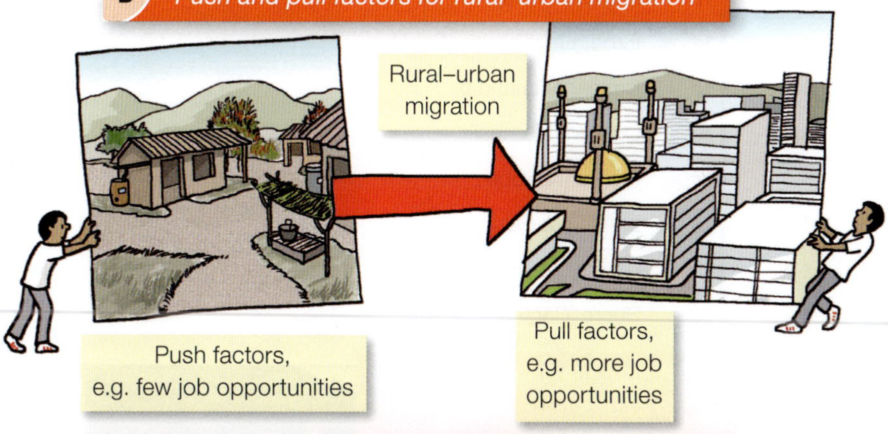

E *Lagos in Nigeria, now the largest city in Africa*

Activities

1 Look at map B.
   a Name the five largest cities in Africa. Which countries are they in?
   b Estimate the population of these cities in 2010 and 2025.
2 Read the push factors for rural–urban migration in panel C and then look at photo E. Think of a list of pull factors for rural–urban migration that would attract people to the city. List at least six factors.
3 Imagine you are an African farmer. Predict whether life in the city would live up to your expectations.
   a Write down your predictions of what the city will be like. Think of at least two positive and two negative aspects of living in a city.
   b Then turn over to read spread 5.5. Which of your predictions were correct and which predictions were wrong?

Practice question

a Explain how the population of cities in LIDCs and EDCs grows by rural–urban migration. (2 marks)

b Explain how the population of cities in LIDCs and EDCs grows by internal growth. (2 marks)

5.5 Slumming it

On this spread you will investigate how slums develop as a consequence of urbanisation.

Slums

Cities in EDCs and LIDCs are growing faster than they can be planned and their **infrastructure** – things like transport, power and water supply – is unable to cope with the growing numbers of people. Another consequence of rapid urban growth has been the development of **slums**. Over 1 billion people in the world live in slums. Most of those people are in cities in LIDCs or EDCs in Africa, Asia and South America (table A).

A slum is a crowded urban area where people live in inadequate housing and poor living conditions (photo B). Slums vary from **squatter settlements** or shanty towns, where people build their own homes on any available land, to older parts of the city where buildings have deteriorated.

Most slums lack basic services, including:

- a clean water supply
- reliable sanitation, including a sewage system
- reliable electricity
- waste collection
- roads and street lighting.

Rank	Slum (city)	Population (millions)	Rank	Slum (city)	Population (millions)
1	Neza/Chalco/Izta (Mexico City)	4.0	11	Cape Flats (Cape Town)	1.2
2	Libertador (Caracas)	2.2	12	Pikine (Dakar)	1.2
3	El Sur/Cuidad Bolivar (Bogota)	2.0	13	Imbaba (Cairo)	1.0
4	San Juan de Lurigancho (Lima)	1.5	14	Ezbet El-Haggana (Cairo)	1.0
5	Cono Sur (Lima)	1.5	15	Cazenga (Luanda)	0.8
6	Ajengunle (Lagos)	1.5	16	Dharavi (Mumbai)	0.8
7	Sadr City (Baghdad)	1.5	17	Kibera (Nairobi)	0.8
8	Soweto (Johannesburg)	1.5	18	El Alto (La Paz)	0.8
9	Gaza City (Palestine)	1.3	19	City of the Dead (Cairo)	0.8
10	Oranji Township (Karachi)	1.2	20	Sucre (Caracas)	0.6

A *Top 20 slums in the world, by population, according to Mike Davis in* Planet of Slums

Why slums grow

- **Rural–urban migration:** Most people move to the city for jobs or education, and housing is not their main priority. They are prepared to live wherever they can afford.
- **Unemployment:** People are forced to live in cheap housing if they can't find a job or if they have to work on low wages.
- **Poor urban planning:** Lack of money or poor organisation mean that cities do not build enough affordable houses for new residents.
- **Poor infrastructure:** Poor roads and lack of transport force people to live in a place where they will be able to travel to work easily.

B *Ajengunle, the largest slum in Lagos*

C *Hazards of living in a squatter settlement*

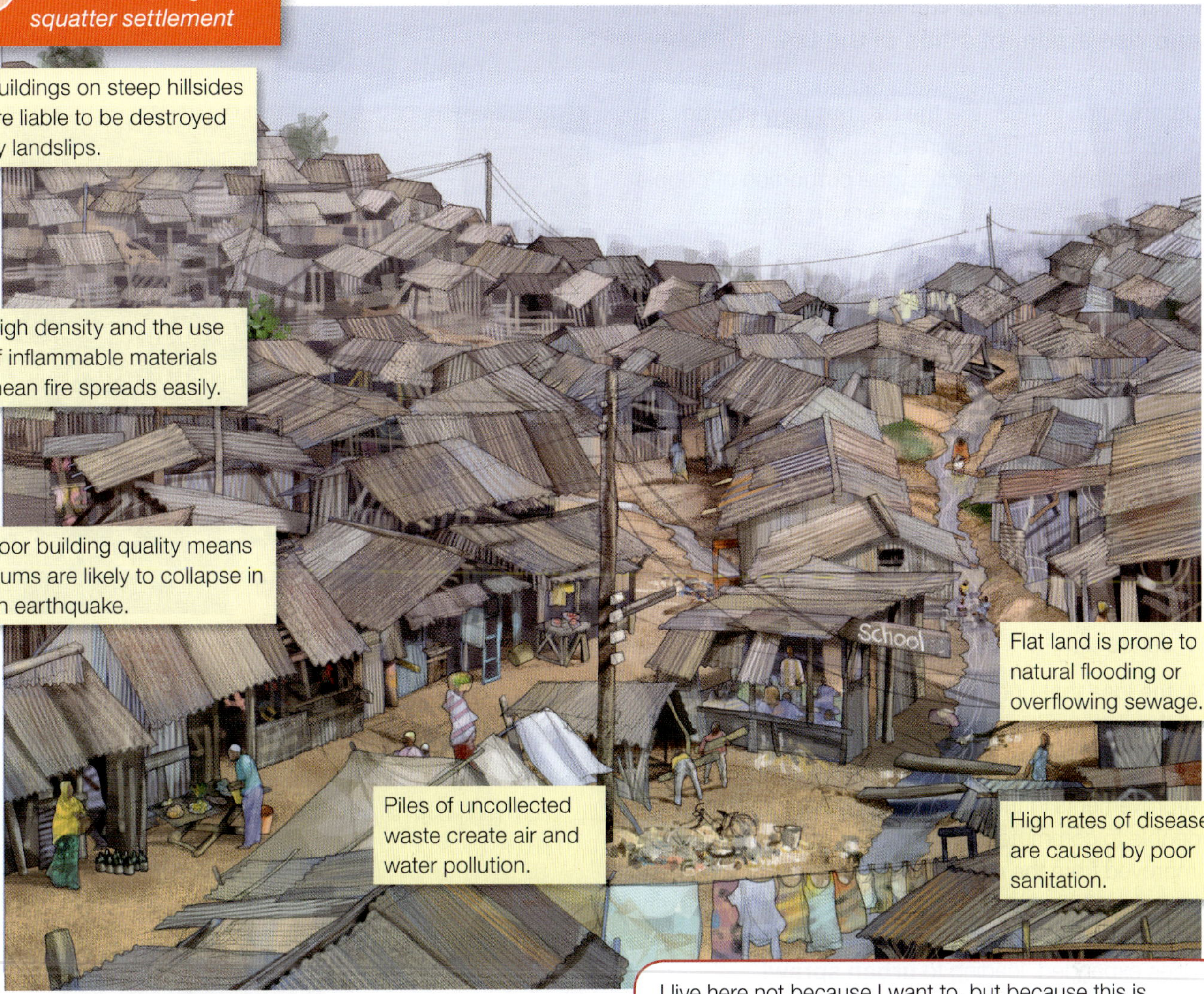

I live here not because I want to, but because this is where I can afford to live for now. It is suitable for the level of income I make. It is not the best place for children to be raised, I admit, but I have to make do with what I have.

D *The views of Mrs Chidi Nmaneke, an Ajengunle resident*

Activities

1. Study table A. Use the data to create a map showing the world's top 20 slums.
   - **a** Find out which country each city is in. Label the countries on an outline map of the world.
   - **b** Show each slum as a circle on the map in the correct country. Give the map a scale and draw each slum in proportion to its population.
2. Use the map you created in activity 1.
   - **a** Describe the pattern on your map. Which parts of the world have most slums? Which have few or none?
   - **b** Try to explain the pattern on your map.
3. Read the quote from Mrs Nmaneke. Imagine how she came to live in Ajengunle. Write her story, explaining why she is prepared to tolerate conditions in the slum. (Look back at spread 5.4 to help you.)

Go further

Choose one of the slums in table A. Find out more about the slum and the city in which it is located, including its physical geography. Identify the hazards in drawing C that might pose a particular threat for your chosen slum. Give reasons why these hazards might be a threat.

Practice question

Explain how the growth of squatter settlements is connected with the rapid growth of cities. (4 marks)

5.6 Urban trends in the UK

On this spread you will follow the rise and fall, and rise again, of cities in the UK.

Urbanisation in ACs, like the UK, has now slowed down (see graph B on spread 5.1). With over 80% of the population living in cities, the proportion of people living in urban and rural areas is quite stable.

Suburbanisation

During the 20th century in the UK, the main process in cities was not urbanisation, but **suburbanisation**. People moved out of the city centre to new residential areas, or suburbs, on the edge of the city (photo A). Suburbanisation was driven by:

- high population density and congestion in city centres
- pollution from industry and, later, road traffic
- cheaper land and house prices further from the city centre
- the idea that quality of life was better closer to the countryside.

Suburbanisation became possible with transport improvements in the 20th century, which allowed people to commute to and from work. First, railways improved from the mid-19th century onwards. Then, from the 1950s, there was an increase in car ownership and improvements in the road network. As a result, cities expanded, leading to **urban sprawl**.

Green belts to hold back urban sprawl

There was concern that cities were growing too large and eating up the countryside. In 1947 the government introduced the idea of a **green belt** – open space around cities on which no further development is allowed. There are 14 green belts in England, covering 13% of the land area (map B). While the green belt has been successful in preserving rural areas, it has also reduced the amount of land available for building new homes. With a shortage of housing in the UK, there is pressure to relax laws about the green belt.

A *Suburban housing in the UK*

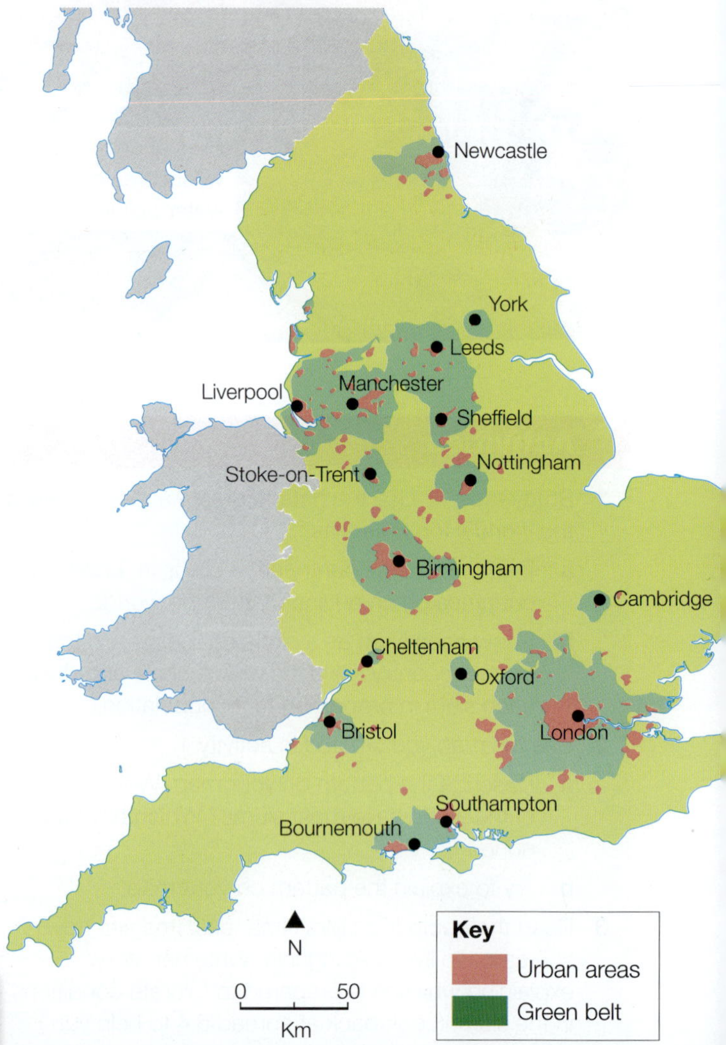

B *The main urban areas and green belts in England*

Counter-urbanisation

During the second half of the 20th century, many UK cities went into decline. People moved out of cities altogether to the countryside or smaller towns. Older people retired to the country or seaside (see spread 7.5), while younger people moved out of the city to raise their families. So old urban areas were neglected and allowed to deteriorate.

Re-urbanisation

Since the 1980s, many cities in the UK have been brought back to life by **urban regeneration** (photo C). This aims to:

- bring derelict land and buildings back into use
- build new homes
- encourage investment in new jobs
- improve the quality of life for local residents
- improve the quality of the environment.

Regeneration has led to **re-urbanisation**, with people moving into cities and the population of most UK cities growing once again (table D). Some cities are growing quicker because their economies are stronger and there are more jobs (see spread 7.9).

C *Urban regeneration in Cardiff*

Rank	City	Population change, 2004–13	Change (%)
1	Milton Keynes	36 200	16.5
2	Peterborough	24 900	15.2
3	Swindon	27 600	14.8
4	Luton	24 400	13.3
5	Cambridge	14 300	12.7
6	London	1088 400	12.6
7	Northampton	22 000	11.3
8	Ipswich	13 500	11.1
9	Cardiff	34 600	10.9
10	Bournemouth	37 600	10.8
55	Glasgow	28 300	2.7
56	Liverpool	20 100	2.6
57	Rochdale	4 800	2.3
58	Hull	4 600	1.8
59	Birkenhead	5 200	1.7
60	Middlesbrough	4 300	0.9
61	Grimsby	1 300	0.8
62	Burnley	700	0.4
63	Blackpool	500	0.2
64	Sunderland	– 4 000	– 1.4

D *Fast- and slow-growing cities in the UK, 2004–2013*

Activities

1 Look at photo A.
   a What are the attractions of living in a suburb? Give at least four.
   b What are the drawbacks? Give at least two.

2 Look at map B.
   a Describe the location of green belt land in England.
   b Explain why the green belt was introduced.
   c Was this a good idea? Give reasons.

3 Look at table D.
   a Mark and label the cities on an outline map of the UK.
   b Describe the pattern on your map. Where are cities growing fastest? Where are they growing slowly?
   c Suggest reasons for the pattern on your map.

4 Complete a large table like the one below.

	Causes	Consequences
Suburbanisation		
Counter-urbanisation		
Re-urbanisation		

Practice question

Compare the causes and consequences of recent urban trends in ACs with those in LIDCs and EDCs. (8 marks)

The future of cities

On this spread you will consider the consequences of living in cities and the prospect of future urbanisation.

Cities are both good and bad

By the year 2050, 70% of the world's population will live in cities. Globally, there is no sign of urbanisation coming to an end yet, even though it has started to slow down. Cities often have challenges and attract negative publicity, but urbanisation offers opportunities too (panels A and B).

A *Opportunities from urbanisation*

- **Quality of life:** Education, healthcare and transport are better in cities as it is easier to provide these services on a large scale.
- **Life expectancy:** People live longer in cities than in rural areas, probably because healthcare is better.
- **Employment:** There are more jobs in cities. Even where unemployment is high, people can find work in informal jobs (unofficial work or self-employment).
- **Squatter settlements:** People can build their own housing on any available land to provide much-needed accommodation in fast-growing cities.
- **Sustainability:** Although cities use resources and create waste, they are more efficient because everything is within easy reach.

B *Challenges of urbanisation*

- **Inequality:** Cities have extreme wealth and poverty, with high rewards for those in top jobs but low pay in informal jobs and no pay for the unemployed.
- **Segregation:** Rich and poor live in separate areas of the city, often with the rich in exclusive gated communities and the poor in slums.
- **Slums:** The development of large squatter settlements creates risks from disease, fire or environmental disasters such as floods and landslides.
- **Overpopulation:** Population growth in the city can exceed the rate at which jobs and homes can be provided.
- **Environmental damage:** Unplanned development can damage the natural environment through air and water pollution and the accumulation of solid waste.

Metacities

A new name – **metacity** – now describes a city with a population of more than 20 million. Tokyo in Japan is currently the world's largest city with 37 million people, but most of the world's future megacities and metacities will be in China.

China has plans to merge a number of cities in the Pearl River Delta, including Hong Kong, Shenzhen and Guangzhou, to create a metacity of 41 000 km², 26 times larger than Greater London. It will have a population of 120 million – almost twice the population of the whole of the UK!

C *The world's predicted urban population in 2050*

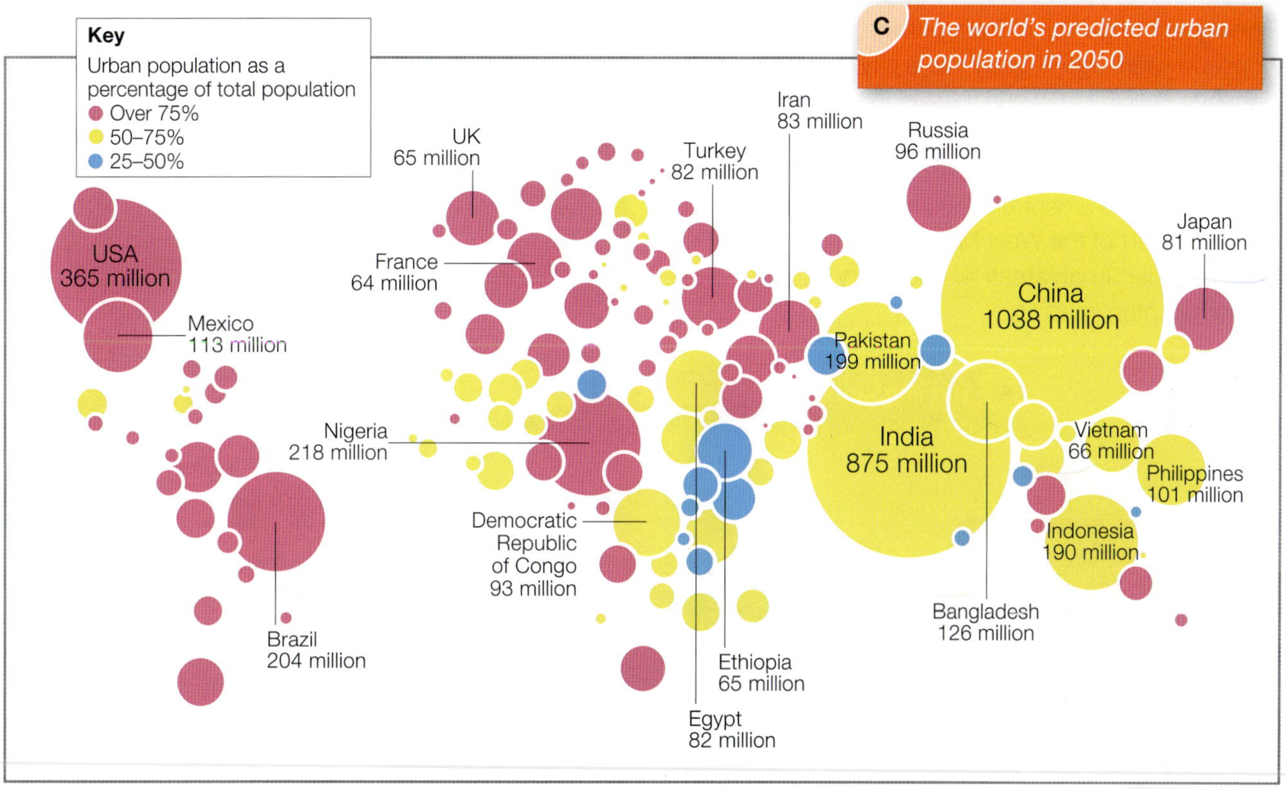

Activities

1 Read the positives and negatives of urbanisation in panels A and B. Think what each of these people's views on urbanisation might be:

- **a** slum resident
- **b** migrant from a rural area
- **c** wealthy business owner
- **d** city mayor.

Choose the most important aspects of urbanisation – both positive and negative – for each person. In each case, give a reason.

2 Look at map C. Describe what the map tells you about:

- **a** how urbanised each continent will be by 2050
- **b** what the urban population in each continent will be by 2050.

Your enquiry

Why do more than half the world's population live in urban areas?

Imagine an alien from outer space visits Earth. The alien asks, 'Why do so many of you humans live in urban areas?' How would you answer?

Break the question down into smaller questions. Write a sentence or short paragraph to answer each one.

- **a** Why did cities first appear?
- **b** Why did they grow bigger?
- **c** Where do most people live in cities? Why?
- **d** Where are cities growing now?
- **e** Why do cities still attract people?

5.8 Birmingham on the map

On this spread you will explore the location and importance of Birmingham, using maps.

Case study

Introducing Birmingham

Birmingham is the UK's second largest city, after London (map A). Its population is about 1 100 000, and growing.

Birmingham began as a Saxon village over a thousand years ago. It grew into a city during the Industrial Revolution, when it developed large-scale manufacturing industries. In recent years, manufacturing has declined and the city has re-invented itself as a centre of culture and shopping. Today it is part of the West Midlands region, which includes other urban areas like Coventry and Wolverhampton (map B).

A Birmingham's location in the UK

B Birmingham's location in the West Midlands

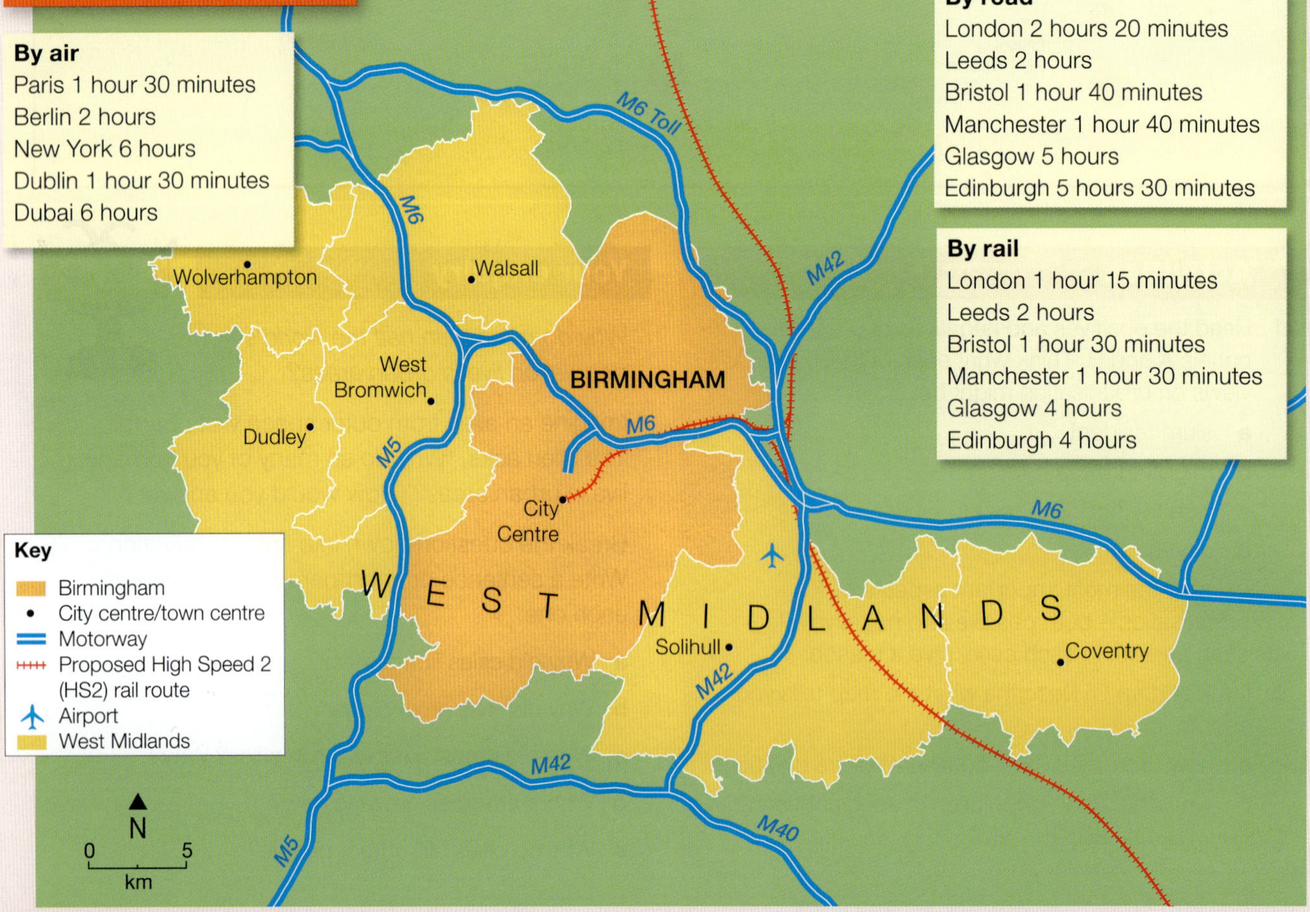

Birmingham's global connections

Birmingham is a young, **multi-cultural** city. Its population has come from all around the world, including people from Asia, the West Indies, Eastern Europe and elsewhere. It has one of the largest Pakistani communities in the world and the largest Irish community in England.

Over the years, while people migrated to the city for work, its factories exported their manufactured products around the world. Birmingham was once known as 'the workshop of the world'. It still has many global connections today (map C).

C *Birmingham's global connections*

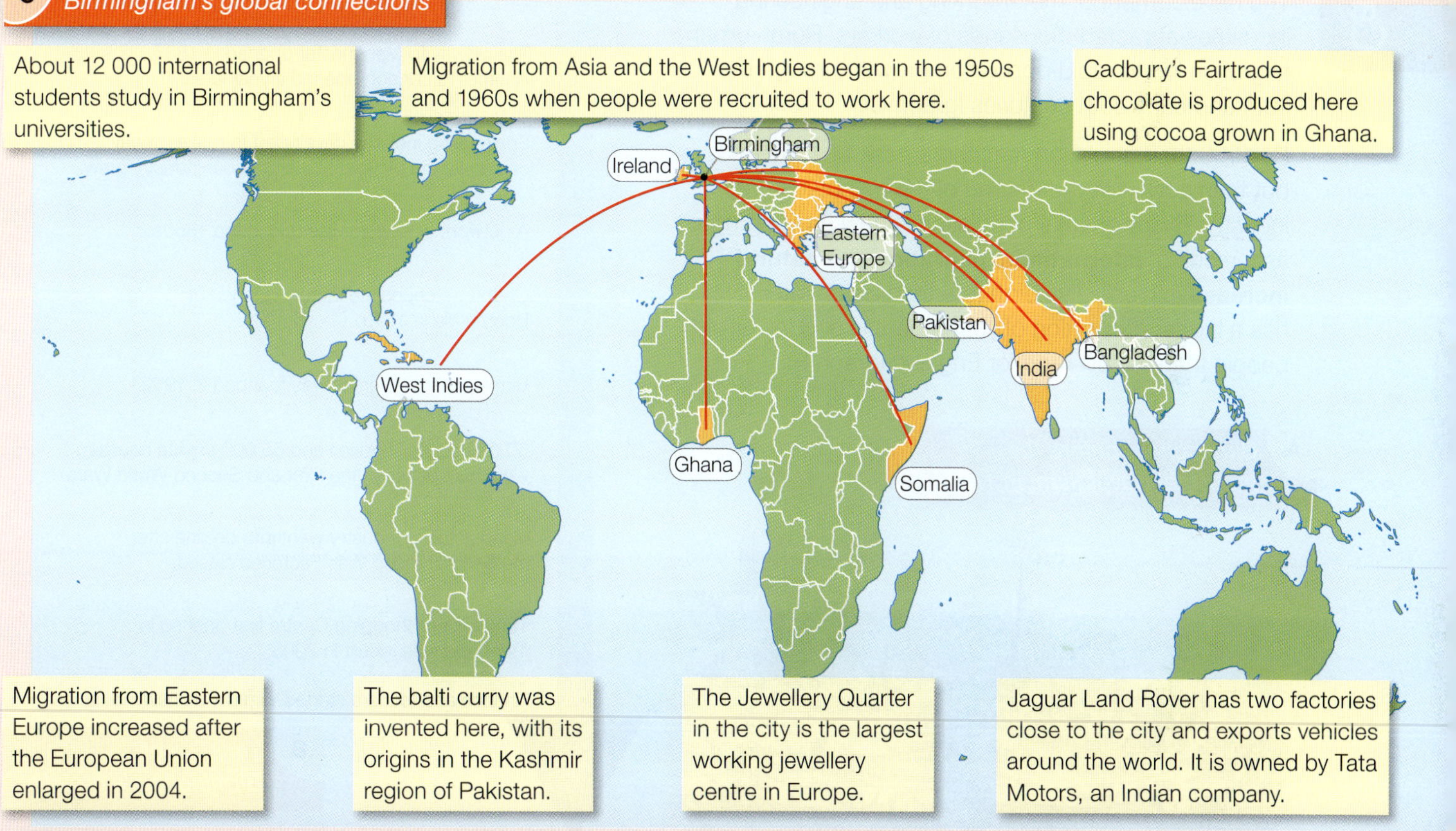

Activities

1 Look at maps A and B. Why might Birmingham be a good location for:
- **a** a manufacturing company that sells products around the UK?
- **b** a retail chain to open a new shop?
- **c** a student to study at university?
- **d** a migrant from Eastern Europe?

2 Draw a large copy of the diagram opposite. List all the places with which Birmingham is connected in the correct circle, e.g. put 'Coventry' in the West Midlands circle. If you can, add the travel time to each place by road, rail or air.

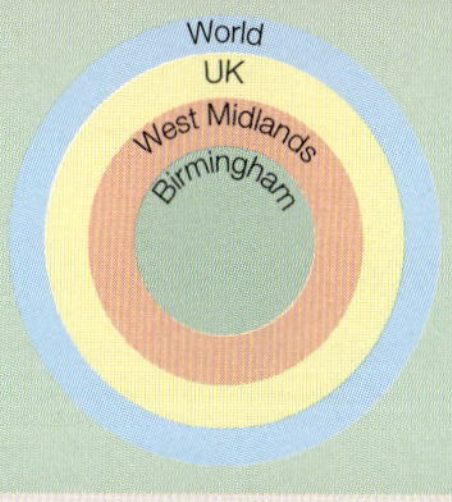

Your enquiry

What are the challenges and opportunities for cities today?

As you work through this section, make notes about anything that will help you to answer the question. You will bring your ideas together to do an activity at the end of spread 5.19.

5.9 Come to Birmingham

On this spread you will find out how past and present migration changes the character of the city.

Case study

Birmingham's growth

In 1700 Birmingham was a small market town of 10 000 people (drawing A). During the Industrial Revolution it grew into a large industrial city. By 1900 it had a population of over half a million people. The development of metalworking and engineering industries attracted thousands of workers. Rural–urban migration led to rapid population growth through the 19th and early 20th centuries (graph C).

By 1950 the population reached a peak of 1 112 000, but then it fell with the decline of the manufacturing industry. Birmingham's population is now growing again as a result of **international migration** and **natural increase** (a higher live birth rate than death rate). It has a higher proportion of young people and fewer old people than the average for England (graph E).

A *Birmingham before the Industrial Revolution, in 1732*

Metalworking in Birmingham began before the Industrial Revolution.

The first canal to Birmingham was built in 1769. The canal network grew.

The city's first General Hospital was built in 1779.

The railway to Manchester opened in 1837. Another to London opened a year later.

Engineering industry developed to make railway carriages and bicycles. Later, the car industry grew.

A network of sewers was dug in the 1850s.

Cadbury's chocolate factory opened at Bourneville in 1879.

Birmingham University was founded in 1909.

50 000 council houses and 65 000 private houses were built between the First and Second World Wars.

Manufacturing industry went into decline after the Second World War. Factories closed.

The Bullring Shopping Centre first opened in 1964 and was rebuilt in 2003.

Finance and tourism started to grow as new industries.

B *Timeline of Birmingham's growth*

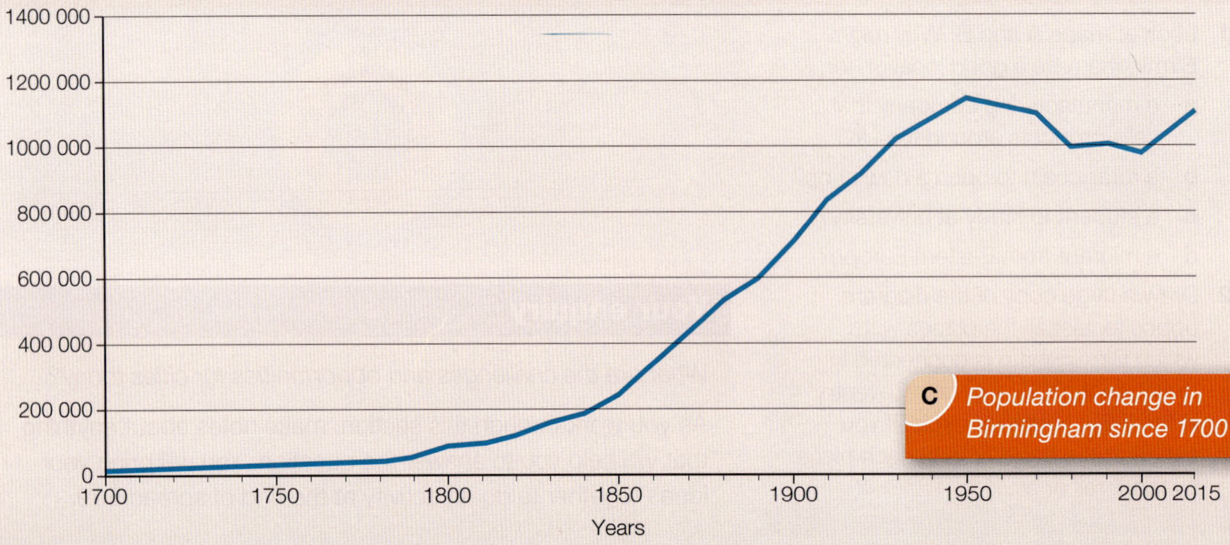

C *Population change in Birmingham since 1700*

Diversity in Birmingham

Birmingham's population is one of the most diverse in the UK outside London (photo D). During the 1950s and 1960s, most immigrants came from south Asia and the West Indies. Since 2000, more people have come from Eastern Europe, Africa and the Middle East. In 2011, only 53% of Birmingham's population were white British (map F). Migrants bring their own culture, language and religion with them. The character of the city is changing with new types of shop and places of worship, for example. However, 86% of the city's population, regardless of **ethnicity**, see themselves as being British.

D Birmingham's multi-ethnic population

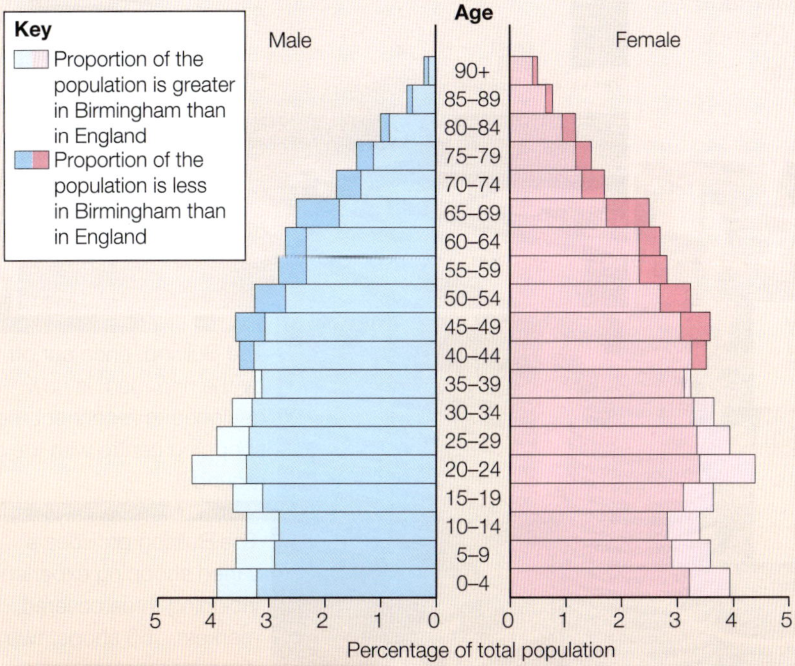

E Birmingham's population compared with England's

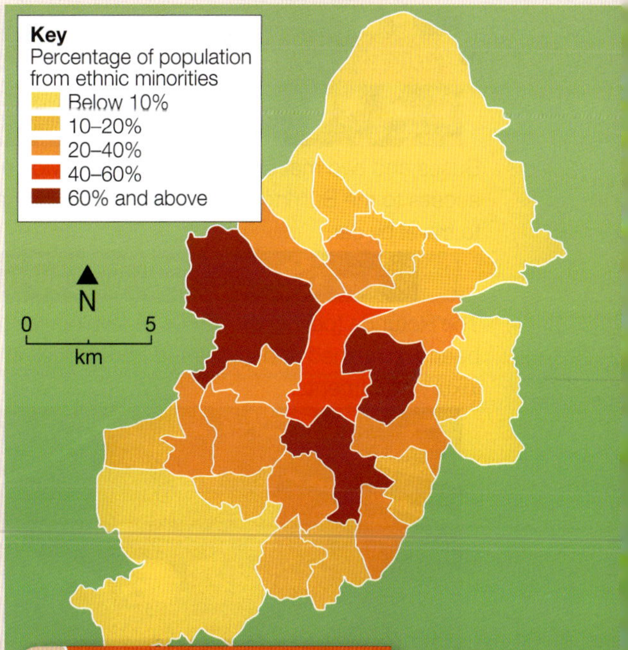

F Ethnic minority population in Birmingham

Activities

1. Study timeline B and graph C.
   - **a** On a copy of the graph, label at least five events from the timeline at the correct point on the line.
   - **b** Explain the connection between each event and the population change at the time.
2. Look at graph E. Which age groups have:
   - **a** a higher proportion in Birmingham than in England?
   - **b** a lower proportion in Birmingham than in England?
3. Suggest reasons for the differences in the age of Birmingham's population from the rest of England. (Clue: think about the people who have come to Birmingham and why they have come.)
4. Look at map F. Write a paragraph to describe the distribution of ethnic minorities in Birmingham.
5. Are you surprised that 86% of Birmingham's population see themselves as British? Would you expect the percentage to be higher or lower? Why?

Practice question

For a named city in an AC [Birmingham], explain how migration is changing the character of the city. (3 marks)

5.10 Explore Birmingham

On this spread you will explore new shopping and leisure opportunities in Birmingham.

Case study

Birmingham's Bullring

Birmingham is changing its image. At the heart of the city is the Bullring, Birmingham's historic market place. In the 1960s the Bullring was redeveloped into a large, concrete shopping centre surrounded by an urban ring road (photo A). By the 1990s, it had become a run-down, depressing place, choked by traffic that made life difficult for pedestrians. Birmingham City Council decided to demolish it and **regenerate** the site (photo B).

A *The Bullring in the 1960s*

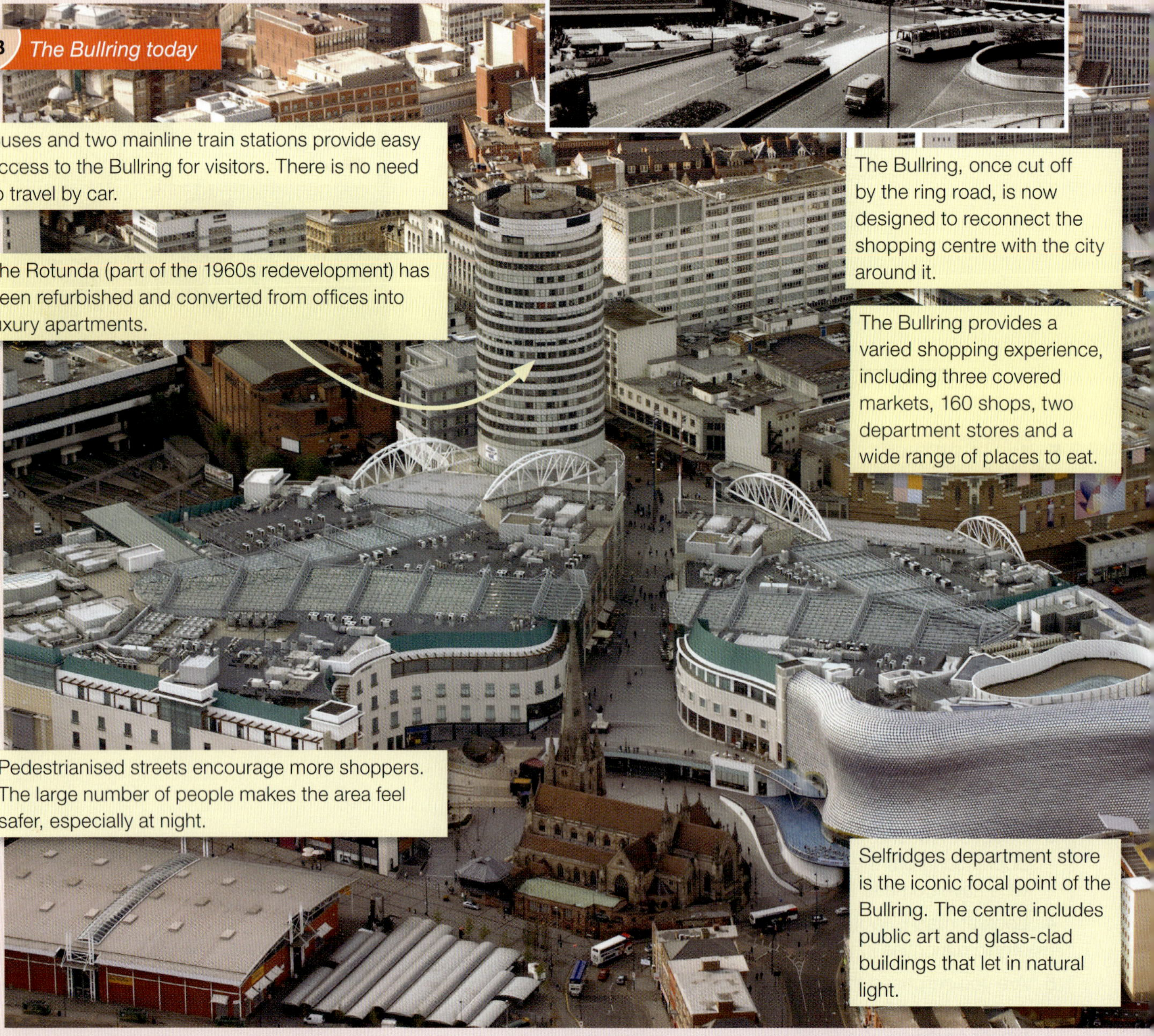

B *The Bullring today*

City-centre regeneration

The Bullring reopened in 2003 as Europe's largest city-centre, retail-led, urban regeneration project. Since then, Birmingham's ranking as a retail destination in the UK has risen from twelfth to third. During the 1980s, shoppers travelled to out-of-town shopping centres like the Merryhill Centre in nearby Dudley. Many of them have now been drawn back to shop in Birmingham city centre.

The Bullring was the first of many similar projects in the 21st century in the UK. It is not the only regeneration project in Birmingham. Other projects include:

- **Brindleyplace:** an area around the old canals in the city centre, including the National Indoor Arena and the International Convention Centre.
- **Gateway Plus:** redesigned city-centre train station with a shopping centre above it.
- **Library of Birmingham:** opened in 2013 as the largest library in the UK (see spread 5.12).
- **High Speed 2 (HS2):** a new high-speed railway linking London and Birmingham, planned to open in 2026. It will help to regenerate Birmingham city centre.

C *Opinions about the new Bullring*

Percentage of visitors who identified improvements there:

- shopping in the city centre (29%)
- access to the city centre (24%)
- quality of life in Birmingham (19%)
- community wellbeing (e.g. health) (20%)
- safety (18%).

A local shop owner said:

> Without doubt, the Bullring has dramatically improved the city, helping us become the regional shopping capital. It attracts new brands and is a fantastic experience, especially with its extensive food and drink offer.

D *An impression of the planned new HS2 station in Birmingham*

Activities

1. Look at photos A and B. Compare the old and new Bullring. Mention at least three differences. What are the similarities, if any?
2. Suggest why each of these people might be attracted to the new Bullring:
   - **a** a visitor to Birmingham
   - **b** a businesswoman hoping to open a shop in Birmingham
   - **c** a local family who used to shop at the Merryhill Centre
   - **d** a young professional couple who want to live close to their city-centre jobs.
3. Study the data in panel C. Explain how the Bullring might have improved:
   - **a** shopping in the city centre
   - **b** access to the city centre
   - **c** quality of life in Birmingham
   - **d** community wellbeing
   - **e** safety.

   Write a sentence about each.

Go further

Find out more about a new regeneration project either in Birmingham or in your own city. Explain how it could change the way of life for different groups of people.

Practice question

a For a named city in an AC [Birmingham], describe efforts to improve the way of life for people in the city. (4 marks)

b Evaluate how successful you think these efforts have been. (2 marks)

Unequal Birmingham

On this spread you will investigate inequality in Birmingham, comparing two parts of the city.

Case study

Urban inequality

Like other cities in the UK, one of the main challenges in Birmingham is **inequality**. While some areas of the city enjoy greater wealth and a good quality of life, other areas suffer **deprivation** and a poor quality of life.

Inequality is often linked to levels of unemployment. Birmingham is divided into **wards** (small areas within the city) with different unemployment levels (map A). Unemployment varies between wards close to the city centre and those in the suburbs, close to the edge of the city.

Unemployment, in turn, is connected with other indicators of deprivation, like poor educational achievement, child poverty and low household income. Table B compares indicators of deprivation for two wards in Birmingham – Sparkbrook near the city centre and Sutton Four Oaks in the suburbs.

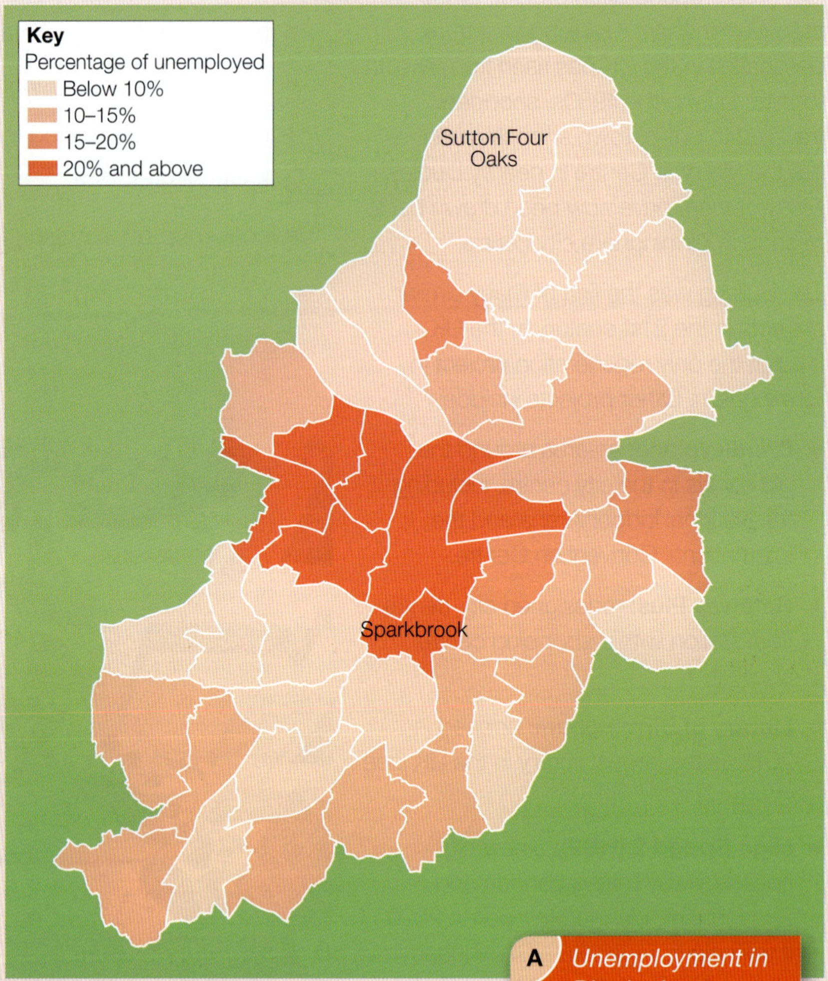

A *Unemployment in Birmingham*

B *Comparing wealth and deprivation in Sparkbrook and Sutton Four Oaks*

	Sparkbrook	Sutton Four Oaks	Birmingham
Unemployment (%)	24.5	3.1	12.0
Economically active or at work (%)	48	81	68
Working age population with no qualifications (%)	49.7	20.9	37.1
Pupils with five GCSEs, A*–C (%)	51	74	58
Children living in poverty (%)	49	7	34
Average household income (£)	21 000	40 000	31 000
Households with income less than £15 000 (%)	46	12	27
Households with income over £35 000 (%)	12	47	27

Housing provision

Other indicators of deprivation include:

- poor-quality housing
- lack of access to good services like health and education
- high crime rates
- poor-quality physical environment.

House prices help to keep inequality going. Only wealthy people with high incomes can afford the highest house prices, so they move to the least deprived areas (advert C). Meanwhile poorer people, including the unemployed, are forced to live in more deprived areas (advert D). More of the houses in deprived areas are for rent, either from the local council, housing associations or private landlords. And so the pattern of inequality is repeated.

C *Advert for a house in Sutton Four Oaks*

Four-bedroom detached house

- Two large reception rooms and large fitted kitchen
- Two bathrooms, one en-suite
- Garage and large garden

£525 000

Two-bedroom terraced house

- Two reception rooms and kitchen
- First-floor bathroom
- In need of some renovation

£99 950

D *Advert for a house in Sparkbrook*

Activities

1 Look at map A. Write a paragraph to describe the pattern of unemployment in Birmingham.

2 **a** Study table B. Choose four indicators of deprivation. For each one, draw a bar chart like the one opposite to compare Sparkbrook, Sutton Four Oaks and Birmingham, using data from the table.

b Is a bar graph the best way to show this data? Would a line graph, scatter graph or pie chart be better? Give reasons for your ideas.

3 Look at the graphs you have drawn for activity 2.

a Write a few sentences to compare levels of deprivation in Sparkbrook and Sutton Four Oaks.

b How do the two wards compare with the rest of Birmingham? Justify your answer from the data.

4 Look at house adverts C and D.

a Do you think one house is really worth more than five times the other? Give reasons for your answer.

b What other factors affect the price of a house, apart from the size and quality of the building?

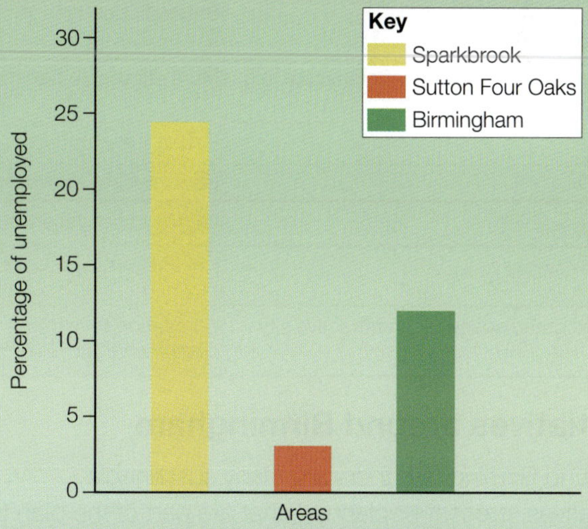

Practice question

For a named city in an AC [Birmingham], outline one challenge that it faces [inequality]. (4 marks)

5.12 Sustainable Birmingham

On this spread you will consider Birmingham's initiative to be a more sustainable city and explore one example of how this is being achieved.

Case study

Why be sustainable?

The world faces a huge challenge from the impacts of climate change (see Chapter 2). Cities consume 75% of the world's energy and produce 80% of greenhouse gases. If they are to be more sustainable, cities need to change.

As the UK's second largest city, Birmingham has an important contribution to make. It was once one of the world's leading manufacturing cities, producing more than its share of carbon emissions. It now aims to be a more sustainable city.

Birmingham's plan

Birmingham has developed a plan to be 'a leading green city' (panel A). The main challenge is to reduce the amount of resources the city consumes and the amount of carbon emissions it produces. However, becoming more sustainable also brings opportunities:

- The city could develop a new, greener economy, specialising in low-carbon technology such as manufacturing solar panels.
- It could become a 'smart city' – one that, through better planning, uses fewer resources and has more efficient transport.
- People could become fitter and healthier by living a more sustainable lifestyle.

A *Birmingham's plan to be 'a leading green city'*

Economic	Social	Environmental	Sustainable themes
	Birmingham aims to be...		
more prosperous	**healthier and fairer**	**resource-efficient**	
Education, skills and knowledge for a low-carbon, green economy	Lower energy costs and reduced fuel poverty	Low-carbon energy generation, e.g. solar panels	Energy
'Smart city' using fewer resources	More efficient, well-insulated buildings	Reduced energy demands Alternative energy sources, e.g. ground-source heat	
	Better air quality	Better waste management and conversion of waste into energy	Resources
Low-carbon transport infrastructure, e.g. metro system	More affordable transport		
	Better public transport		Transport
	Better cycling and walking options		
	Improved natural environment		
		Planned and managed ecosystems	Green living
More green space for an attractive city	Adaptation to extreme weather and climate change		Climate change adaptation

Initiatives around Birmingham

Around Birmingham there are many sustainable initiatives, present or planned, that are part of the plan to be 'a leading green city',

- **Birmingham Energy Savers:** This includes plans to fit older houses with better insulation, double-glazing and new boilers to make them more energy-efficient.
- **Midland Metro:** This light rail/tram network has been extended into Birmingham city centre, improving transport across the city centre and creating new jobs.
- **District energy schemes:** The city centre is supplied with energy from a combined heat and power plant that saves money and reduces CO_2 emissions (see spread 11.5). Another scheme plans to burn waste to generate electricity for 40 000 homes.
- **High Speed 2:** HS2 is planned to improve rail connections to Birmingham from London by 2026, providing a quicker alternative to road journeys and helping to regenerate the city centre (see spread 5.10).

The Library of Birmingham

The Library of Birmingham is another sustainable initiative. It opened in 2013 and is the largest public library in the UK (photo B). It was built on a brownfield site that was once a multi-storey car park and is part of a wider plan to regenerate the city centre.

The building reflects Birmingham's past but also a more sustainable future. The design of the metal exterior is a reminder of the city's industrial past, while its function as a library points to Birmingham's aim to develop a more educated workforce for a new, knowledge-based economy.

As part of Birmingham's plan to be 'a leading green city', the library:

- recycled 95% of waste material from the site during construction
- employed over 250 people during construction to reduce local unemployment
- minimises carbon emissions by using energy-efficient systems
- reduces energy consumption by making use of natural daylight and ventilation
- uses less water by harvesting rainwater and recycling water
- has a roof garden to attract wildlife and improve biodiversity
- promotes sustainable transport use and provides cycle storage space
- attracts about 2.5 million visitors a year, helping to raise educational achievement.

B *The Library of Birmingham*

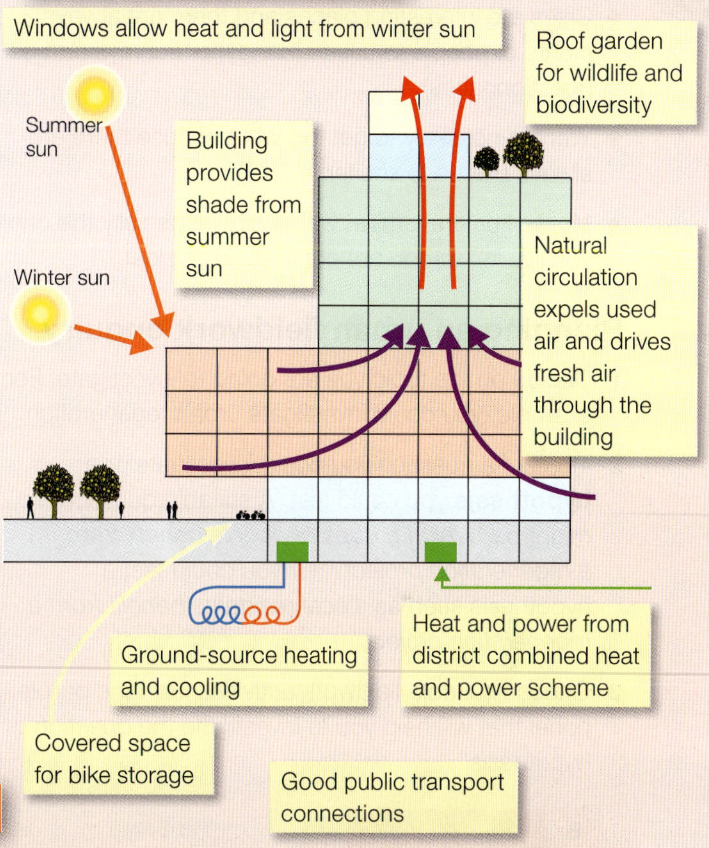

C *Sustainable features of the library's design*

Activities

1 Describe the challenges and the opportunities for Birmingham as it plans to become a more sustainable city.

2 Look at photo B. Explain how the building reflects:
- a the city's past
- b a more sustainable future.

3 Study panel A. For each sustainable initiative below, list the economic, social and environmental aims that the initiative helps Birmingham to achieve.
- a Birmingham Energy Savers
- b District energy schemes
- c Midland Metro
- d High Speed 2.

4 Assess the contribution of the Library of Birmingham to the city's plan to be more sustainable.
- a How does it contribute to economic, social and environmental sustainability? Write at least three sentences.
- b Do you think its most important contribution is to economic, social or environmental sustainability? Give reasons for your answer.

Practice question

For a named city in an AC [Birmingham], describe one initiative you have studied and how it will help to make the city more sustainable. (6 marks)

5.13 Urban fieldwork – 1

On this spread you will plan an urban fieldwork investigation to do in your own city.

Fieldwork

Why do urban fieldwork?

Unless you live in Birmingham it is unlikely that you will do any fieldwork there. However you may live in another city or within travelling distance of one (most of us in the UK do!). You will be able to do urban fieldwork there.

Urban fieldwork has several advantages:

- Cities are interesting places and there are always plenty of people to talk to (handy for surveys and questionnaires).
- You are not likely to get too wet because there are always places to shelter.
- Most of us are familiar with a city (especially the one you live in), so you are not likely to get lost.

A *Doing a survey in a shopping centre*

Planning an urban fieldwork enquiry

1. Decide on the topic you are going to investigate. For example, you could investigate urban regeneration.
2. Think of a question you could investigate or a **hypothesis** you could test. A suitable question might be *'Which groups of people benefit from a regeneration project?'* Or you could test a hypothesis such as *'Local residents benefit from a regeneration project.'*
3. Choose suitable fieldwork activities from the purple panel that will help you to investigate the question or hypothesis.

B *Urban role play*

You are:
a teenager in Birmingham

You live: in Sparkbrook

You want:

- a safe place where you can meet friends in the city centre
- things to do, like going shopping or eating out
- good public transport so you can travel around easily.

Urban fieldwork activities

- **Urban role play:** Look at an area through the eyes of different types of people.

- **Annotated sketch/photo:** Draw a sketch or take a photo and annotate it.

- **Environmental quality survey:** Give a score to different environmental features.

- **Questionnaire:** Ask people about their experiences and opinions.

- **Traffic/pedestrian survey:** Count the number of vehicles or pedestrians going past.

- **Land use mapping:** Complete an outline map to show how each area is being used.

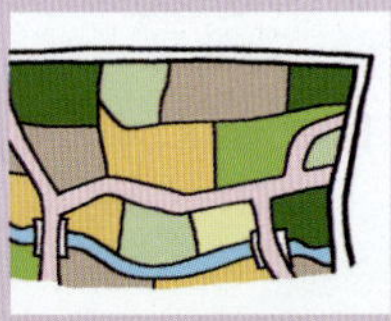

C *An annotated photo of the Bullring, Birmingham*

URBAN ENVIRONMENTAL QUALITY SURVEY
Name of site: Birmingham Bullring
General description: Large, modern shopping centre built in 2003

	Quality being assessed	High +2	OK +1	Av 0	Poor –1	Bad –2	
General	Vibrant, interesting place	/					Empty, uninteresting place
	Variety of natural features					/	No natural features
	Feels safe, light and busy	/					Feels unsafe, dark and quiet
	Clean environment	/					Dirty environment
Buildings	Well designed	/					Poorly designed
	In good condition		/				In poor condition
	Large, spacious buildings		/				Small, cramped buildings
	No vandalism	/					Extensive vandalism

D *Urban environmental quality survey*

Activities

1 Think of a question or hypothesis about an urban area. It should be one that you could investigate through fieldwork.

2 Look at the suggested fieldwork activities. Select three activities you could use to investigate your chosen question or hypothesis.

3 Look at role play B.
  - **a** Write a similar role-play card for one of these people – an older resident, a student or a local business owner.
  - **b** Look at photo C. Give at least three benefits of the Bullring for your person.

4 Copy and complete survey D, adding space for two more sections.
  - **a** Design two more sections on shopping and transport. List four positive and four negative qualities for each section.
  - **b** Complete a survey for the Bullring. Work out a total score between +32 and –32, by adding up the scores you give.

Practice question

For a fieldwork investigation you have done in an urban area:

- **a** State the question or hypothesis you investigated. (1 mark)
- **b** Describe one fieldwork technique you used and why you chose it. (4 marks)
- **c** Evaluate how successful the fieldwork was. (3 marks)

5.14 Urban fieldwork – 2

On this spread you will use primary and secondary data to investigate urban change.

Primary and secondary data

Urban fieldwork involves the collection of **primary data** – information that you collect yourself during the fieldwork. Photos, sketches, completed maps and surveys are all examples of primary data.

However, sometimes you will also need to use **secondary data** – information that you can obtain from other sources, like the internet, newspapers, even textbooks. For example, to investigate urban change, you may need secondary data to tell you what the city used to be like.

Questions about urban change

- How is the population of the urban area changing?
- Why is the land use in the urban area changing?
- How is the urban area becoming more sustainable?

Hypotheses about urban change

- The proportion of ethnic minorities in the urban area is increasing.
- Land use in the urban area is changing because more housing is needed.
- Transport improvements are making the urban area more sustainable.

Investigating urban population change

How is the population of the urban area changing? To investigate this question, you could find out about the population now by doing a population survey on the street (survey B). However, to find out how the population is changing you will also need secondary data that give you information about population in the past. You could use a past census (graph C).

A *Doing a survey on the street*

B *Population survey*

	Count or tally Make a mark for each person who passes, up to a total of 100. Turn your total into a percentage.
White, including white British, Irish and European	𝍸 𝍸 𝍸 𝍸 𝍸 𝍸 𝍸 𝍸 𝍸 𝍸 𝍸
Asian, including Pakistani, Indian, Bangladeshi and Asian British	𝍸 𝍸 𝍸 𝍸 𝍸
Black, including West Indian, African and Black British	𝍸 ////
Chinese and British Chinese	//
Mixed race and other ethnic groups	𝍸 ////

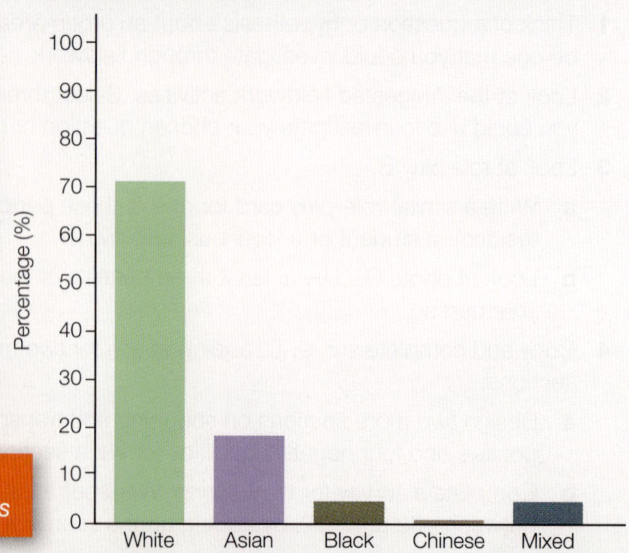

C *Population of Birmingham by ethnic group from 2001 census*

Evidence of population change

In urban areas with a large or growing ethnic minority population, there are often specialist shops and services that meet their needs. New shops open and old ones close as the population changes. You can walk around an area to classify shops and services by the ethnic group they are aimed at (map E). However, to find out how shops and services are changing, you will also need to use secondary data. Use an old directory to find out what shops and services were there before (extract F).

D *Asian butchers selling halal meat*

E *Classifying shops and services*

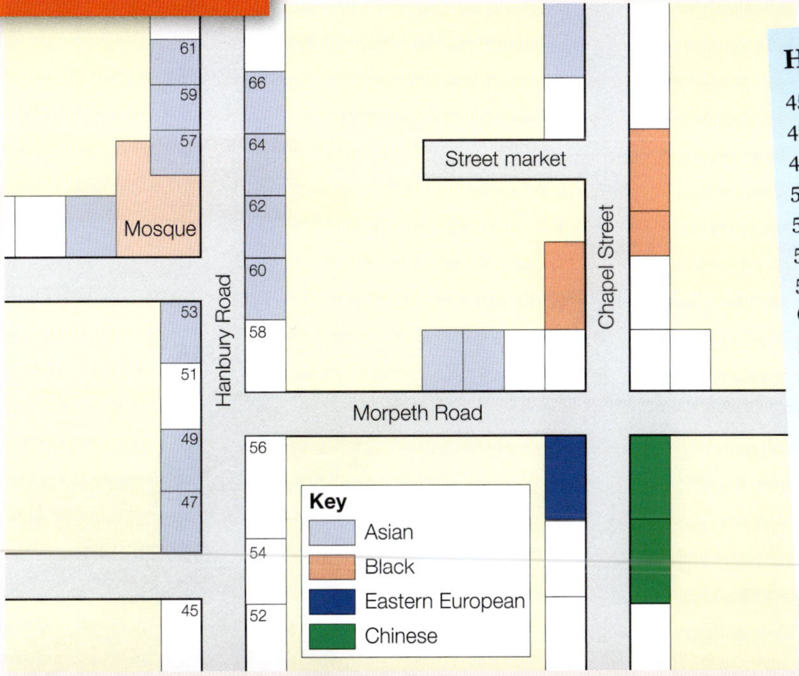

Hanbury Road

45 Brothers, fish and chips
47 Walkers, newsagents
49 Sun City, tanning and beauty
51 Curls and Locks, Afro-Caribbean hair salon
53 Corner Shop, Asian grocery
57 Greggs, bakers
59 Wakefields, ironmongery
61 Comtessa, womens' clothing
52 Bits and Bobs, fabrics
54 Chicken Licken, fast food
56 William Hill, betting shop
58 Travel Far, bags and luggage
60 Clean and Go, dry cleaners
62 Sagor, Indian restaurant
64 Tour de Force, bicycle shop
66 Kamal's Kebabs, Turkish takeaway

F *Extract from a 2001 street directory*

Activities

1. Look at survey B. Count the population in each category. Use the percentage figures to draw a bar chart showing the ethnic composition of the population now.
2. Look at graph C.
   - **a** Compare the graph you drew in activity 1 with graph C. How is the population changing?
   - **b** Explain why it is necessary to use primary and secondary data together to show how the population is changing.
3. Look at map E and extract F.
   - **a** What evidence do they provide of the change in population?
   - **b** Explain why it is necessary to use primary and secondary data together.
4. Choose one of the other enquiry questions in the panel. Suggest what fieldwork activities and secondary data you could use to investigate urban change.

Go further

Plan an urban fieldwork enquiry you could carry out in your own city.

- **a** Decide on the topic you would like to investigate, e.g. regeneration or population change.
- **b** Think of an enquiry question or hypothesis that you could investigate through fieldwork.
- **c** Select three suitable fieldwork activities to investigate the question or hypothesis. They could be your own ideas.
- **d** Think of one or more locations in your city where you could carry out your investigation. Now try to persuade you teacher that your plan will work!

5.15 Istanbul on the map

On this spread you will explore the location and importance of Istanbul, using maps.

Case study

Introducing Istanbul

Istanbul has a unique location, straddling Europe and Asia. It is the only city in the world to span two continents (map A). Between the two sides of the city lies the Bosphorus, a 30-km long strait that links the Black Sea in the north with the Sea of Marmara in the south.

Modern Istanbul is one of the fastest-growing cities in the world. It now extends along the coastline of the Sea of Marmara on both sides of the Bosphorus. It is the largest city in Turkey, with a population of about 15 million. But Istanbul is not even the capital of Turkey – that is Ankara (map B).

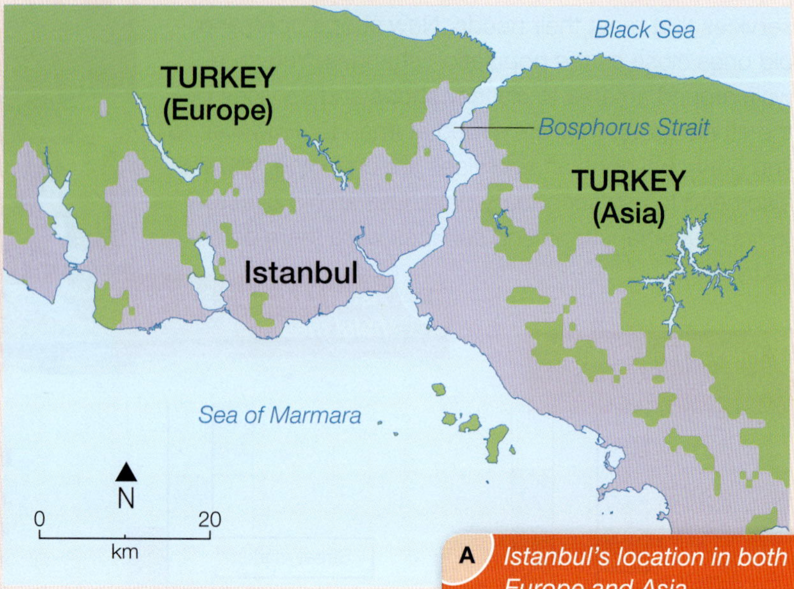

A *Istanbul's location in both Europe and Asia*

B *Turkey and neighbouring countries*

A brief history of Istanbul

Few cities in the world can boast a more interesting history than Istanbul. Over the centuries it has had three names and been at the heart of three empires.

It began as the city of Byzantium in the Greek Empire in the 7th century BC. Later it became the capital of the Roman Empire under the Emperor Constantine in AD 300 and changed its name to Constantinople. Constantine was the first Christian emperor.

In 1453 the city was conquered by the Ottomans, Muslims from the Anatolian peninsula (now Turkey).

The Ottoman Empire, with Constantinople as its capital, grew to control much of south-east Europe, west Asia and north Africa (map C). The Empire declined during the 19th century and the new country of Turkey was established in 1923. Constantinople changed its name to Istanbul.

C *Growth of the Ottoman Empire*

Activities

1. Look at map A. Write a paragraph to describe the location of Istanbul. Mention all the labels on the map.
2. Look at maps A and B. Suggest why Istanbul was a good location for the capital of an empire. (Clue: think about how people used to travel before the days of cars, trains and planes.)
3. Look at map C.
   - **a** Describe how the Ottoman Empire grew from 1451 to 1683. Mention the areas of land it occupied and the seas it surrounded at each date.
   - **b** Compare the map with a modern atlas. Name the modern countries that were once part of the Ottoman Empire and ruled from Constantinople.
4. Today Istanbul has 143 Christian churches and 3190 Islamic mosques (some converted from churches). How can you explain this from Istanbul's history?

Practice question

a For a named city in either an LIDC or EDC [Istanbul], explain its importance within the country. (2 marks)

b For the same city, explain its importance within the wider world. (2 marks)

Expanding Istanbul

On this spread you will find out how Istanbul's population and the size of the city have grown.

Case study

A growing population

Throughout history, Istanbul has been ranked among the largest cities in the world (photo A). However, its most dramatic growth has happened over the past 60 years or so (table B). In 1950 it was a city of just under 1 million people. By 2015 it had grown to almost 15 million and it is expected to reach 20 million by 2025.

Most of Istanbul's growth has been due to **national migration** from other parts of Turkey. Migrants come to the city for employment and better living conditions. Initially, many of them find work in the **informal sector**, doing jobs as street vendors, porters (see photo D on spread 5.17) or waste recyclers. The fastest growth was during the 1980s when its population doubled in less than ten years. That is like some African cities today (see spread 5.4).

A *Modern Istanbul*

A city of migrants (and tourists!)

Istanbul is home to most of Turkey's ethnic minorities, including 3 million Kurds from south and east Anatolia. It is the largest Kurdish population in the world. There are also smaller Armenian, Greek and Jewish communities. Only 28% of Istanbul's population were actually born in the city – a sign of how fast it has grown.

There is not much international migration to Istanbul but it is growing as a tourist destination. Turkey is now among the world's most popular tourism destinations. And the Turkish city that most tourists visit is Istanbul!

Year	Population	Growth rate over 5 years
1950	983 000	–
1955	1 258 000	28%
1960	1 466 000	16%
1965	1 742 000	19%
1970	2 132 000	22%
1975	2 547 000	19%
1980	2 772 000	9%
1985	5 475 000	77%
1990	7 620 000	39%
1995	9 260 000	21%
2000	10 923 000	18%
2005	12 061 000	10%
2010	13 256 000	9%
2015	14 377 000	8%

B *Population growth in Istanbul, 1950–2015*

Coastal expansion

Istanbul's location is not ideal for urban growth. The city is divided between two peninsulas – one in Europe, the other in Asia – separated by the Bosphorus. Inland are steep hills covered by forest and reservoirs that supply most of the city's water. There is little space for the city to expand inland.

In order to grow, the city has spread along the coastline of the Sea of Marmara and on either side of the Bosphorus. It now stretches 100 km from west to east – an unusual shape for a city.

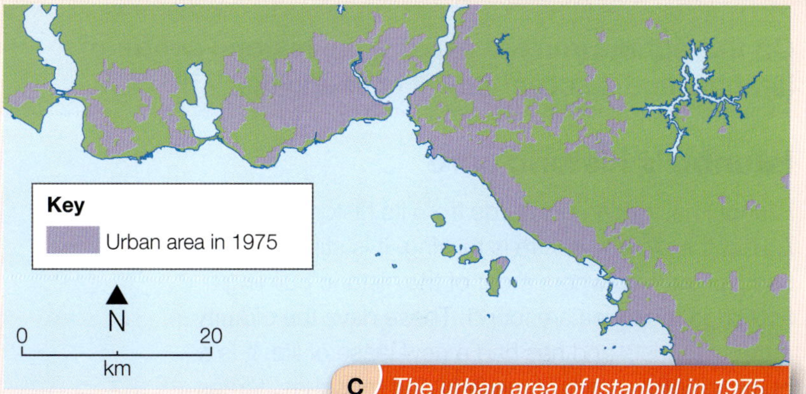

C *The urban area of Istanbul in 1975*

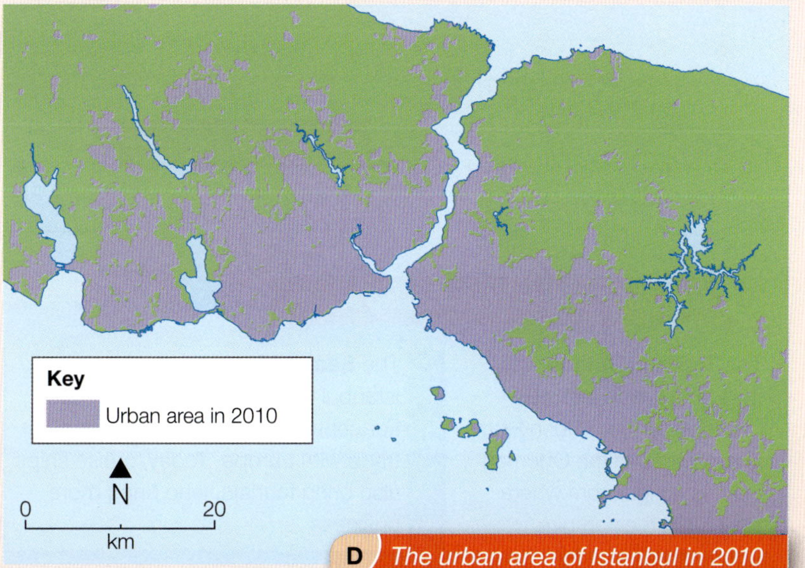

D *The urban area of Istanbul in 2010*

Activities

1. Look at photo A. Write two sentences to give your first impressions of Istanbul. Mention:
   - natural features, including water
   - size and density of the city.
2. Look at table B.
   - **a** Draw a line graph to show population growth in Istanbul from 1950 to the present. You could find the latest population on the internet to bring the graph up to date.
   - **b** Work out the mean population growth rate over a five-year period from 1950 to 2015. To work out the mean annual growth rate, divide your figure by 5.
3. Look at maps C and D.
   - **a** Describe the shape of Istanbul in 1975.
   - **b** Describe how the city had grown by 2010.
   - **c** Explain why the city is this shape and why it has grown in this way.

Go further

Look back at population growth in Birmingham on graph C on spread 5.9. Compare it with the graph you drew in activity 2.

a How does Istanbul's growth compare with Birmingham's? When did both cities grow? What has happened since 1950 in both cities?

b Explain the difference between the two graphs. (Clue: it will help to look back at spread 5.1).

Practice question

For a named city in either an LIDC or EDC [Istanbul] describe how the population has grown. Refer to data in table B in your answer. (3 marks)

Explore Istanbul

On this spread you will explore the historic centre of Istanbul and discover how it is changing.

Case study

Istanbul's historic core

Istanbul has grown outwards from its historic core (map A), but the old city centre is thriving. It is still the cultural heart of the city, where most of the historic buildings and large mosques are found. These days the old city attracts tourists and has had a new lease of life. It is a bustling hotch-potch of markets and mosques, ferries and fishermen, tourists and trams (photo B).

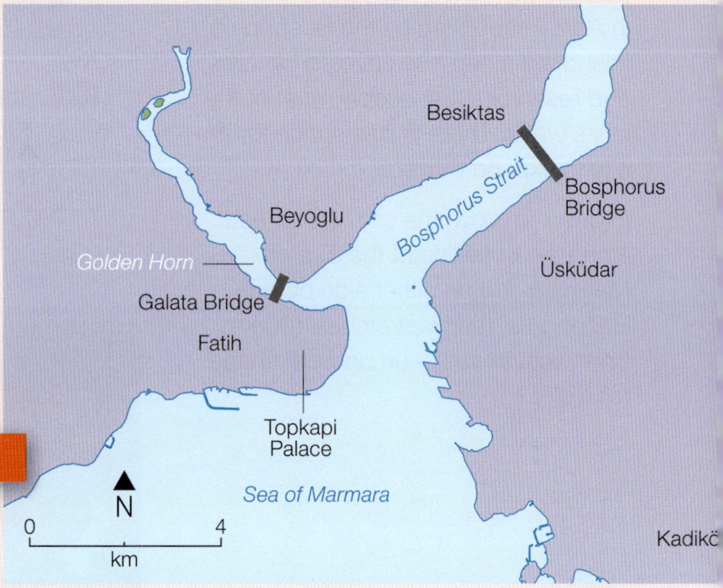

A *Istanbul city centre*

B *Istanbul's historic core, looking south*

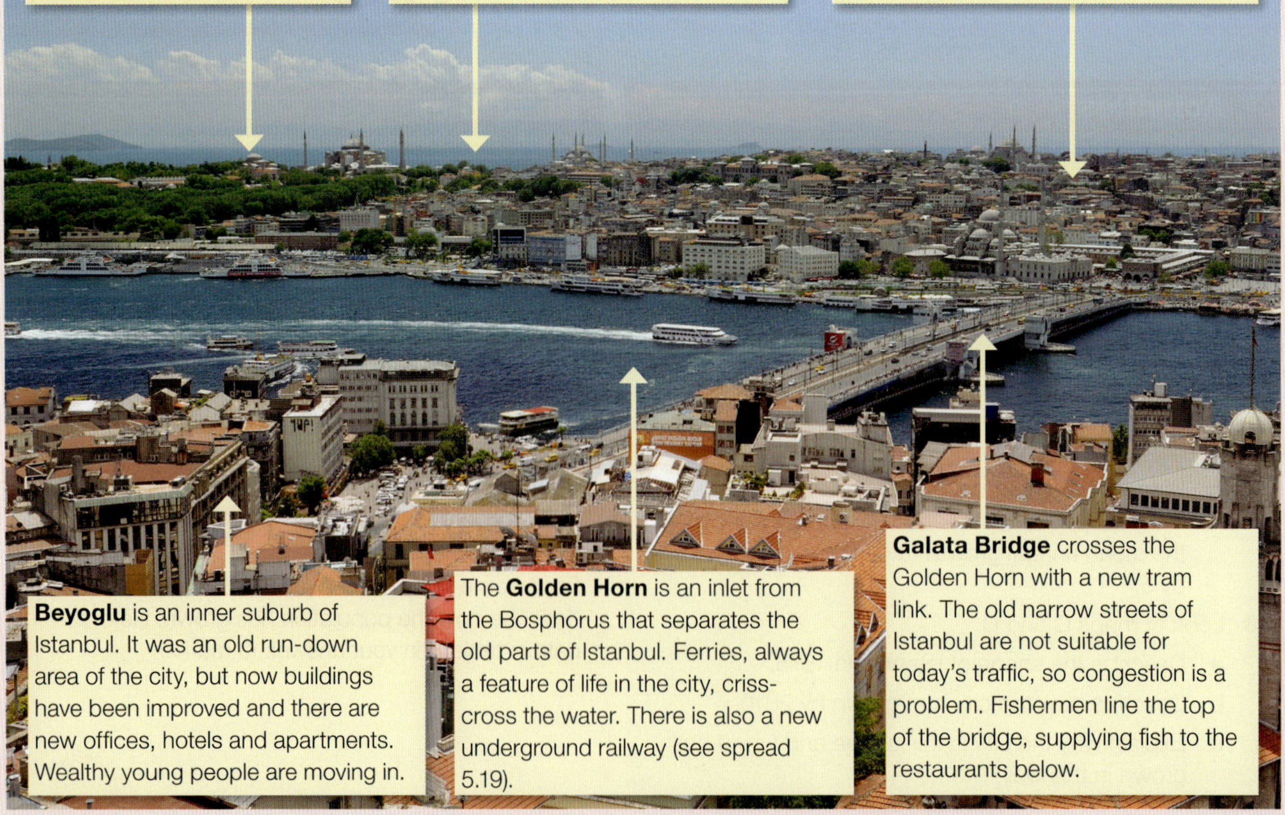

The **Topkapi Palace** was built on a prominent site overlooking the Golden Horn. It was home to the Ottoman sultans (kings) from where they ruled their empire.

The **Sea of Marmara** connects Istanbul with the Mediterranean Sea. For centuries, boats have brought trade with Europe. Today, cruise ships also bring tourists, who bring more business.

Fatih is the oldest district in Istanbul and site of the original city. Today it is the market centre of Istanbul. At its heart is the **Grand Bazaar** dating from 1461. Here you can still buy anything from a handmade rug to Turkish delight.

Beyoglu is an inner suburb of Istanbul. It was an old run-down area of the city, but now buildings have been improved and there are new offices, hotels and apartments. Wealthy young people are moving in.

The **Golden Horn** is an inlet from the Bosphorus that separates the old parts of Istanbul. Ferries, always a feature of life in the city, criss-cross the water. There is also a new underground railway (see spread 5.19).

Galata Bridge crosses the Golden Horn with a new tram link. The old narrow streets of Istanbul are not suitable for today's traffic, so congestion is a problem. Fishermen line the top of the bridge, supplying fish to the restaurants below.

Istanbul's changing role

Istanbul is home to just 20% of Turkey's population, yet the city generates almost half the country's wealth. It plays a vital role in the manufacturing production, creativity and culture of the country. The city is surrounded by agricultural areas producing cotton, fruit, olive oil and tobacco. Two of Istanbul's main industries are textile manufacturing and food processing. However, the fastest-growing sectors are finance and tourism.

As the city has grown, so its population has shifted from the core to the suburbs. From 1985 to 2012, the population of Fatih – the historic core of the city – fell, while the population of the suburbs has more than doubled (graph C).

This suburbanisation has happened at the same time as jobs, services and shopping have also moved out of the centre. New high-rise commercial centres have appeared in other parts of Istanbul with the growth of the financial and service industries.

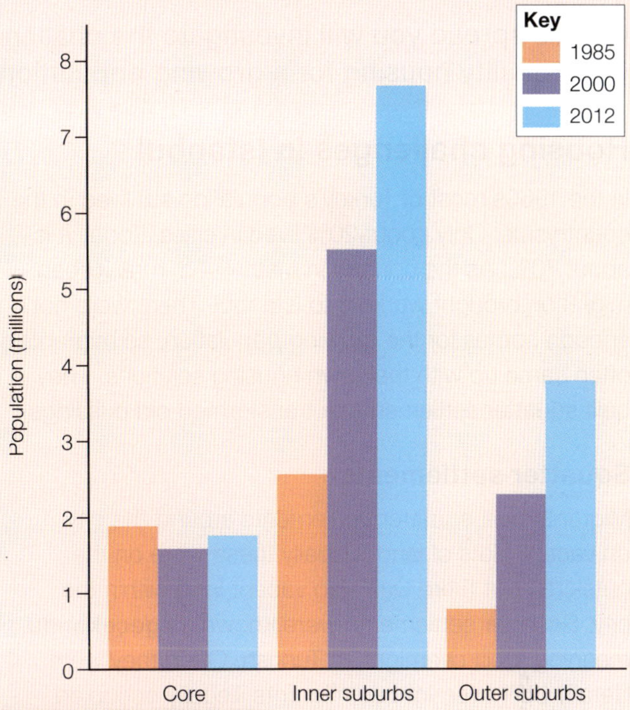

C *Shift of population in Istanbul, 1985–2012*

Activities

1 Look at map A and photo B.
  - **a** Suggest why Istanbul was a good site on which to build a city. Give at least three reasons.
  - **b** Do you think it is still a good site for a modern city? Give more reasons.

2 Look at photo B.
  Would each of the people or businesses below want to be in the historic core of the city, or would they want to leave? Give reasons.
  - **a** A tourist from a cruise ship
  - **b** A large bank employing hundreds of people
  - **c** A modern supermarket
  - **d** A shop selling traditional Turkish rugs

3 Look at graph C.
  - **a** Write three sentences to describe how Istanbul's population has changed in each of the core, inner suburbs and outer suburbs. Estimate figures from the graph.
  - **b** Try to explain these changes.

4 Look at photo D.
  - **a** Suggest how a street porter's life might have changed in the past 30 years. Mention his work, who he meets and where he lives.
  - **b** How do you think he might feel about these changes?

D *A street porter in the old city centre*

Practice question

For a named city in either an LIDC or EDC [Istanbul], explain how and why ways of life are changing. (4 marks)

At home in Istanbul

On this spread you will investigate the challenge of providing good quality housing for a growing population.

Case study

Housing challenges in Istanbul

In the 1960s most of Turkey's population still lived in the countryside. Only about 20% lived in cities. Today it is about 70%. As the country industrialised, rural–urban migration brought workers to Istanbul. There were not enough homes for the growing population, so migrants often came up with their own housing solutions. They built squatter settlements or transformed old buildings.

Squatter settlements

Migrants built squatter settlements around the city on vacant plots of land. Usually these were on the outskirts, but there was also vacant land within the city. Squatter settlements were known as **gecekondu**, meaning 'built overnight' in Turkish. Often they were literally built overnight as migrants illegally occupied vacant land. This is similar to many African and Asian cities today.

The gecekondu lacked basic services like water supply, sanitation, electricity and street lighting. Residents often devised their own solutions, like adding water tanks and pumps. Over time, services have improved and gecekondu have been rebuilt to become part of the city.

A *An old residential building*

Old buildings

Some migrants moved closer to the city centre and transformed large old family homes into apartment blocks (photo A). The buildings were often in poor repair and in need of improvement. Some buildings were expanded upwards by adding new storeys. Four and five-storey buildings became common in Istanbul. But, like the gecekondu, they weren't built to very high standards – a serious problem in an earthquake zone!

B *Districts in Istanbul*

Modern Istanbul

Today, with increasing wealth, Istanbul is investing in better infrastructure and housing for residents. Strategies vary from one district to another (map B).

Beyoglu

Beyoglu was once a fashionable neighbourhood close to the city centre. But it had fallen into disrepair by the 1990s. Most of the inhabitants were poor and property prices were low. Since 1990, young professional people have moved in and improved the homes. The area has been transformed and, as a result, property prices have increased. Poor people can no longer afford to live in Beyoglu. This is called **gentrification**, a process that is also common in UK cities.

Esenler

As the city grew in the 1980s the area was transformed from a village just outside Istanbul to part of the growing city. Most of the newcomers were poor and built squatter settlements made of low-quality concrete. The homes were built at very high density, with little space or light between them.

Esenler is now being redeveloped. The gecekondu are being demolished and replaced by new high-rise apartments with more space in between.

C A street in Beyoglu

D A redeveloped housing estate in Esenler

Activities

1 Look at photo A. What are the advantages and disadvantages of migrants building their own squatter settlements or transforming old buildings? Make two lists.

2 Look at map B here and at maps C and D on spread 5.16.
   - **a** Describe the locations of Beyoglu and Esenler in 1975 and 2010. Refer to their positions in relation to the whole city.
   - **b** Explain how the locations of the two districts affected the types of homes that people lived in.

3 Look at photo C.
   - **a** What views about gentrification might the following people have?
     - A poor resident?
     - A young professional?

     Why would they have these views?
   - **b** On balance, do you think gentrification is a good or bad thing? Give reasons.

4 Look at photo D. Suggest what advantages or disadvantages the redeveloped estate would have compared to a squatter-settlement.

Go further

a Research housing conditions EITHER in a UK city in the 19th century OR in an African city today.

b What similarities or differences can you find with Istanbul?

Practice question

For a named city in either an LIDC or EDC [Istanbul], assess how it is meeting one of the challenges it faces. (6 marks)

Congested Istanbul

On this spread you will find out how an integrated transport system would help to make Istanbul more sustainable.

Case study

Traffic congestion

Istanbul is one of the most congested cities in the world (photo A). At peak hours the average traffic speed is as little as 8–10 km/hour. Most commuters spend more than two hours daily in traffic. This is despite the fact that car ownership in Istanbul is still quite low. Only about one-third of households own a car, but that is expected to rise to two-thirds by 2025.

One of the main obstacles to transport in Istanbul is the Bosphorus. There are three bridges linking the European and Asian sides of the city. 420 000 vehicles cross the Bosphorus each day and this is where some of the worst congestion occurs.

A *Congestion in Istanbul*

Transport solutions

One obvious solution to the congestion problem would be to reduce the number of vehicles using the roads. And the best way to do this would be to improve other forms of transport.

- **Taxis, minibuses and dolmus:** These were traditionally the most common way for people to travel in Istanbul, although they are decreasing with the growing number of cars. Dolmus are shared taxis for up to eight people. Like taxis or minibuses, they have fixed routes but can stop anywhere.
- **Company** and **school buses:** Many companies and schools provide bus services for their workers or students. This is a popular means of transport in Istanbul, but the buses only operate on fixed routes at set times of day.
- **Rail:** The underground is a relatively small network in Istanbul compared to other cities like London, but it is growing.
- **Ferries:** These are a quick way of getting from one side of the Bosphorus to the other side, but they don't always link with other forms of transport.
- **Bicycles:** Very few people in Istanbul risk their lives by cycling! It is too dangerous and seen as a 'poor person's transport'.

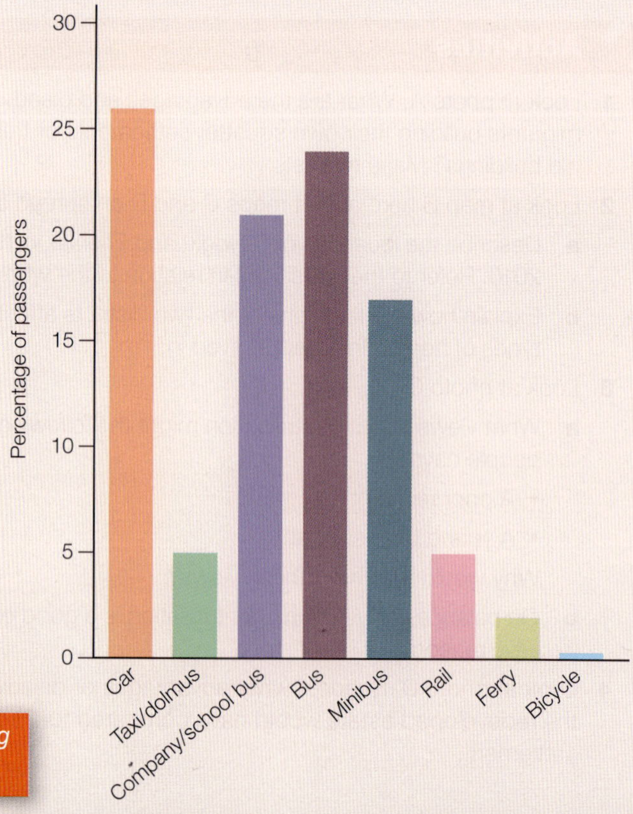

B *Percentage of passengers using different forms of transport*

Integrated transport system

The most sustainable solution to Istanbul's congestion problem would be an **integrated transport system** where all forms of transport link together better. There are signs that this is starting to happen.

In 2013 the first railway tunnel under the Bosphorus opened as part of the Marmaray Rail Project to link the European and Asian sides of the city. When complete, the line is expected to increase rail travel from 5% to 30% of passengers. That should help to take a lot of cars off the road and reduce air pollution and carbon emissions.

Another part of an integrated system could be one of the oldest forms of transport in Istanbul – ferries. They are more efficient and produce lower carbon emissions per passenger journey than cars.

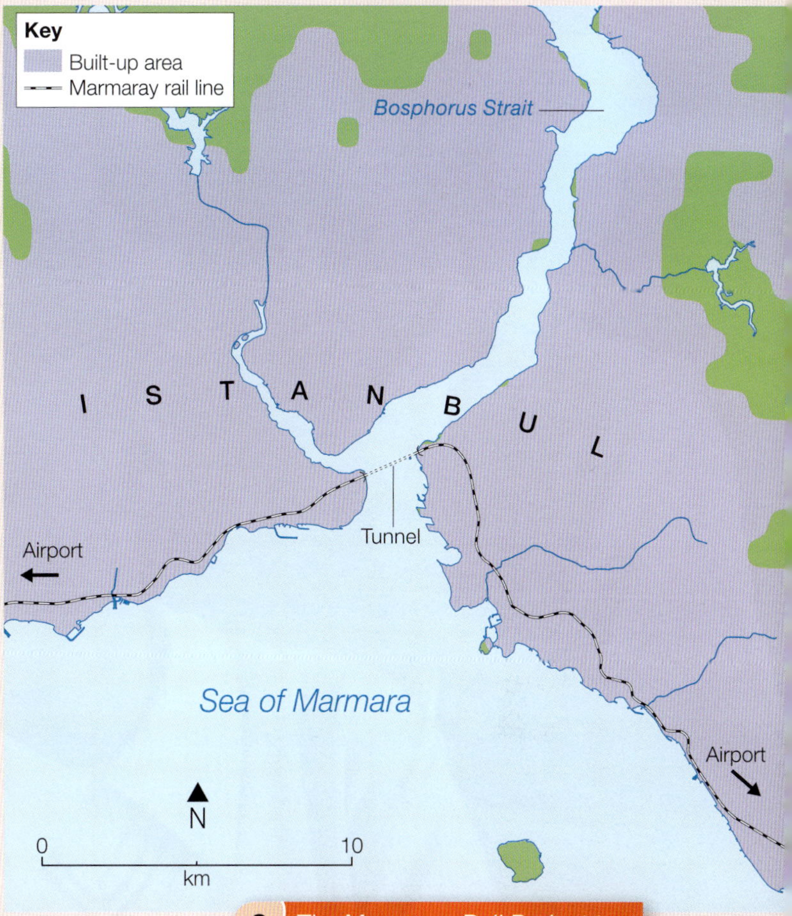

C *The Marmaray Rail Project to link both sides of Istanbul*

D *Ferries are a part of Istanbul's transport system*

Activities

1. Look at photo A. Why is traffic congestion bad for:
   - **a** drivers?
   - **b** the environment?
   - **c** the economy?
2. Look at graph B.
   - **a** Rank the means of transport in Istanbul in order of percentage of total passengers.
   - **b** Now try to rank them in order of sustainability. Think about their carbon emissions, number of passengers and efficiency.
   - **c** What do you notice about the two rank orders? How sustainable is Istanbul's transport system?
3. Plan a new integrated transport system for Istanbul.
   - **a** Outline the main problems with the existing transport network.
   - **b** Suggest at least one solution to each of the problems.

Your enquiry

What are the challenges and opportunities for cities today?

You have studied two cities – Birmingham and Istanbul. How do they compare? Do you find the same challenges and opportunities in both cities, or do they differ?

Draw a large table, like the one below, to compare the two cities. List the challenges and opportunities found only in one city or in both cities.

	Challenges	Opportunities
Birmingham		
Istanbul		
Both cities		

Dynamic Development

What is this topic about?

The world is forever developing. Even in your lifetime there has been **development**. On average, worldwide, we are wealthier and healthier than we were at the start of the 21st century. And, over the same period, technology has changed our lives. Until the year 2000, even in the UK, it was unusual to own a mobile phone or home computer, let alone a smartphone or laptop. Now look at us! That's development for you.

It's not just the UK that's developing. Changes in other parts of the world are happening even faster than here. The most rapid development so far this century has been in China. You already know how fast cities are growing there (see spread 5.7). But other countries are developing fast too – countries like Mexico, Indonesia, Nigeria and Turkey.

Your enquiry

In this chapter you will investigate the specification's two key questions.

Why are some countries richer than others? You will:

- Define the term 'development' in your own words.
- Describe the global spread of ACs, EDCs and LIDCs.
- Compare economic and social development for some countries.
- Understand the links between measures of development.
- Learn how uneven development between countries has come about.
- Explore the obstacles to development for countries today.

Are low-income developing countries likely to stay poor? You will:

- Study Zambia as an example of a low-income developing country.
- Investigate how the country has developed over the past 50 years.
- Consider whether it has met its targets for development.
- Explore trade, aid and investment in Zambia.
- Decide whether these have been good for the country or not.
- Compare the impact of two development strategies in Zambia.

Look at the photo opposite.

- What has this school in Africa got to do with development?
- How does education help a country to develop?
- What would be your top three priorities for development?

6.1 Defining development

On this spread you will come up with your own definition of development.

Development varies in different countries

In geography, **development** means an improvement in living standards through better use of resources. We often talk about a city, a country or even the world developing.

Development can mean different things, depending on where you live. If you live in a rural area in Africa, development could mean building a new road so you can reach the nearest town. If you live in the UK, it could mean building a new high-speed railway to get from one city to another more quickly. The places in photos A–D are all in need of development.

Classifying countries

Geographers try to classify countries according to their level of development. However, they don't always agree! In this book, countries are classified into three groups:

- **Advanced countries (ACs):** countries that are wealthy and have a wide range of jobs and many services.
- **Emerging and developing countries (EDCs):** countries that are in transition from being low-income developing countries to becoming ACs.
- **Low-income developing countries (LIDCs):** countries that are poor and have a narrow range of jobs and few services.

A *Rochdale, UK*

B *Beijing, China*

C *Damascus, Syria*

D *A rural area in Zambia*

Development is about more than wealth

Our understanding of development is also developing! In the past it was all about wealth, but now we think about other aspects of development too:

- **Economic development:** an increase in the total goods and services a country produces.
- **Social development:** an improvement in human welfare to meet people's needs in a country.
- **Sustainable development:** economic, social and environmental development to meet people's needs now without compromising the ability of future generations to meet their needs (diagram E).

E *Sustainable development*

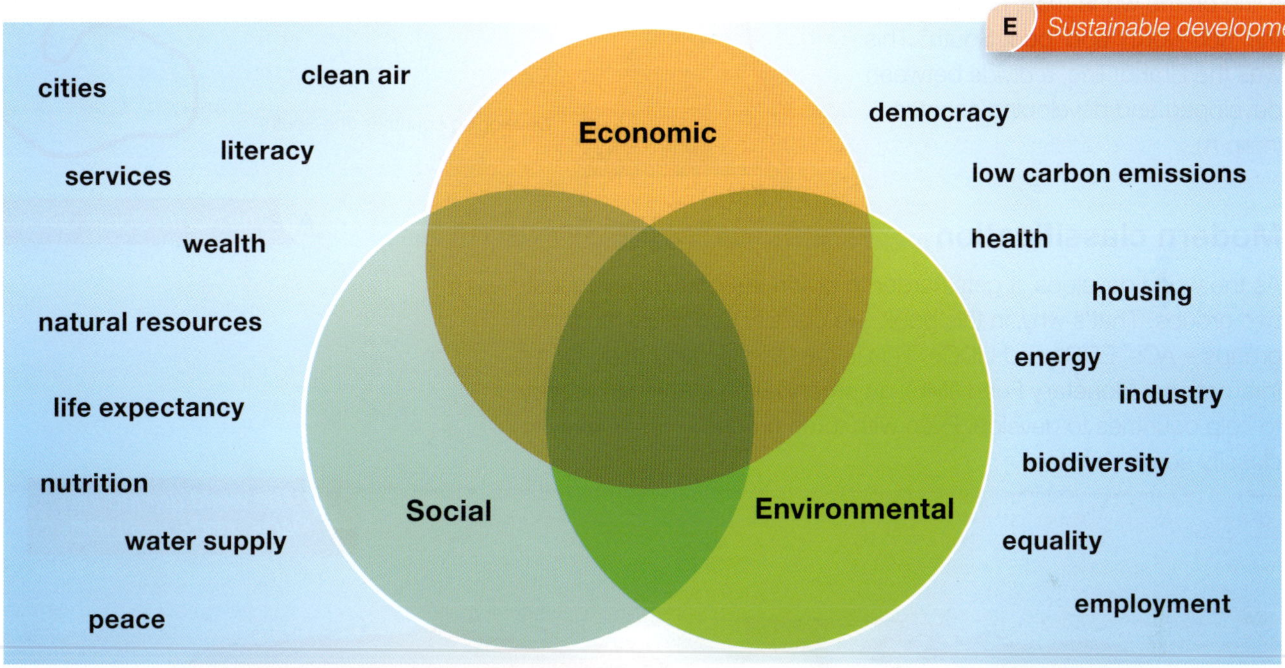

Activities

1. Look at the photos A–D.
   - **a** What problem can you identify in each photo?
   - **b** What might development mean in each of these places?
   - **c** Try to decide whether each place is in an AC, EDC or LIDC. Give reasons for your choices. (You can check your answer on map B on spread 6.2.)
2. What would development mean in the area where you live? List at least five things you think the area needs to develop, e.g. more jobs for young people.
3. Look at diagram E.
   - **a** Make a large copy of the Venn diagram and give it the overall title, 'Sustainable development'.
   - **b** Put each of the aspects of development around diagram E in the best space on your diagram. For example, 'cities' could go in the space where economic and social development overlap.
4. Write your own definition of development to summarise what you have learned. Try to avoid copying from the book. Your definition should be no longer than a Tweet (140 characters).

Your enquiry

Why are some countries richer than others?

As you work through this section, make notes about anything that will help you to answer the question. You will bring your ideas together to do an activity at the end of spread 6.8.

The global development divide

On this spread you will consider how the world is divided into countries at different stages of development.

North–South divide

For a long time, geographers divided the world into 'developed' and 'developing' countries. Put simply, developed countries were rich and developing countries were poor. In the 1980s an international report on development drew a line on the world map, dividing it into the 'rich North' and 'poor South'. This was the Brandt line, a divide between developed and developing countries (map A).

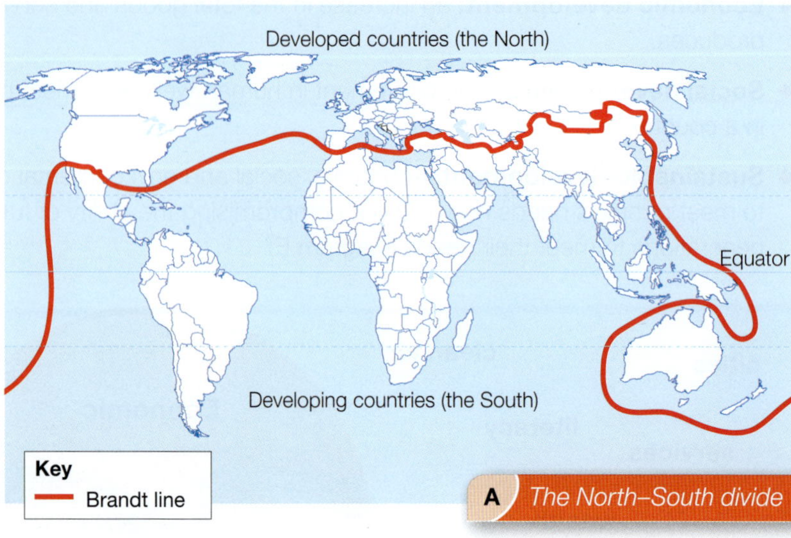

A *The North–South divide*

Modern classification

As the world develops, it gets harder to divide the countries into just two groups. That's why, in this book, we classify countries into three groups – ACs, EDCs and LIDCs. This is the classification used by the International Monetary Fund (IMF), an organisation that lends money to help countries to develop. Even with three groups, it is not easy to classify some countries.

B *The world divided into ACs, EDCs and LIDCs*

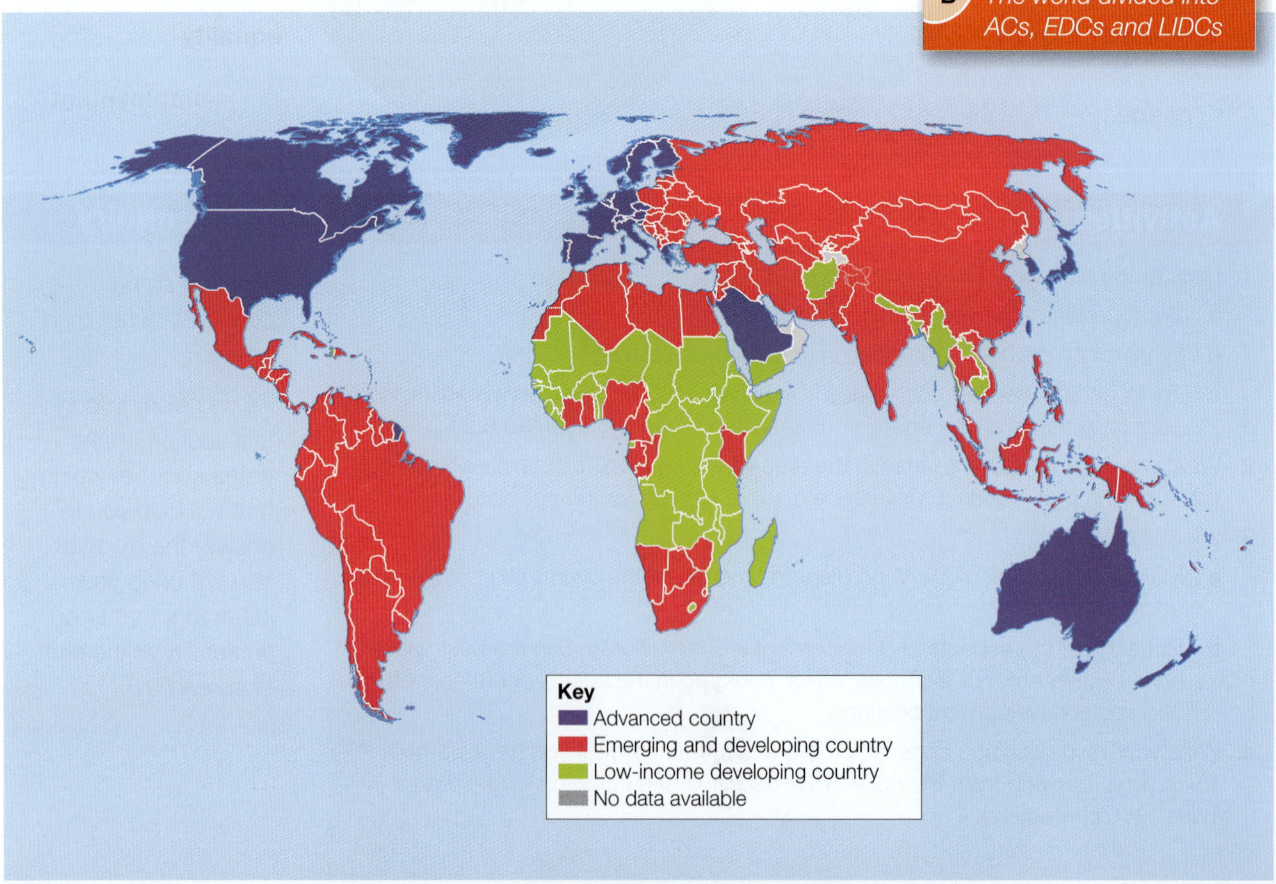

BRICs

In 2001, a small group of countries – Brazil, Russia, India and China – was identified and called **BRICs**. These are all large countries with rapidly growing economies, as measured by **gross domestic product (GDP)** – the total value of what the country produces. They have the potential to be the world's largest economics by 2050 (graph C).

MINTs

More recently, in 2014, another group of countries was identified, following in the footsteps of the BRICs. These are **MINTs** – Mexico, Indonesia, Nigeria and Turkey. They also have potential to become major world economies by 2050.

C *The world's largest economies in 2012 and 2050*

GDP, 2012 ($ trillion)		Rank		Estimated GDP, 2050 ($ trillion)
16.24	USA	1	China	52.62
8.23	China	2	USA	34.58
5.96	Japan	3	India	24.51
3.43	Germany	4	Brazil	9.71
2.61	France	5	Russia	8.01
2.47	UK	6	Japan	7.37
2.25	Brazil	7	Mexico	6.95
2.01	Russia	8	Indonesia	6.04
2.01	Italy	9	UK	5.69
1.84	India	10	France	5.36
1.82	Canada	11	Germany	5.23
1.53	Australia	12	Nigeria	4.91
1.32	Spain	13	Turkey	4.45
1.18	Mexico	14	Egypt	3.61
1.13	South Korea	15	Canada	3.47
0.88	Indonesia	16		
0.79	Turkey	17		
0.26	Nigeria	39		

Activities

1 Look at map A.

a Name three countries that were in the 'rich North' and three in the 'poor South'. You could look in an atlas.

b Explain why the North–South divide curves and does not follow the Equator.

2 Look at map B.

a Describe the distribution of the three groups of country – ACs, EDCs and LIDCs. Mention the continents where each group is found.

b In which groups are the six countries you listed in activity 1?

3 Look at graph C.

a List the climbers and fallers in the rank order of GDP from 2012 to 2050. Which continents are each of these countries in?

b What does this say about the rate of development in each continent up to 2050?

Go further

Choose one of the BRIC or MINT countries. Do some research to find out more about its development.

- How is the population changing and how could this affect its development?
- What does it produce?
- How much money do people earn (GDP/person)?

Practice question

What is the evidence from graph C that BRICs and MINTs are EDCs? Include data from the graph in your answer. (4 marks)

6.3 Measuring development

On this spread you will compare countries using economic and social measures of development.

Measuring by wealth

It is important to be able to measure development. Otherwise, how would we know if a country is making progress or how countries compare with one another?

One way to measure development is to focus on economic development, using a country's wealth. Two common measures are GDP and **gross national income (GNI)** – the sum of all the money earned in a country. The country with the largest GNI is the USA.

However, to find the real wealth of a country, we need to divide GNI by population. This gives us GNI/capita, or average wealth per person. But this does not account for the cost of living between countries, because the same amount of money may buy more in one country than in another. GNI/capita at **purchasing power parity (PPP)** does take this into account. The country with the largest GNI/capita at PPP is Qatar (photo A and table B). By this measure, the USA comes eighth.

Criticisms of using wealth alone as a measure of development

- GNI/capita is only an average measure of wealth. Many people are richer or poorer than the average.
- Not all wealth is included in GNI. For example, food grown by farmers to feed their own families is not included.
- Wealth is not the only measure of development. Social development also needs to be included and development should be sustainable.

A *Qatar, the world's richest country*

Rank	Country	GNI/capita ($)
1	Qatar	123 860
2	Kuwait	88 170
3	Singapore	76 850
4	Norway	66 520
5	Luxembourg	59 750
6	United Arab Emirates	58 090
7	Switzerland	56 580
8	USA	53 960
9	Saudi Arabia	53 780
10	Oman	52 170
24	UK	35 760

B *The world's top ten richest countries measured by GNI/capita at PPP*

PPP means purchasing power parity, a way of saying that the cost of living in each country has been taken into account. That's because a US dollar buys more in some countries than in others.

Human Development Index

Another way to measure development is to use the **Human Development Index** (**HDI**), calculated using the following three economic/social measures of development:

- **Wealth:** measured by GDP/capita at PPP, that is, adjusted to take account of the cost of living in each country.
- **Health:** measured by **life expectancy**, the average age to which people in a country can expect to live.
- **Education:** measured by the **adult literacy rate** (the percentage of adults able to read) and also the percentage of children in education.

Scores for HDI range from 0 (the lowest level of development) to 1 (the highest level) (table C and map D).

Rank	Country	HDI
1	Norway	0.944
2	Australia	0.933
3	Switzerland	0.917
4	Netherlands	0.915
5	USA	0.914
6	Germany	0.911
7	New Zealand	0.910
8	Canada	0.902
9	Singapore	0.901
10	Denmark	0.900
14	UK	0.892

C *The world's top ten most developed countries ranked by HDI*

D *HDI by country*

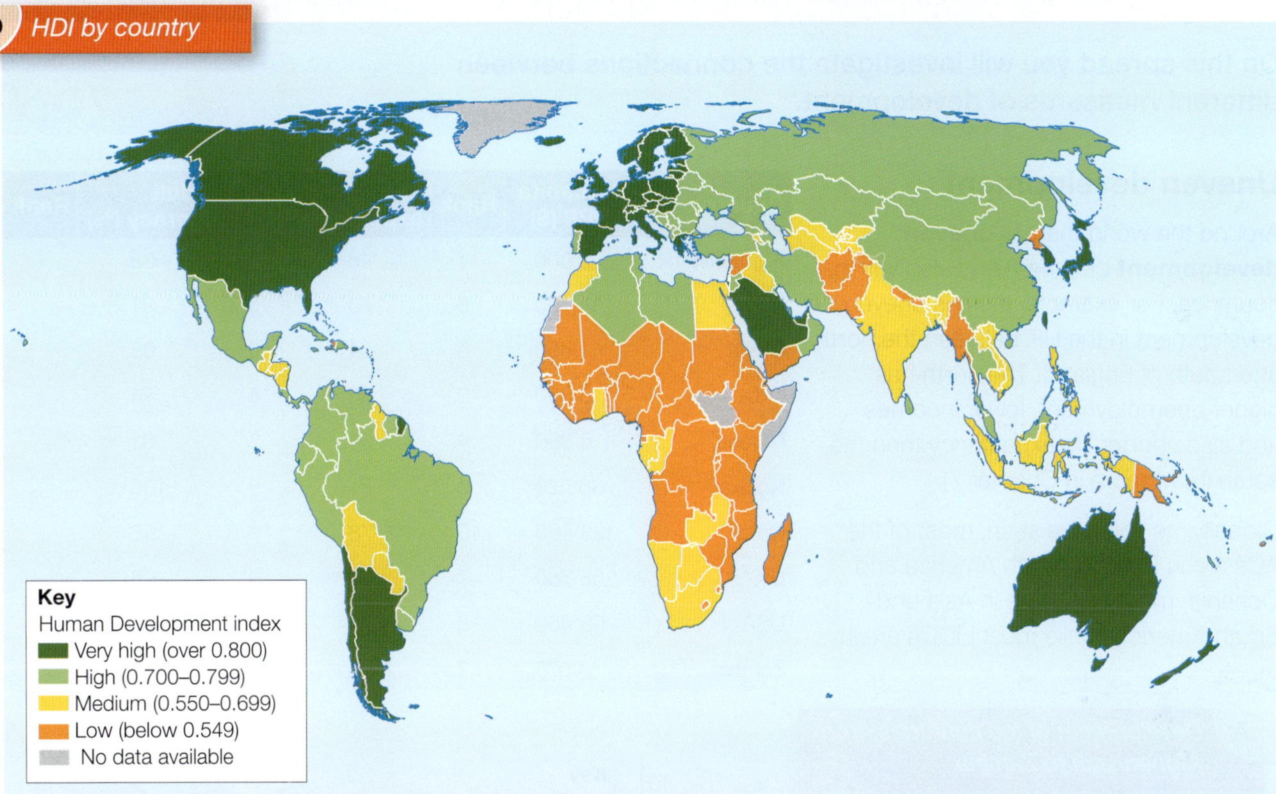

Activities

1. **a** What connects Qatar (photo A) and many of the countries in table B? (Clue: think about what they produce.)
   b Is wealth a good way to measure development? Give reasons for your answer.
2. Compare tables B and C.
   a Where in the world are most of the countries in each list located? You could check in an atlas.
   b The UK is ranked higher by HDI than by GNI/capita at PPP. How could you explain this?
3. Look at map D.
   a Write a few sentences to describe the distribution of countries grouped by HDI.
   b Compare map D with map B on spread 6.2. What do you notice about the HDI for ACs, EDCs and LIDCs?
4. **a** Draw a scatter graph like the one below to show the relationship between GNI/capita at PPP and HDI. On your graph, plot the countries that appear in both tables B and C. Use the rank order to plot them.
   b Is it possible to plot a line of best fit? What does this show about the relationship between GNI/capita at PPP and HDI?

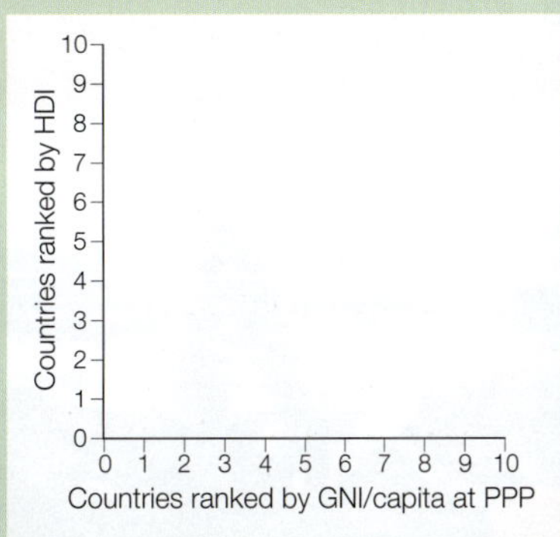

Go further

Find out more about Qatar.

- Why is it so wealthy?
- How does it rank using other development measures?
- Which parts of the UK does Qatar own?

Practice question

a Locate and label the top ten countries, ranked by their HDI in table C, on a world map. (2 marks)

b Explain the location of these countries on the map. (2 marks)

6.4 Development changes lives

On this spread you will investigate the connections between different measures of development.

Uneven development

Around the world there is **uneven development** between, and also within, countries. For example, there is uneven development in the UK between the north and south of England. The north has higher unemployment, lower incomes and also shorter life expectancy than the south (read more in Chapter 7).

Globally, as you have seen, most of the ACs are in Europe, North America and Oceania, most EDCs are in Asia and South America, while most LIDCs are in Africa.

Country	GNI/capita at PPP ($)	Rank	Life expectancy	Rank	Adult literacy (%)	Rank
Afghanistan	2 000	1	48	1	38	
Brazil	14 750	6	73	5	91	
China	11 850	5	74	6	95	
India	5 350	3	65	4	74	
Nigeria	5 360	4	54	3	61	
Norway	66 520	9	81	9	100	
Qatar	123 860	10	78	7	96	
UK	35 760	7	81	9	99	
USA	53 960	8	79	8	99	
Zambia	3 076	2	52	2	80	

A *Development of ten countries, from Afghanistan to Zambia*

Making connections between measures

Often, there is a connection between the measures of development (table A and graph B). Wealthy countries are likely to have more money to spend on health and education. So social measures like life expectancy and adult literacy should improve with wealth.

The reverse is also true. Poor countries are less likely to have money to spend on health and education. That is one reason why the ebola epidemic spread in West Africa in 2014, killing thousands of people (photo C).

However, wealth is no guarantee of development. You may notice in table A that Qatar, the richest country in the world, has a lower life expectancy than some countries with less wealth.

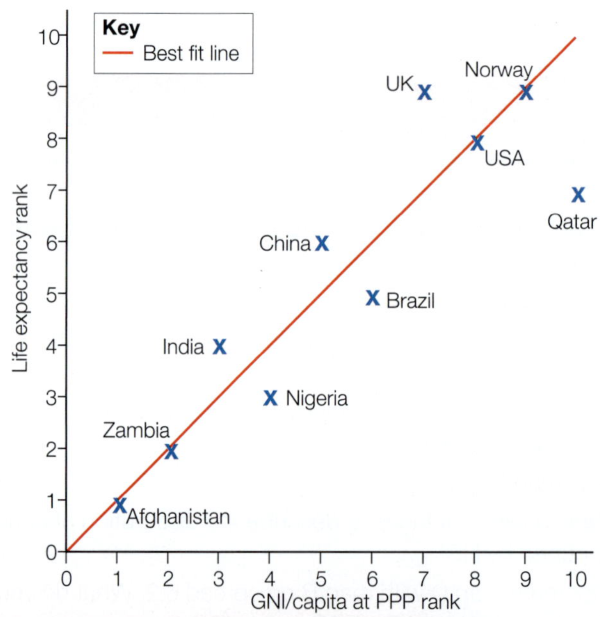

B *Wealth and health of ten countries, shown by GNI/capita at PPP and life expectancy*

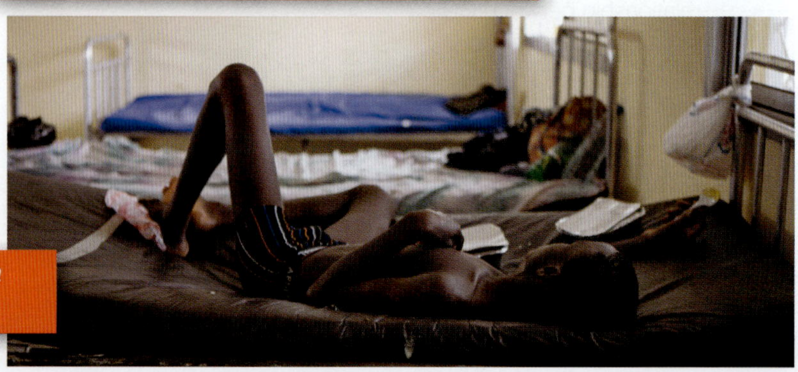

C *An ebola clinic in Sierra Leone, West Africa*

Qatar

Qatar is a small country in the Middle East with a population of just 2.1 million. It has become extremely wealthy from its resources of oil and gas, which are the third largest in the world.

There has been a big construction boom in Qatar in recent years (photo A, spread 6.3), not least to build the football stadiums it needs for the World Cup in 2022. Qatar does not have a large workforce, so **migrant workers** have come from countries like Nepal, India and Bangladesh (map D). It is estimated that, by 2022, a million migrant workers will have flown to Qatar.

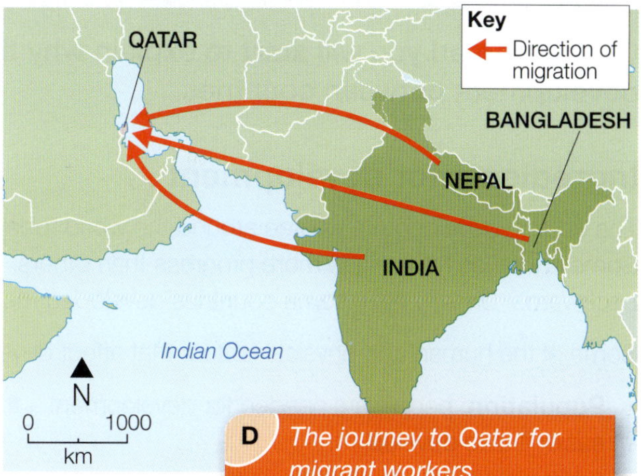

D *The journey to Qatar for migrant workers*

The workers building Qatar

Most of these workers live in poor, overcrowded conditions in camps (photo E), far from the bright lights of the capital city Doha. They claim they are treated like slaves, paid less than £1 an hour. They claim that employers take away their passports so they can't go home. By 2014, about a thousand workers had died, mainly from heart failure, working outdoors in temperatures over 40°C.

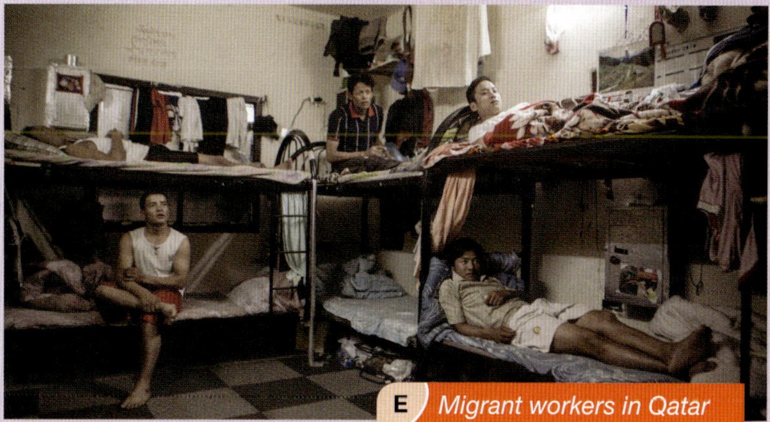

E *Migrant workers in Qatar*

Activities

1 Complete a large copy of table A. In the last column, rank the countries from 1 to 10 by their adult literacy rate, where 1 is lowest and 10 highest. Give equal rank to countries with the same adult literacy rate.

2 Look at scatter graph B. It shows some data from table A.

- **a** Describe the pattern you can see on the graph. Mention the overall trend and the countries that don't fit the pattern.
- **b** Explain the pattern on the graph. How are wealth and health connected?

3 a Draw a scatter graph to show the connection between wealth (GNI/capita at PPP) and education (adult literacy). Use axes like those on the right. Follow the rank orders in your table to plot the countries on your graph.

- **b** Draw a line of best fit on your graph (like the one on graph B), with an equal number of points above and below.
- **c** Describe and explain the pattern on your graph.

4 Imagine that you come from a poor family in India.

- **a** Why might you want to go to work in Qatar?
- **b** What might happen once you got there? Write a letter home to your family to describe your experience in Qatar.

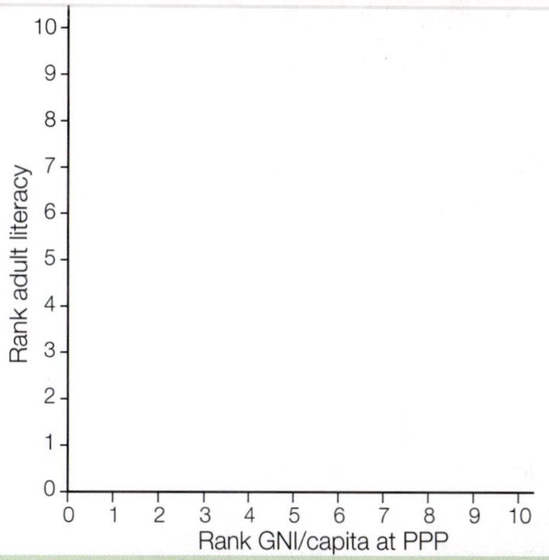

Practice question

The ebola outbreak in West Africa in 2014 killed thousands of people. Explain how this could be related to the level of development in those countries. (3 marks)

6.5 How uneven development happened

On this spread you will start to explain why there is uneven development between countries.

Ingredients for development

The uneven development we see around the world today has not always existed. Some countries have made more progress than others, while a few have actually gone backwards. So why have some countries developed more than others?

Some of the human and physical factors that affect development are:

- **Population:** people are needed for development – the larger the population, the easier it is to develop.
- **Climate:** a warm, moist climate, good for growing crops and for people to live.
- **Land:** flat, fertile land, suitable for growing crops and building cities.
- **Natural resources:** abundant minerals, like oil or coal, that can be traded or used for manufacturing.

Unfortunately, none of these factors fully explains why some countries developed more than others. Some LIDCs, like Bangladesh, have large populations. They can also have warm, moist climates, flat fertile land and abundant natural resources (photo A). The opposite can be true for most ACs. If we want to find out why some countries like Bangladesh developed less than others, history can help us (cartoon B).

A *Bangladesh is an LIDC*

B *A history of colonialism*

Historically, the development gap we see between countries today did not exist. Countries once traded with each other as equal partners.

From the 15th century, Europeans began to explore Asia, Africa and America. They went in search of gold and other valuable resources.

Europeans traded with people in other continents. But, if negotiation didn't work, they conquered the land and enslaved people.

Over the next few centuries, Europeans colonised much of Asia, Africa and South America. The **colonies** grew crops like sugar that were sent back to Europe.

Colonies supplied Europe with natural **commodities** and a market for manufactured goods. But by the 20th century the colonies wanted to be independent.

Post-independence, the previous pattern of trade continued. Manufactured goods grew more expensive than commodities, so some countries got into debt.

The effects of colonialism

The period of **colonialism** lasted about 400 years. During that time, European countries controlled much of the world.

The 'triangular trade' between Europe, West Africa and the Caribbean is an example of how colonialism affected later development (map C). Europe traded goods for African slaves, who were transported to the Caribbean, depriving Africa of its working population. In the Caribbean, land was taken from native people and turned into large plantations, growing crops for Europe. This trade made the merchants very rich and helped them to invest in new industries in Europe.

Colonialism helped Europe develop, but slowed down development in the colonies. The impact can still be seen in those countries now (table D).

The patterns of trade established during the colonial period continue today (see spread 6.7). Debt and political unrest also hold back development in ex-colonies (see spreads 6.7 and 6.8) and climate change may add to the problem of uneven development (see spread 6.6).

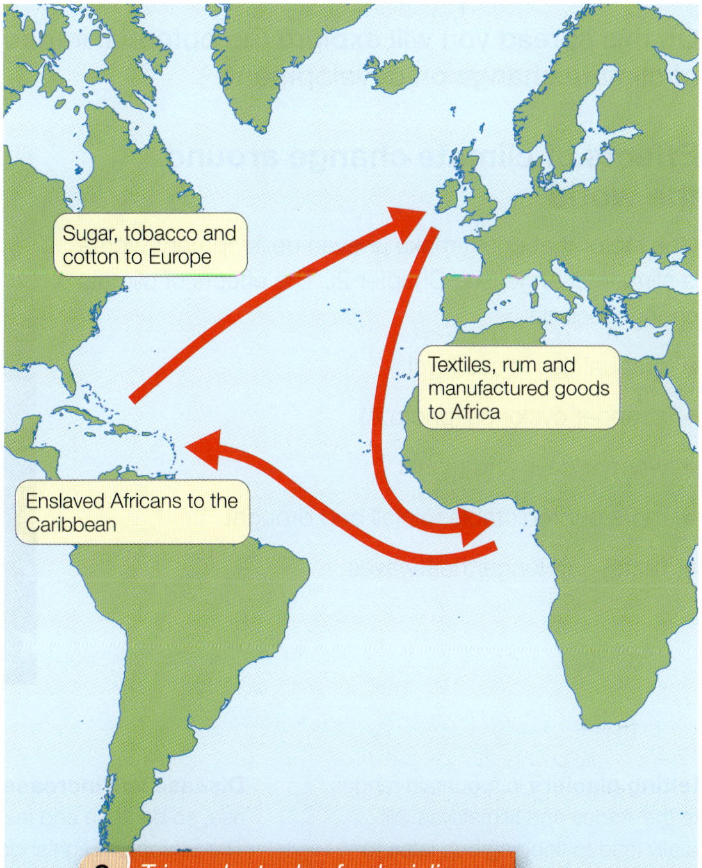

C *Triangular trade of colonialism*

Country	GNI/capita	Life expectancy	Adult literacy (%)	HDI	Main exports
UK	36 590	80	99	0.892	financial services, military arms, motor vehicles
Nigeria	5 360	54	61	0.459	oil (95%), cocoa, rubber
Jamaica	7 450	71	88	0.727	tourism, bauxite (for aluminium), sugar

D *Development today in countries involved in the triangular trade*

Activities

1. Look at photo A. What factors do not explain why Bangladesh is an LIDC?
2. Study cartoon B and map C. In what ways did colonialism:
   - **a** benefit Europe? Mention three ways.
   - **b** disadvantage Asia, Africa and America? Mention three ways.
3. Look at table D.
   - **a** Identify which country is now an AC, an EDC or an LIDC. (Check on map B on spread 6.2.)
   - **b** How might each country's history help to explain its level of development today? Refer to map C to help you.

Go further

Find out more about the British Empire.

- Name some other countries Britain colonised?
- When did each one gain its independence?
- What is the level of development in each of those countries today?

Practice question

History rather than geography offers a better explanation of why some countries are more developed than others today. Do you agree or disagree? Discuss. (6 marks)

6.6 Uneven development and climate change

On this spread you will explore the potential impact of climate change on development.

Effects of climate change around the world

One factor that could make uneven development worse is climate change (see Chapter 2). The effects of climate change include:

- gradual rise in sea level
- stronger cyclones (photo A)
- warmer days and nights
- more unpredictable rainfall and drought
- hotter and longer heatwaves.

A *Typhoon Haiyan killed over 6000 people in the Philippines in 2013 (see spreads 1.8–1.9)*

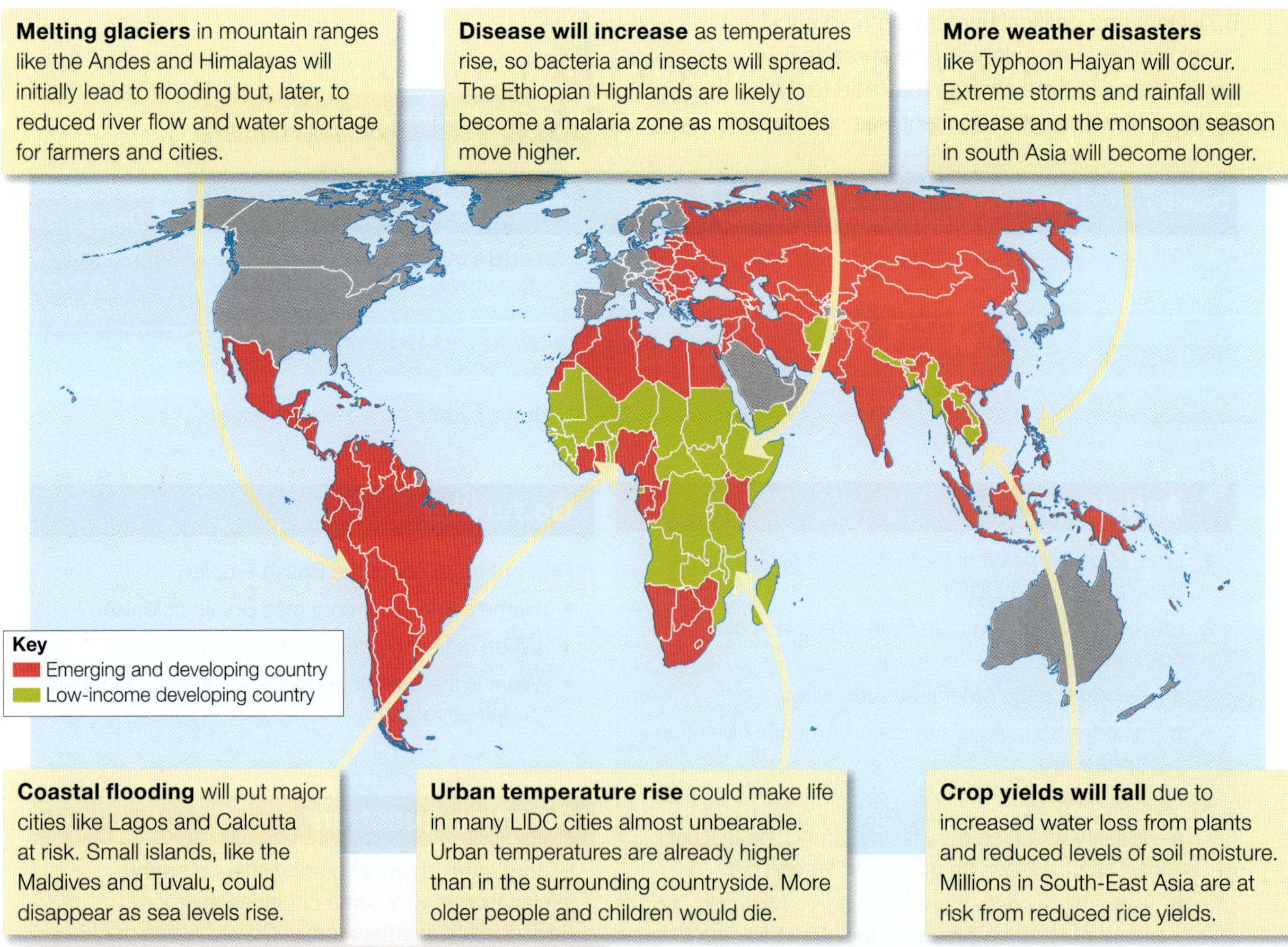

B *The impact of climate change in LIDCs and EDCs*

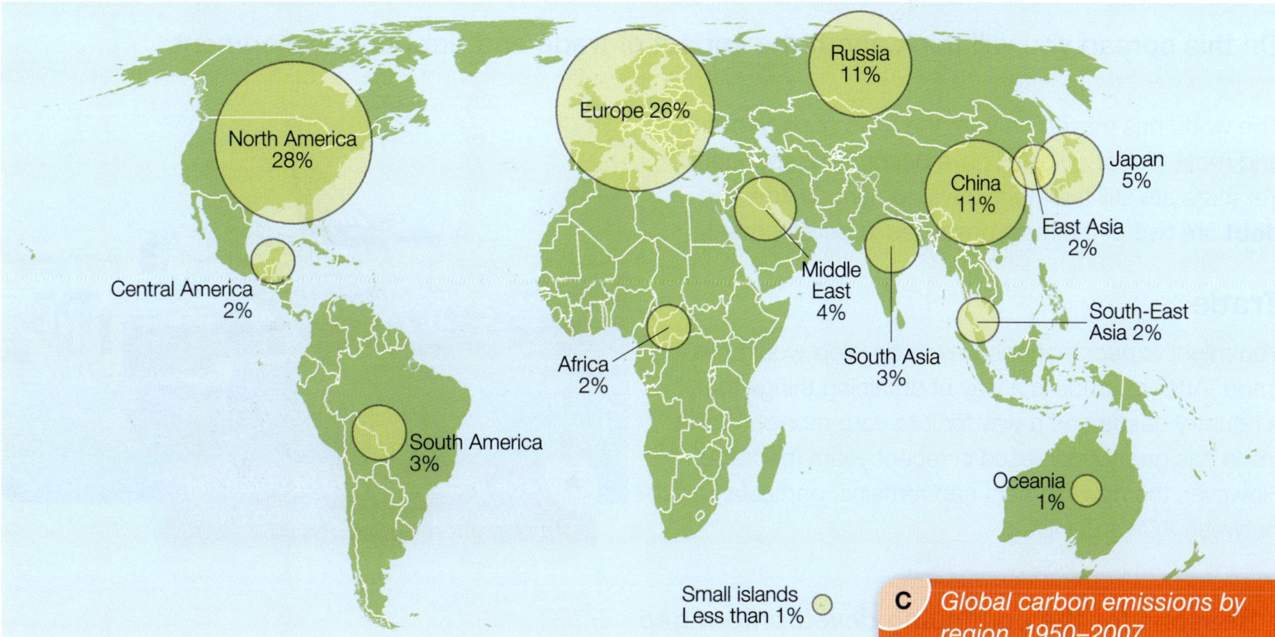

C *Global carbon emissions by region, 1950–2007*

Global carbon emissions

ACs like the UK have produced carbon emissions (map C) for over two centuries. So they have made a much greater contribution to global climate change than other countries. Although EDCs like China are now burning more fuel, their overall contribution has been smaller.

LIDCs emit small amounts of carbon dioxide and other greenhouse gases. This has relatively little effect on the world's climate. Yet these countries are most vulnerable to the impact of climate change.

Effects of climate change on LIDCs

The consequences of climate change are likely to be worse for LIDCs because they lack the resources necessary to adapt. Possible consequences include:

- **increased hunger** as crop yields fall (reversing the recent trend of reducing hunger)
- **more conflict** between countries as they fight over scarce water supplies
- **migration** from areas that become uninhabitable, creating climate refugees
- **worse poverty** as the gains made from development are lost.

Activities

1 Look at map B. Why might each of these people be concerned by climate change?
  - a A farmer in the Andes
  - b A slum dweller in Manila, in the Philippines
  - c The president of the Maldives

2 Look at map C.
  - a Which regions of the world have made:
    - the greatest contribution to global carbon emissions (over 10%)?
    - the smallest contribution to global carbon emissions (2% or less)?
  - b Why might a person living in an LIDC think this is unfair?

3 a How does climate change increase the problem of global uneven development? (Clue: think about the effects of climate change on ACs and LIDCs.)
  - b Is climate change a human or physical factor influencing uneven development? Argue the case that it is either human or physical, or both. (Clue: look back at Chapter 2).

Practice question

How could climate change add to the problem of uneven development around the world? Refer to named countries and how they are affected by climate change. (6 marks)

Obstacles to development – trade and debt

On this spread you will think about the impact of trade and debt on development.

The world has moved on from the days of colonialism and most former colonies have become independent. Yet there are still threats to development. **Trade** and **debt** are two of the main obstacles to development.

Trade

You might expect that one way to develop would be trade. After all, trade is a way of obtaining things that a country needs and a way for it to earn money. World trade has greatly increased in recent years (photo A). However, the development gap remains, particularly between ACs and LIDCs.

A The increasing size of container ships is a sign of growing trade

Reasons for the continuing development gap

- Most of the increase in trade has been between ACs. Europe still dominates world trade, even though its population is relatively small. China now plays an important role too (map B).
- The types of goods traded by ACs and LIDCs differ. ACs produce more manufactured goods and services while most LIDCs still rely on the export of natural commodities.
- The price of natural commodities fluctuates with supply and demand, while the price of goods and services has generally increased.
- Much of world trade is now controlled by large **trans-national companies (TNCs)**, mainly based in ACs (table C). Although many TNCs also operate in LIDCs, the profits each TNC makes comes back to the country where it is based.

B World trade – imports and exports by region

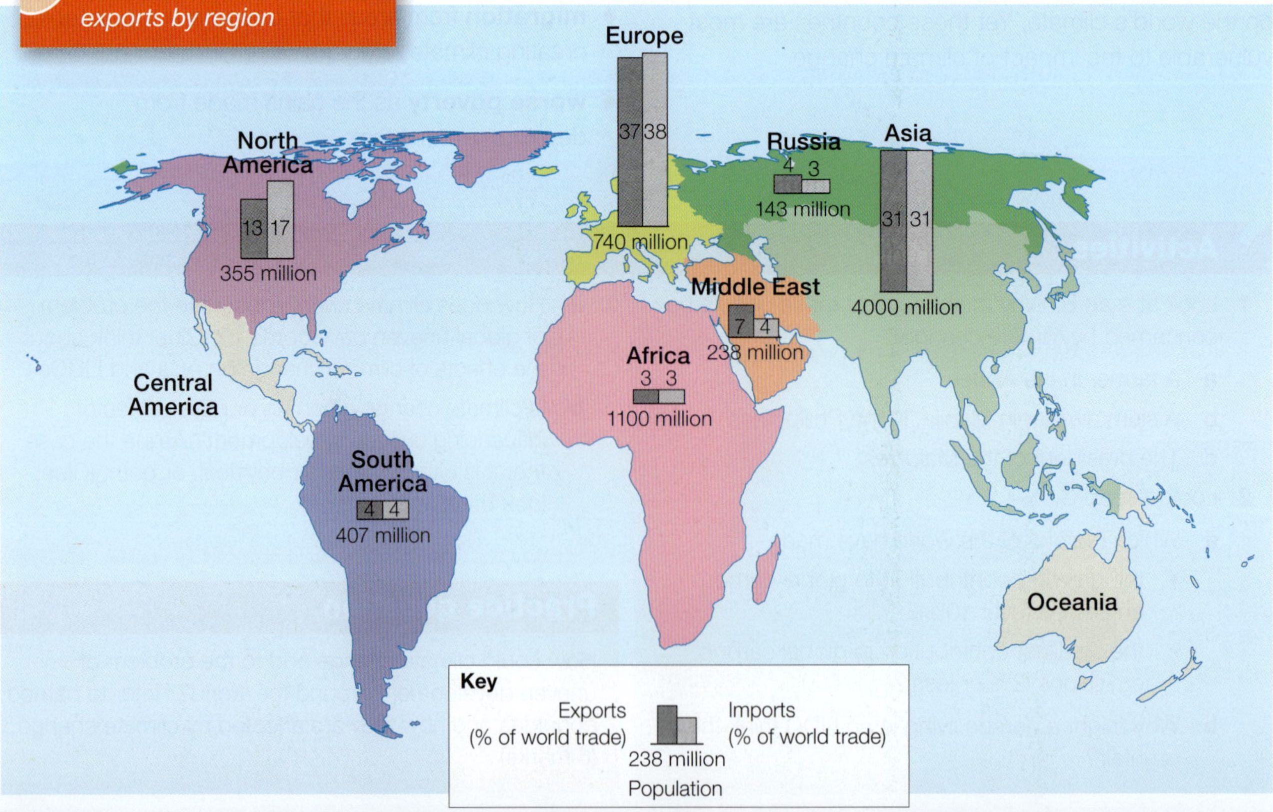

Debt

Debt is a problem for any country that borrows money to pay for goods and services (diagram D). However, the impact of debt is worse for LIDCs with a smaller GNI. For example, it can make the difference between having enough money to pay for health and education, or not. The IMF, which lends money for development, does so on strict conditions that countries repay their debts and cut spending.

Since 2000, 30 of the world's poorest countries have benefited from international **debt relief**, so they can now spend more on essential services.

C *The world's top ten TNCs in 2014*

Rank	Company	Country	Industry	Revenue (US$ billion)
1	Walmart	USA	Retail	476
2	Royal Dutch Shell	Netherlands/UK	Oil	459
3	Sinopec	China	Oil	457
4	China National Petroleum Corporation	China	Oil	432
5	ExxonMobil	USA	Oil	407
6	BP	UK	Oil	396
7	State Grid Corporation of China	China	Power	333
8	Volkswagen	Germany	Cars	261
9	Toyota	Japan	Cars	256
10	Glencore	Switzerland	Commodities	232

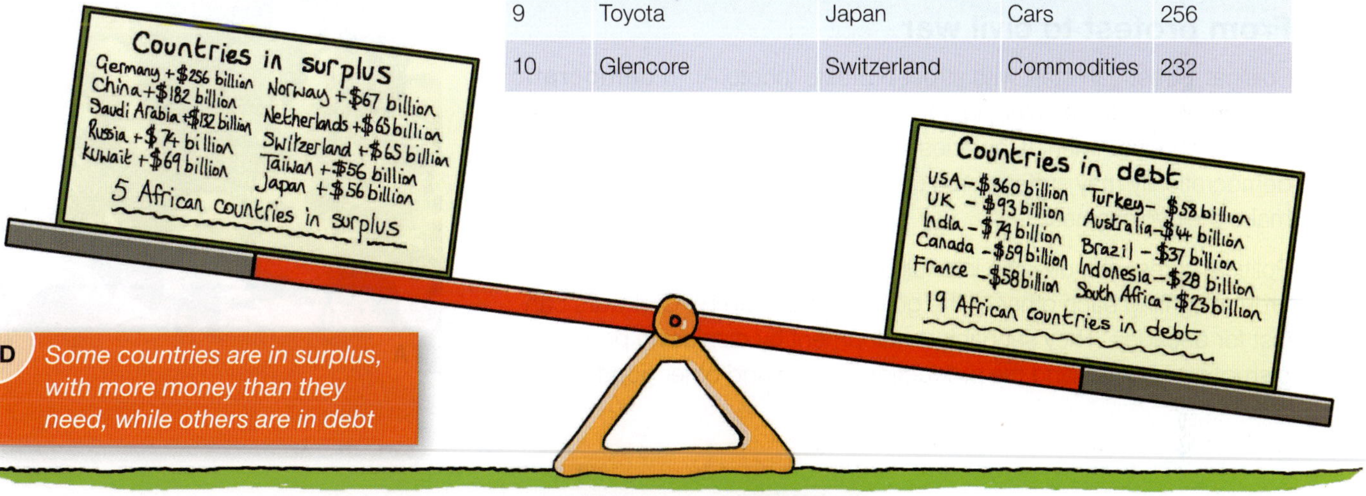

D *Some countries are in surplus, with more money than they need, while others are in debt*

Activities

1 Look at photo A.

a What do you think the benefits of trade are for:
- a consumer (like you!) in an AC?
- a TNC?

b Why is trade less likely to benefit LIDCs?

2 Look at map B. Which regions have:

a most trade?

b least trade?

c more exports than imports?

d more imports than exports?

3 Look at table C.

a In which regions are most TNCs based? And which have none?

b Why is this important for regions with and those without TNCs?

4 Look at diagram D.

a The UK is in debt. Why is that a problem for us?

b Most African countries have a smaller debt than the UK. Why is debt a bigger problem for them than for us?

Practice question

Explain how trade and borrowing money could help an LIDC to develop. What are the disadvantages of each strategy? (6 marks)

Obstacles to development – political unrest

On this spread you will find out why political unrest is also an obstacle to development.

Reasons for political unrest

Political unrest is another obstacle to development. It happens where there is widespread dissatisfaction with the government in a country. Underlying most political unrest is one or more of these factors:

- social inequality
- poor government or corruption
- high taxes
- lack of government spending of services such as schools and roads
- rising prices of essential items like food
- conflict over ownership of resources such as water, land and fuel.

From protest to civil war

Political unrest can start with street protests (photo A). In **democratic** countries, people have the right to vote and unrest can be resolved by a change of government. Countries with governments that are not elected democratically may suffer worse unrest if people resort to violence to make their voices heard.

Political unrest can develop into full-scale **civil war**, with different groups fighting each other or the government (map B). Civil wars can last for decades. They set back development by displacing people, and disrupting services and the production of food and other goods.

A *Protests in 2011 led to the fall of the Egyptian government*

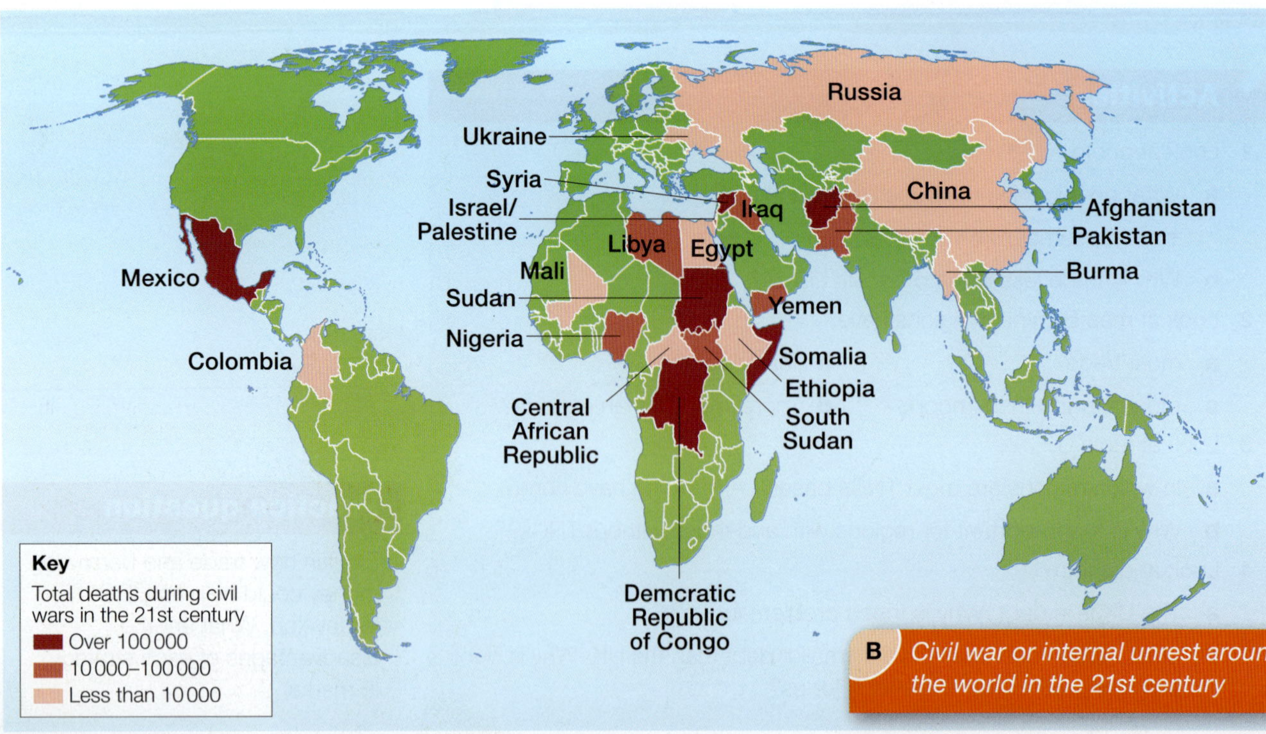

B *Civil war or internal unrest around the world in the 21st century*

Political unrest in South Sudan

South Sudan is the world's newest country and also one of the poorest. It gained independence from Sudan in 2011 after a peace deal in 2005 ended Africa's longest-running civil war between the north and south of the country. An overwhelming majority of people in South Sudan voted in favour of independence.

There were many reasons for the civil war:

- Until 1956, Sudan was under British control. It was made up of regions, each with its own identity.
- After independence from Britain, the north of Sudan controlled the south and forced an Islamic identity on it.
- Sudan was rich in oil, most of it in the south of the country. Under the peace deal, money from oil was split 50:50.

After independence from Sudan

Unfortunately, the story did not end there. In 2013 a new civil war broke out between the government in South Sudan and rebels within the country. Thousands were killed and nearly a million were displaced. Production of oil, the country's main export, almost came to a halt.

Despite its oil wealth, most of South Sudan's population are farmers or herders who depend on what they produce to survive. This produces further conflict. Rival ethnic groups fight each other over the right to graze their cattle or farm the land.

C *Location of South Sudan*

D *A cattle herder in South Sudan*

Activities

1 Look at photo A.

a Do you think people should protest if they are unhappy with their government? Give reasons.

b How can democratic government help to reduce the likelihood of political unrest?

2 Look at map B.

a Describe the distribution of civil war around the world in the 21st century. Which regions experience most conflict? Which experience the least?

b Name five countries that have had the worst conflict.

3 Look at photo D. Suggest what impact civil war in South Sudan has on:

a a cattle herder

b a foreign oil company hoping to invest in the country

c a family displaced and forced to become refugees

d the country as a whole.

Your enquiry

Why are some countries richer than others?

Ask a few of your friends (who don't study geography) this question. You may hear some of these answers:

'That's the way it's always been.'

'Rich countries have more technology.'

'Poor countries are all in Africa.'

'The countries with oil get richer.'

a Which answers are right and which are wrong? Are any partly right? Explain.

b Now give your own answer to the question, using what you have learned in this section. Write up to 300 words. You should be able to do better than your friends!

Zambia – a low-income developing country

On this spread you will get an overview of economic development in Zambia.

Introduction to Zambia

Zambia is a country in central southern Africa that was once a British colony. Today, it is classified as an LIDC. Life expectancy is low and almost half the country's population lives in poverty. Most people still live in rural areas, although there are also cities, including Lusaka, the capital (photo A). Zambia is a **land-locked country**, surrounded by other countries and without a coastline of its own (maps B and C).

A Lusaka, capital of Zambia

A brief history of Zambia

Originally, the country was inhabited by different groups of African people, each with their own culture and language. Many were farmers or herders. The land was taken over as a British colony in 1888 and named Northern Rhodesia after the British business tycoon, Cecil Rhodes. His company discovered copper in the north of the country and claimed the rights to mine it.

The country gained independence from Britain in 1964, when it was renamed Zambia. Despite being rich in minerals like copper, Zambia has struggled to develop over the past 50 years and remains one of the world's poorest countries.

B Zambia's location in Africa

C Zambia

Key facts

Zambia

Population	14.5 million
Area	752 614 km^2
Population growth rate	2.9%
Birth rate	43 per 1000
Death rate	13 per 1000
Average life expectancy	52
Literacy rate	80%
GNI/capita	$3070
Human Development Index	0.43 (ranked 164/187 in the world)
Urban population	36%
Economic growth rate	6.5% per year
Exports	copper (64%), cobalt, tobacco, flowers, cotton
Imports	machinery, transport equipment, oil, fertiliser, food, clothing

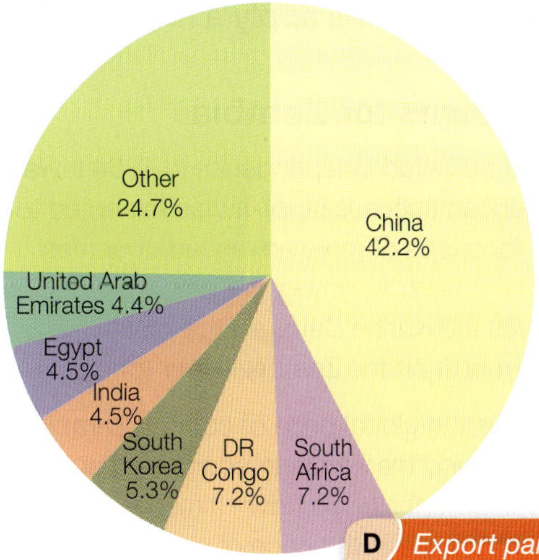

D *Export partners*

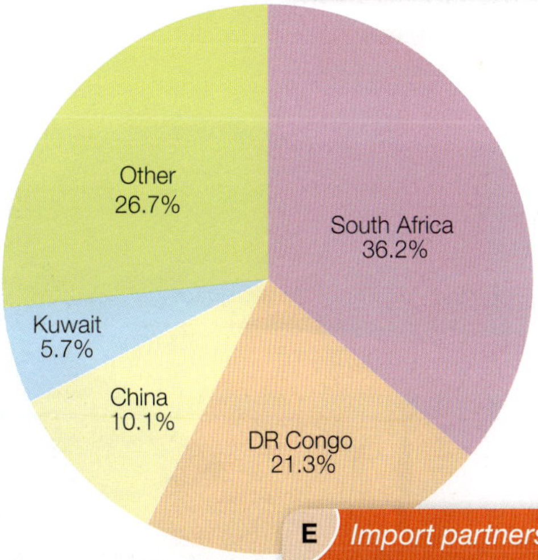

E *Import partners*

Activities

1 Look at map B.
- **a** Describe the location of Zambia in Africa.
- **b** What could be the disadvantage of this location for trade?

2 Look at map C.
- **a** Name the eight countries that have a border with Zambia.
- **b** What could be the advantage of this location for trade?

3 Study the key facts about Zambia, including charts D and E.
- **a** Which data show that Zambia is an LIDC? Select five bits of data and explain why you chose each one.
- **b** One bit of data suggests Zambia is getting richer. Which one? Explain why.

4 Complete a world map to show Zambia's export and import partners.
- **a** Label Zambia on the map. Colour and label the countries Zambia trades with. Use one colour for export partners and another for import partners.
- **b** Draw arrows to and from Zambia on the map linking it with its trading partners. Make sure the arrows go in the correct direction! Make the width of the arrows proportionate to the amount of trade.

Your enquiry

Are LIDCs likely to stay poor?

Through this section, make notes about anything that will help you to answer the question. You will bring your ideas together to do an activity at the end of spread 6.17.

6.10 Zambia's zigzag path to development

On this spread you will apply a model of economic development to Zambia.

Case study

Ups and downs for Zambia

When Zambia achieved independence in 1964 it was seen as an African success story. It was expected to follow in the footsteps of more advanced countries towards development. A symbol of Zambia's ambition at the time was the Kariba Dam, a large hydro-electricity dam built on the Zambezi River (photo A).

But in the 1970s the global price of copper, Zambia's main export, fell and the Zambian economy went into decline. Development slowed down (timeline B). Only since 2000, with a rise in the price of copper, has the economy begun to grow again (graph C).

A *The Kariba Dam*

1964 Zambia gains independence. However, few Zambians are trained to run the country. Most of the power and wealth is still in the hands of Europeans.

1970 The global price of copper falls. It remains low for the next 30 years. Zambia has to borrow money to develop.

1975 The Kariba Dam starts to generate hydro-electric power for Zambia. This is important for the copper industry.

1980 HIV/AIDS spreads across Africa. The death rate in Zambia increases and life expectancy falls.

1990 Zambia's debt is now very high. Food is expensive to buy because the government cannot afford to subsidise it and there are riots.

2000 The global price of copper starts to rise again, increasing the amount of money Zambia can earn.

2006 The IMF cancels Zambia's debt, enabling the government to spend more on services such as health and education.

2010 Zambia begins to develop new industries like tourism, farming and hydro-electric power to reduce its reliance on copper.

B *Zambia's development timeline*

C *Zambia's economic growth, 2001–2015*

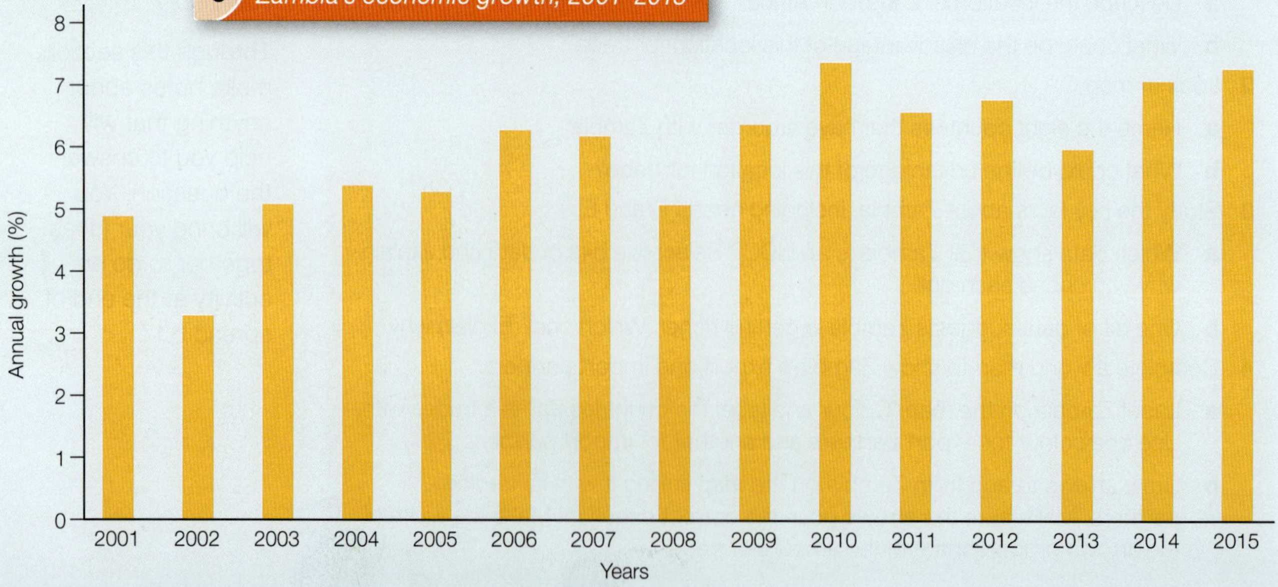

Rostow's model of economic development

In 1960 the American economist Walt Rostow devised a model of economic development (graph D). He identified five stages that all countries must go through as they develop. The model was based on the experience of ACs like the USA.

Can the same model be applied to all countries, including LIDCs? The experience of Zambia since 1964 suggests that some countries' paths to development are more of a zigzag than a straight line.

D *Rostow's model of economic development*

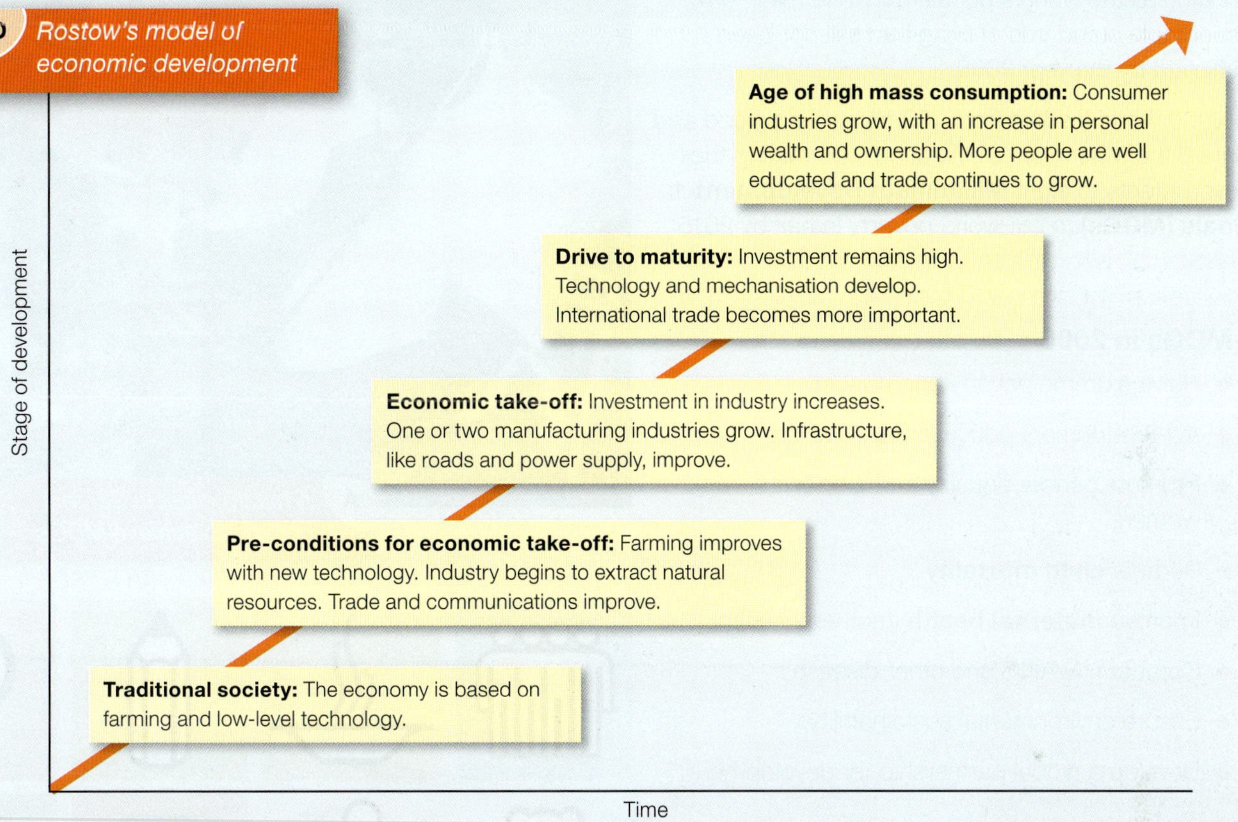

Activities

1 Study photo A and timeline B. Identify these steps on Zambia's path to development:

- **a** three steps forward
- **b** three steps backward.

In each case, explain why it was a step forward or backward.

2 Look at graph C.

- **a** Describe economic growth in Zambia from 2001 to 2015. Mention high and low points, average growth and the overall trend.
- **b** Suggest what effects economic growth might have had on Zambia over that time. (Clue: average economic growth in the UK over the same time was about 2% per year.)

3 Look at graph D. What stage on Rostow's model of economic development do you think these countries are at now?

- **a** The UK
- **b** Zambia

Give reasons for your answers.

Go further

Find out more about stages of development in the UK. Give approximate dates for when the UK reached each stage of development in Rostow's model.

Practice question

Do you think Rostow's model can be applied to LIDCs [like Zambia], or not? Give reasons for your answer. (4 marks)

Zambia's development goals

On this spread you will assess how well Zambia has achieved the Millennium Development Goals.

Millennium Development Goals

Poverty has always been a problem on our planet. Yet despite decades of development, the aim of bringing all the world's population up to minimum acceptable standards of living had still not been achieved by the year 2000.

In 2000 world leaders from 189 countries agreed and signed the Millennium Declaration, committing their governments to eight **Millennium Development Goals (MDGs)** to cut world poverty in half by 2015.

MDGs in 2000

- Halve extreme poverty and hunger.
- Achieve primary education everywhere.
- Promote gender equality and empower women.
- Reduce **child mortality**.
- Improve **maternal health** (mothers' health).
- Combat HIV/AIDS and other diseases.
- Ensure environmental sustainability.
- Develop a global partnership for development.

A *A health clinic in Zambia helping to achieve MDGs*

B *Symbols for the eight Millennium Development Goals*

Progress towards the MDGs

In order to achieve the eight goals by 2015, specific targets were set. Globally, progress on reaching the targets has been mixed. Targets to improve access to safe drinking water, the lives of slum dwellers and access to primary schools for girls were met by 2015. On the other hand, targets to reduce maternal mortality and improve **sanitation** were not.

Zambia's progress towards meeting the MDGs by 2015 was also mixed (photo C). For example, Zambia fell short of the first goal to halve extreme poverty (the number of people living on less than $1.25 a day). Urban areas, like Lusaka, saw much bigger reductions in poverty than rural areas. Zambia remains one of the world's most unequal countries, with large differences between rural and urban areas.

C *Most children in Zambia now attend primary school*

Zambia meets some targets...

The targets on which Zambia did well include:

- most children attend primary school (MDG2)
- there is gender equality in primary education (MDG3)
- the number of people infected with HIV/AIDS has fallen (MDG6)
- the country has obtained debt relief (MDG8).

... but falls short on others

The targets on which Zambia fell short include:

- child mortality is still high (MDG4)
- many mothers still die in pregnancy and childbirth (MDG5)
- the amount of land covered by forest has fallen (MDG7)
- many people still lack proper sanitation (MDG7).

D *Zambia's progress towards meeting Millennium Development Goals*

Extreme poverty

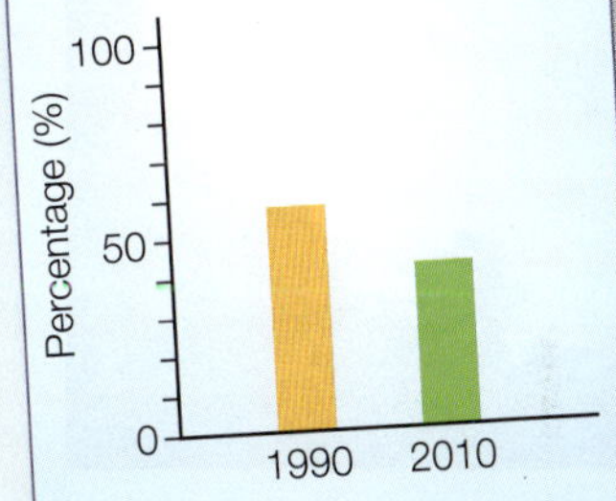

Children in primary education

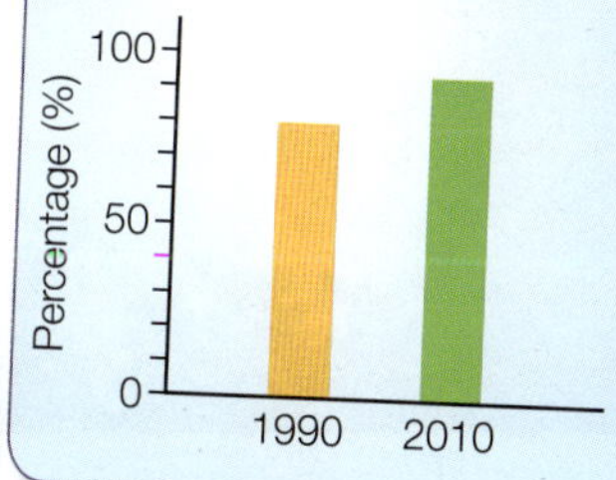

Gender equality

Zambia was expected to achieve equal primary school enrolment for boys and girls in 2015. More has to be done to achieve equal enrolment in secondary schools and universities.

Child mortality

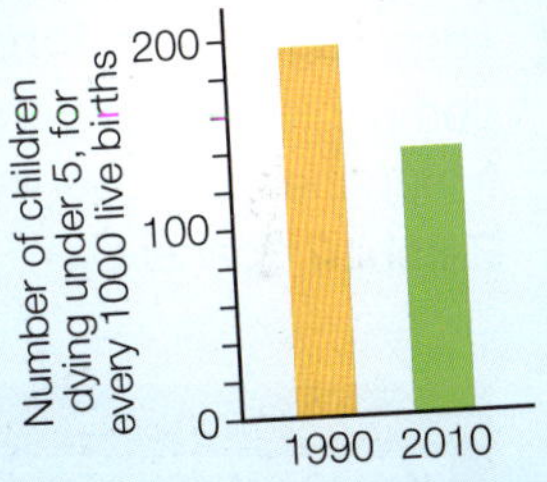

Maternal health

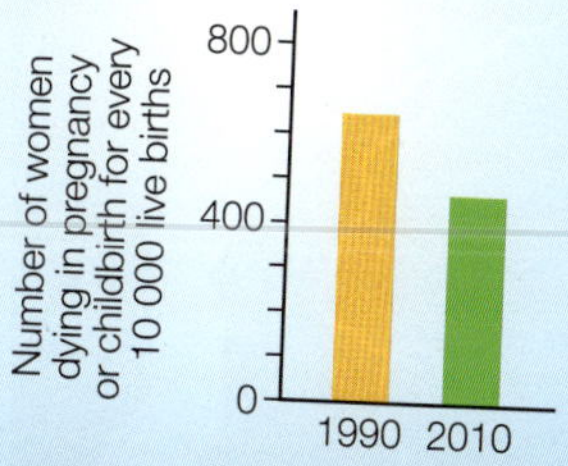

HIV/AIDS

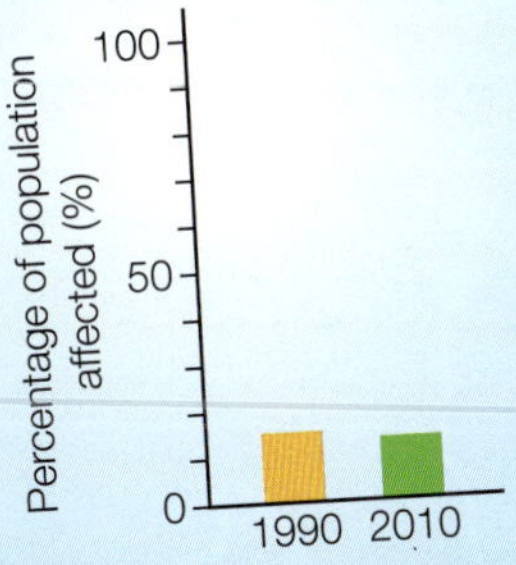

Access to safe water

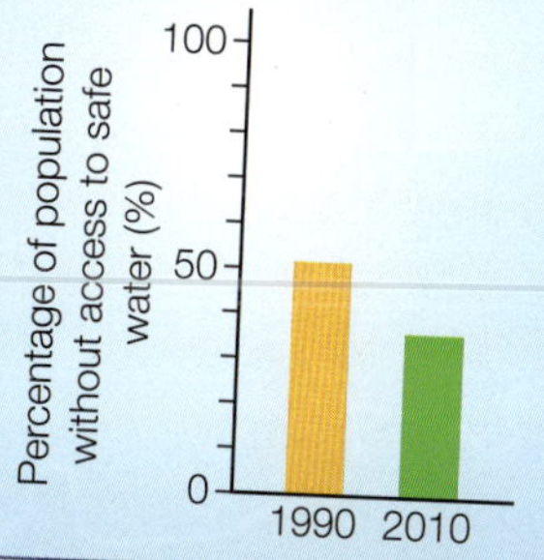

Global partnership

Zambia has obtained international debt relief. It has also attracted foreign investment, particularly from China into mining. Zambia's trade has grown, with more products now exported.

Activities

1. Look at photos A and C. Which MDGs could the following help to achieve?
   - **a** A new health clinic
   - **b** Improved primary school attendance

   Explain how in each case.
2. Look at panel B. For each of the eight MDGs, suggest at least one way that progress towards that goal could be:
   a measured **b** achieved.
3. Study the information in panel D. Use the data to assess the progress made in Zambia towards each MDG.

Go further

In 2015 the UN set 17 new global Sustainable Development Goals.

- Find out and list the 17 new goals.
- How do they compare with the original MDGs?

Practice question

For an LIDC you have studied [Zambia], evaluate how successful it has been in achieving the Millennium Development Goals. (8 marks)

Invest in Zambia

On this spread you will consider the reasons why foreign companies might want to invest in Zambia.

A better Zambia for all?

There has been progress towards development in Zambia since 2000, but the country still faces some challenges. Zambia has one of the fastest-growing economies in Africa – on average about 6% per year (see graph C, spread 6.10). Yet it remains a very unequal country with almost half the population living in poverty.

The Zambian government wants to achieve 'a better Zambia for all', building on the country's rich natural resources (photo A and map B). It aims to become an EDC by 2030. In order to achieve this, the government encourages more **foreign investment**, through large companies bringing their business to Zambia. In 2012, it launched an advertising campaign, 'Why invest in Zambia?' (panel C).

A *Victoria Falls on the Zambezi River – the largest waterfall in Africa and a major tourist attraction*

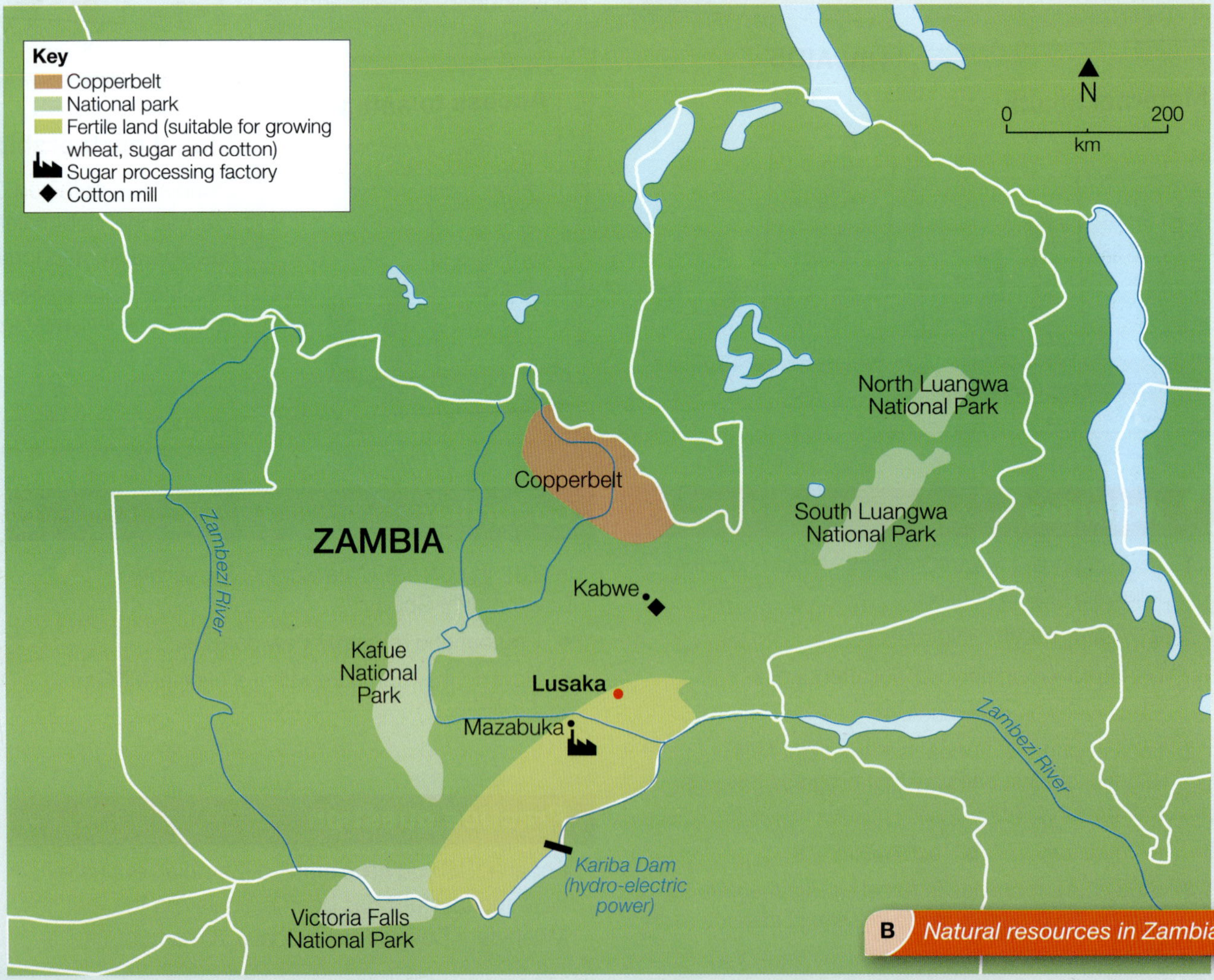

B *Natural resources in Zambia*

C *Why invest in Zambia?*

1 Stable political system

The country has been a peaceful democracy since 1964. It holds elections every few years and there has been little political unrest.

2 Positive economic environment

The economy has grown steadily since 2000. This is due to more businesses and more people earning money.

3 Attractive investment incentives

Companies that invest in Zambia do not pay tax for the first five years. Tax rates increase gradually after that.

4 Good banking, legal and insurance services

Many international banks operate in Zambia, including Citibank and the Bank of China. There are also well-established law and insurance companies.

5 Abundant natural resources

Zambia has extensive resources, including copper, over 50% of land suitable for farming and several large national parks (map B).

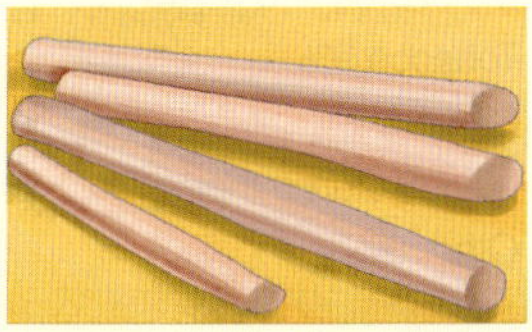

6 Access to other markets

Zambia borders eight other African countries. It is a member of the Common Market for Eastern and Southern Africa (COMESA) and the World Trade Organisation (WTO).

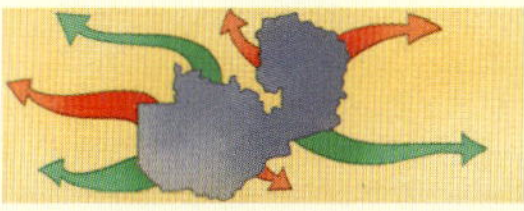

7 Thriving private sector

The Zambian government encourages business. All government-owned business has been sold to private companies (whose owners share the profits).

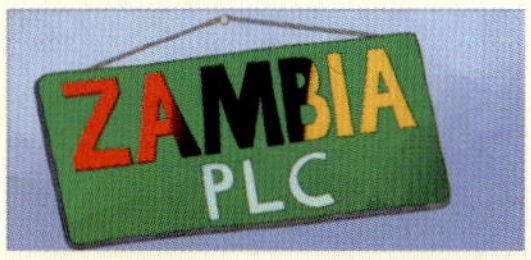

8 Good place to work and live

Zambia is a beautiful country with many national parks to visit (photo D). It is also a very safe country and has good shops and schools for those who can afford them.

D *On safari in South Luangua National Park*

Activities

1. Look at photos A and D. How could tourist attractions help to bring investment to Zambia?
2. Look at map B. In which part of Zambia might a company want to invest in the following?
   - **a** Mining
   - **b** Tourism
   - **c** Agriculture
   - **d** Manufacturing

   In each case, give a reason for your decision.
3. Read panel C.
   - **a** Imagine you are a foreign company in Zambia. Rank the reasons for investing there in order of importance to your company. Choose the three most important.
   - **b** Produce a leaflet to persuade more foreign companies to invest in Zambia.

Practice question

Think of an LIDC you have studied [Zambia]. Suggest ways in which foreign investment could help the country to develop. (3 marks)

Zambia's reliance on copper

On this spread you will explore the importance of copper to Zambia's economy.

Copper in Zambia

From the time Zambia was a British colony, its economy has been based on the copper-mining industry (photo A). Copper was what first attracted Cecil Rhodes (remember him?) to the country. Copper is a good conductor of electricity and today it is vital in the production of wire and cable.

Zambia has come to rely on its international trade of copper for 70% of its exports (chart B). This is not a problem when the global price for copper is high, but it becomes a problem when the price falls. That is what happened in Zambia between 1970 and 2000, causing economic decline in the country. Development slowed down and the country fell into debt.

Since 2000, the global price of copper has risen. Copper production in Zambia has increased and this has helped the economy to grow (graph C). Now the Zambian government wants to diversify the economy so that, in future, the country does not rely so much on copper.

A A copper mine in Zambia

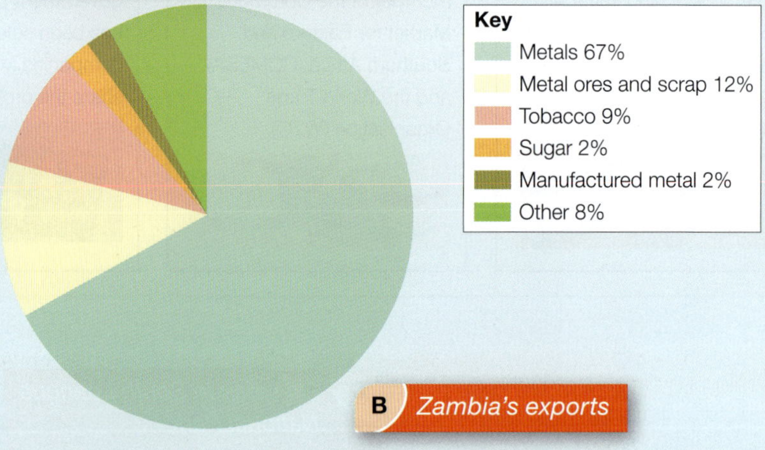

B Zambia's exports

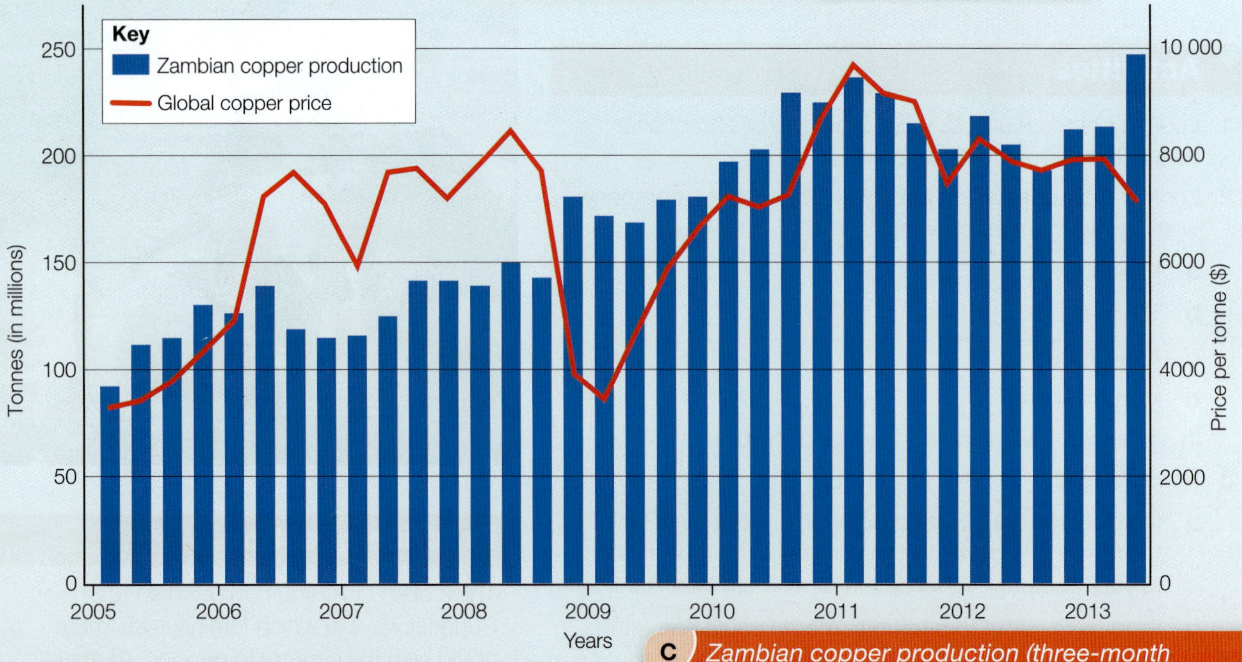

C Zambian copper production (three-month periods) and global price for copper, 2005–2013

Chinese investment in Zambia

China has become the world's largest consumer of copper. This makes Zambia an attractive place for Chinese investment. In 1998, China Non-Ferrous Metals Corporation bought Chambishi mine, one of the largest copper mines in Zambia. The company has invested millions of dollars to improve the mine.

Over 500 Chinese companies now invest in Zambia in businesses ranging from mining to manufacturing, farming to tourism.

Chinese investment in Zambia is not new. Back in the 1970s, the TAZARA (Tanzanian Zambian Railway Authority) railway was funded by the Chinese government. The railway is 1850 km long from the Copperbelt region of Zambia to the port of Dar es Salaam in Tanzania. It makes it easier for land-locked Zambia to export its resources and, at the same time, import other bulky goods.

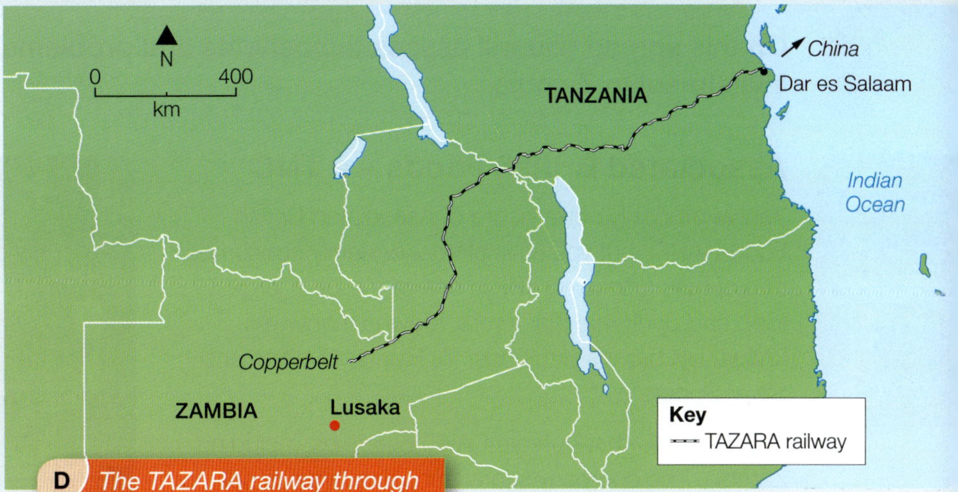

D *The TAZARA railway through Zambia and Tanzania*

Recent Chinese investments in projects in Zambia

- The Kariba Dam hydro-electric power plant has been expanded to generate more electricity for Zambia.
- Link 8000, a project to build 8000 km of new roads in Zambia, will link cities and improve rural connections.
- The TAZARA railway, which had fallen into disrepair in recent years, has undergone improvements.
- 100 000 Chinese people now live in Zambia, working with investment projects.

Activities

1 Look at chart B.

 a What percentage of Zambia's exports are metal-related?

 b Overall, how would you describe the balance of Zambia's exports? (Clue: think what it does not include.)

2 Look at graph C.

 a Describe how the global price for copper changed from 2005 to 2013. Mention high and low points and the overall trend.

 b Describe the connection between the global copper price and Zambia's copper production.

 c Explain the connection.

3 Work with a partner. One of you represents a Chinese company and the other represents the Zambian government. The company wants to take over a Zambian mine. You have to negotiate the best deal for both sides.

 a Individually, write a list of five things you want from the negotiation. For example, the Zambian government might want the company to pay good wages.

 b Together, present your lists to each other. Negotiate the best deal for both sides and come to an agreement. Write down the list of points you agree to.

Practice question

Explain how reliance on a single commodity [copper] in an LIDC you have studied [Zambia] has influenced the country's development. Mention the benefits and the problems of relying on a single commodity. (6 marks)

Trans-national companies in Zambia

On this spread you will assess the benefits and problems of trans-national investment in Zambia.

Case study

Associated British Foods – a TNC

You've probably never heard of Associated British Foods (ABF), but you have almost certainly consumed one of their products. ABF is one of the UK's largest food and drink manufacturers. It owns many well-known UK brands like Kingsmill, Ryvita, Ovaltine and Silver Spoon.

ABF is a TNC that operates in 46 countries, including Zambia. The company's total annual revenue is not much less than Zambia's entire GDP. In 2001, ABF bought Zambia Sugar, the company that produces most of the sugar consumed in Zambia. Its sugar plantations, where sugar cane is grown, cover an area twice the size of London (photo A). It is also Zambia's main sugar exporter.

A *A cane-cutter on Zambia Sugar's plantations*

The sweetest town in Zambia

Mazabuka, south of Lusaka, calls itself the 'sweetest town in Zambia'. It lies at the heart of Zambia Sugar's cane plantations. It is also where one of the company's main sugar-processing factories is located. Zambia Sugar is the main employer in the town, providing seasonal work for cane-cutters and work at the factory.

Apart from jobs, the company also supports the town by providing free healthcare and schools for its workers. Mazabuka is one of the most prosperous towns in Zambia. Even so, it is not free from poverty, especially for people who have no work.

Tax reduction and tax avoidance

The charity Action Aid investigated ABF's operations in Zambia in 2013. It discovered that from 2008 to 2013, Zambia Sugar paid almost no tax in Zambia. It was able to do this partly through the government's tax incentives to reduce tax on a company's new investment. But it was also able to avoid tax by transferring the profit made in Zambia to other ABF companies in countries with lower tax rates (map B).

B *How Zambia Sugar avoids tax*

Investment by TNCs

Zambia, like many other LIDCs, is in a difficult situation. It needs to raise as much income as it can through tax to pay for services like health and education. On the other hand, it also wants to encourage TNCs to invest in the country by offering tax incentives, which means it will raise less income through tax. So investment by TNCs brings both benefits and problems (panel C).

C *Benefits and problems of TNCs*

Benefits of TNCs

- TNCs provide jobs and income for individual workers.
- The company and workers pay taxes, which provide income for the government.
- Investment helps the country to exploit its natural resources.
- TNCs produce goods more efficiently and cheaply, reducing the cost to consumers.

Problems of TNCs

- TNCs put their own profit before the interests of workers or consumers.
- Large companies try to avoid paying tax.
- They dominate and make it hard for smaller, local companies to compete.
- They pay low wages by international standards (a reason for investing in LIDCs).
- Some pollute and damage the environment.

D *Gold diggers – how Africa's resources have been exploited*

Activities

1. You have probably not heard of ABF. Why not, do you think?
2. Look at map B. What are the following people likely to think about Zambia Sugar?
   - **a** A cane-cutter in Mazabuka
   - **b** A director of ABF in the UK
   - **c** A Zambian government minister
   - **d** A Zambian taxpayer

   Explain why for each one.
3. Read panel C. Overall, do you think the benefits of TNCs outweigh the problems, or vice versa? Give reasons for your answer.

Go further

Research one TNC you have heard of, e.g. McDonalds, Coca-Cola.

- Where is it based?
- In which countries does it operate?
- What does it produce or do?

Practice question

What is the message in cartoon D? How does the message apply to the development of an LIDC you have studied [Zambia]? (4 marks)

6.15 Aid and debt in Zambia

Case study

On this spread you will assess the arguments for and against aid and debt relief.

Types of aid

International aid – sometimes just called aid – is help that one country gives to another. Most aid is given by ACs, such as the UK, to LIDCs like Zambia.

Aid comes in all shapes and sizes (diagram A). It can involve individuals, charities, governments or international organisations. A lot of aid is given with conditions attached and is known as **tied aid**. Governments may only give aid to another country if it agrees to purchase goods from the donating country with the aid money. The IMF usually only gives aid on condition that the receiving country makes changes to its economy, such as cuts in spending.

A *Types of aid*

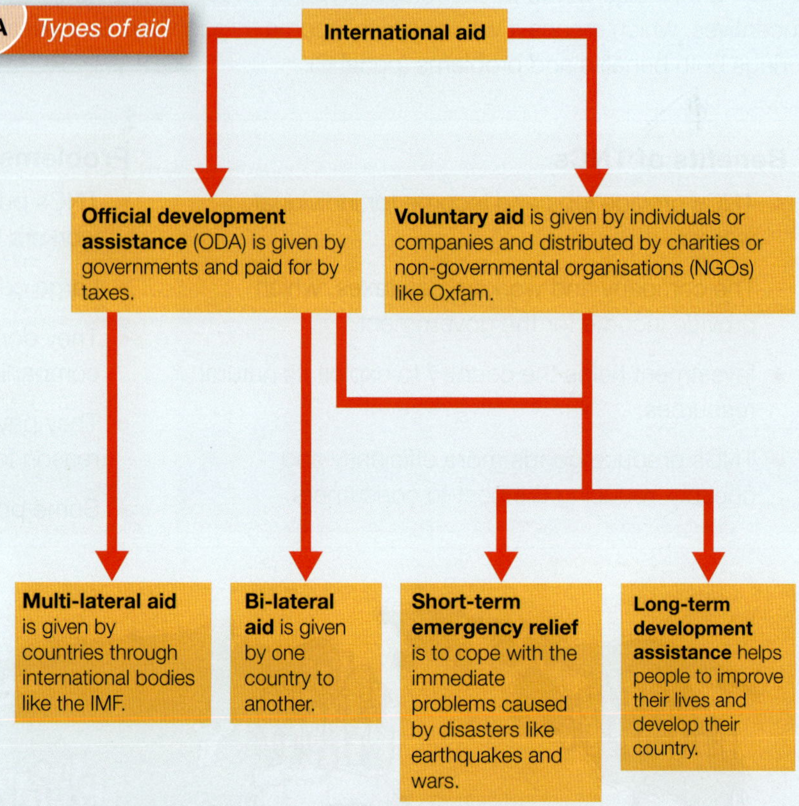

WaterAid

One British charity in Zambia is WaterAid. Aid is needed there because:

- 5.2 million people in Zambia don't have access to safe water
- 8 million people in Zambia don't have access to adequate sanitation
- 5000 children in Zambia die every year from diarrhoea caused by unsafe water or poor sanitation.

The charity works with local Zambian organisations to install simple, low-cost water pumps and toilets that communities can easily maintain (photo B). It also works with the Zambian government to ensure aid goes to the poorest people. In one year, WaterAid provides:

- 54 000 people with safe water
- 42 000 people with improved sanitation.

B *A new community water pump in Zambia*

For and against aid

Aid might sound like good, common sense. After all, who could be against helping? However, there are arguments for and against aid (panel C).

Advantages of aid

- Aid is an attempt to rebalance global inequality.
- It is repayment for the benefits ACs got from colonialism.
- Everyone, wherever they live, should have the right to the essentials of life.

Disadvantages of aid

- It discourages people from trying to look after themselves.
- Donors may decide what aid to give and it may not be what is most needed.
- It may be given to a corrupt government or used to pay for a country's wars.

Aid has not been successful in eradicating poverty because it has not been given on a large enough scale. Much more aid is needed.

Jeffrey Sachs, an American economist

Limitless aid to African governments has fostered dependency, encouraged corruption and, ultimately, perpetuated poverty.

Dambisa Moyo, a Zambian economist

C *Arguments for and against aid*

Debt relief

One problem for LIDCs is international debt. The amount they have to repay can be greater than the aid they receive. That was the case for Zambia until 2006 when the country qualified for debt relief. Until then it was one of 39 **highly indebted poor countries (HIPCs)**. Thirty-three of them are in sub-Saharan Africa.

In 2006, Zambia received $6.5 billion of debt relief from the IMF. As a result it has been able to increase government spending on health and education. For example, it was able to reduce school fees and that helped to increase primary school enrolment. It was also able to buy drugs to help bring the AIDS epidemic under control. However, aid to Zambia dropped from 27% of GDP in 2002 to 6% in 2012.

Activities

1. Look at diagram A and photo B. Into which category of aid would a community water pump fall? Explain why.
2. Which do you think is more effective in helping countries like Zambia – aid or debt relief? Give reasons.
3. Read the arguments for and against aid in panel C.
   - **a** Who do you more agree with – Jeffrey Sachs or Dambisa Moyo? Why?
   - **b** Write a short speech for or against aid. Do not write more than 200 words. You could give your speech as part of a class debate about aid.

Go further

Find out more about WaterAid or another development charity you have heard of (e.g. Oxfam).

- **a** Which countries does it work in?
- **b** What sort of aid does it provide?

Practice question

Name an LIDC you have studied that receives aid [Zambia]. Give three arguments in favour of international aid. (4 marks)

Top-down development in Zambia

On this spread you will find out about one top-down development – the Kariba Dam.

Case study

The Kariba Dam

Top-down development strategies are funded and carried out by government or a large international organisation like the World Bank. The Kariba Dam on the Zambezi River is a good example (photo A). Built in the 1950s, it was one of the earliest large dams in the world, similar to others that have been built since.

The Kariba Dam was a joint project between the governments of Zambia and Zimbabwe – or Northern and Southern Rhodesia as they were then (map B). It was built to produce hydro-electric power for both countries, particularly for major industries like Zambia's copper mines. The Kariba Dam is still one of the largest dams in the world and Lake Kariba lies behind it.

A *The Kariba Dam and Lake Kariba*

Advantages of the Kariba Dam

- The dam generates large amounts of hydro-electric power, which helps the country to develop. The power is vital for Zambia's copper industry. It has also helped Lusaka to develop into a modern city (see photo A on spread 6.9).
- Environmentally, hydro-electric power is a renewable form of energy that does not produce carbon emissions and is a lot cleaner than energy from a coal power station.
- New industries have developed around Lake Kariba, particularly fishing and tourism.

Disadvantages of the Kariba Dam

- 57 000 Tonga people, mostly farming families, were moved from the Zambezi valley. Many were resettled a long way from their land.
- The land they moved to was less fertile than the Zambezi valley, leading to hunger and famine.
- Natural flooding no longer occurs in the Zambezi valley downstream from the dam, leading to the loss of natural ecosystems and farmland.
- Communities north and south of Lake Kariba are cut off from each other – and some communities still have no electricity!

B *The Zambezi River and Lake Kariba*

Concerns about the future

As the Kariba Dam ages there are growing concerns about its safety. Torrents of water from six floodgates in the dam are eroding the bedrock on the riverbed. This is creating a plunge pool and undermining the foundations of the dam (diagram C). It is even possible that the dam could collapse.

The consequences, if the dam were to collapse, are unthinkable.

- A tsunami of water would pour down the Zambezi to reach Mozambique in hours.
- The water would overwhelm the Cahora Bassa Dam and 40% of southern Africa's hydro-electric power would be lost.
- Ecosystems and wildlife along the Zambezi River as far as the Indian Ocean would be lost.
- 3.5 million people's lives are at risk from flooding, including in Lusaka.

Sixty years after the Kariba Dam was built, questions are being asked about whether it was the best development project for Zambia. At the time, top-down development was the only option and there was no consultation with local people about what sort of development they would like to see.

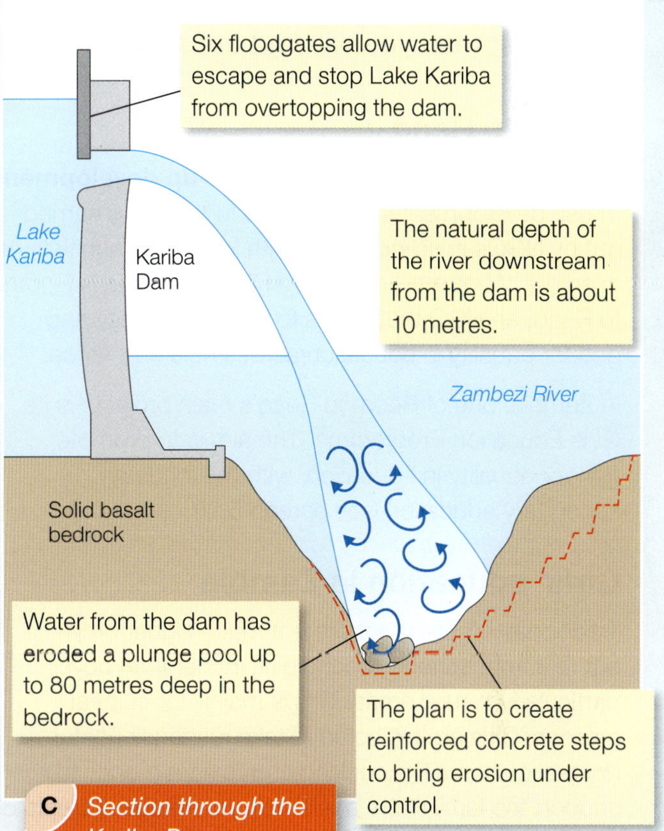

C *Section through the Kariba Dam*

Activities

1. Look at photo A. In what ways is the Kariba Dam an example of top-down development strategy?
2. Look at map B.
   - **a** In the correct order, name the countries through which the Zambezi River flows from source to mouth.
   - **b** Describe the location of the Kariba Dam on the river.
3. List the main advantages and disadvantages of the Kariba Dam for each of the following:
   - **a** a Tonga farmer in the Zambezi valley
   - **b** a mining company in the Copperbelt
   - **c** the Zambian government.
4. Look at drawing C.
   - **a** Explain how the Kariba Dam could collapse.
   - **b** Explain how this would lead to disaster downstream along the Zambezi valley.

Go further

Find out about another large dam project somewhere else in the world. How has it helped or hindered development? Examples you could choose from might be:

- Three Gorges Dam, China
- Narmada Dam, India
- proposed Xayaburi Dam, Laos.

Practice question

Outline the advantages and disadvantages of one top-down development strategy [the Kariba Dam] in an LIDC [Zambia] you have studied. You could use an annotated sketch map to answer the question. (8 marks)

Bottom-up development in Zambia

On this spread you will find out about one bottom-up development – Room to Read.

Room to Read

The opposite of top-down is **bottom-up development**. These development strategies are funded and carried out by NGOs in co-operation with local communities. An example of bottom-up development in Zambia is Room to Read, an NGO that works to improve literacy and gender equality in education across Asia and Africa.

In Zambia, one of Room to Read's main projects is its Girls Education Programme. The aim is to promote gender equality in education, with the focus on secondary education (see spread 6.11).

Girls' education in Zambia

Traditionally in Zambia there is a much higher drop-out rate from schools among girls (rather than boys), particularly in rural areas. This is mainly for cultural reasons. Girls are expected to take responsibility for domestic chores like cooking and to earn money to support the family from a relatively young age. Parents do not value education for girls in the same way they do for boys. So the drop-out ratio for girls to boys is about 2:1.

Girls are sometimes also pushed into early marriage or become pregnant, at which point they leave school. Even in school, class sizes of between 40 and 50 mean that girls (and boys) do not achieve as well as they might in smaller classes. Zambia's educational performance is poor compared to other Africa countries (graph B).

A *Room to Read*

Room to Read®

Kofi Annan, former UN Director-General said:
Literacy unlocks the door to learning throughout life, is essential for development and health and opens the way for democratic participation and active citizenship.

Room to Read's aims

To improve Zambia's educational performance, Room to Read aims to:

- increase the years of schooling for girls
- increase girls' self-awareness and life skills
- increase the girl-friendliness of school environments
- increase family, peer and community support for girls' education
- increase collaboration with government to promote girls' education.

B *Literacy performance for grade 6 pupils (aged 11) in some African countries*

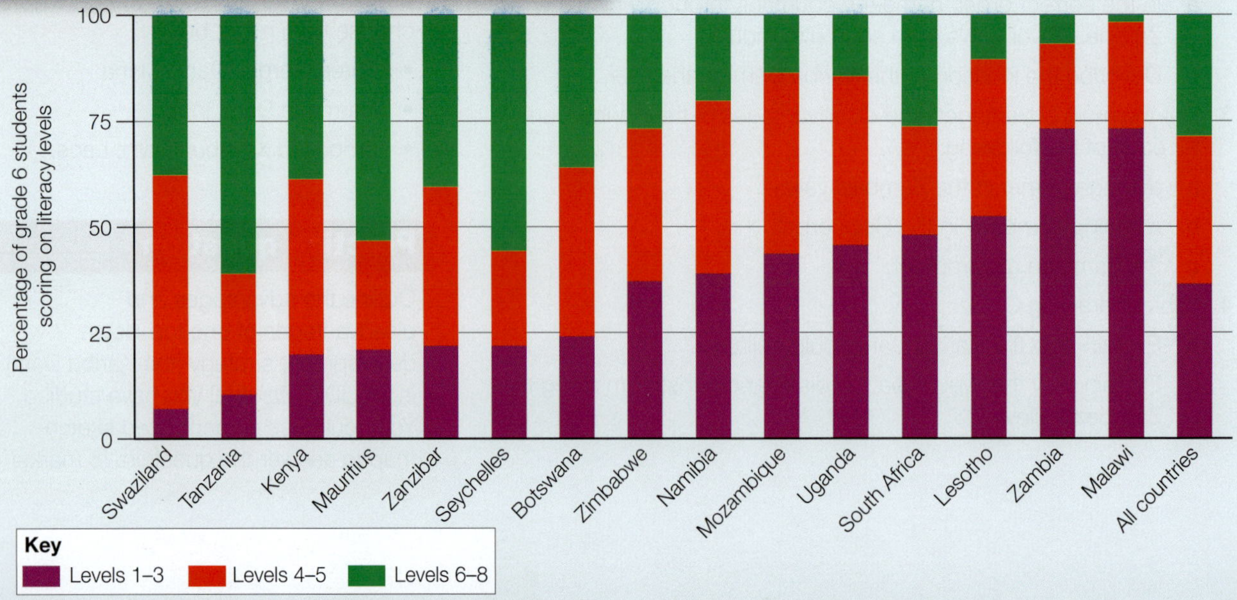

Reading for success

It's not just girls who miss out on education. In rural Zambia, schools are few and far between, and school attendance is often poor. Milimo Muchindu is a nine-year-old boy in Choma, southern Zambia. It takes him half an hour to walk to school from his village. His father works hundreds of miles away in Lusaka, earning money to support the family, and his mother is too busy at home with children to take him to school.

Milimo's school attendance was not good. He preferred to skip school to play football with his friends. Lessons in school did not seem that exciting. That's when Room to Read got involved. They trained a new teacher to work in the school. With proper training, she was able to engage Milimo and get him interested in education. Now, he has learned to read and is even thinking of becoming a teacher himself one day.

C *Milimo with his new reading book*

Activities

1 **a** Describe the difference between top-down and bottom-up development.

b Decide whether each of these developments are top-down or bottom-up:

- a high-speed railway to link major cities
- a water pump to provide safe water for a village.

Explain your reasoning in each case.

2 Look at panel A. Why do you think literacy is essential for the following?

a Development **b** Health **c** Democracy

In each case, explain how literacy might help.

3 Look at graph B. Compare Zambia with other African countries.

a How does the proportion of Level 1–3, 4–5 and 6–8 pupils in Zambia compare?

b How could you explain this from what you know about Zambia?

4 Look at photo C.

a Why might Milimo's chance of getting a good education be less than yours? Give three reasons.

b How might bottom-up development help him?

5 Compare the top-down strategy on spread 6.16 and the bottom-up strategy here. List the advantages and disadvantages of each strategy in a table like the one below.

Top-down strategy		Bottom-up strategy	
Advantages	Disadvantages	Advantages	Disadvantages

Potential disadvantages of bottom-up strategies

Bottom-up strategies may not be planned very systematically and may therefore happen in some places but not in others. They may also rely on volunteers or, if the work is paid, they may rely on aid donations from other countries. If the aid is withdrawn, the scheme may have to close.

Your enquiry

Are LIDCs likely to stay poor?

Zambia, like many other LIDCs, has tried different development strategies over the past few decades. Based on Zambia's experience, which of these strategies would you recommend to another LIDC?

- Encourage more foreign investment.
- Produce more natural resources.
- Ask for more international aid.
- Introduce more government top-down projects.
- Allow more bottom-up NGO projects.

Write a development plan for your country, using one or more of these strategies. Explain why you would recommend each one.

7 UK in the 21st Century

What is this topic about?

The UK is not the same country it used to be. Our country is changing. You will live your entire life through the 21st century (and you might be lucky enough to make it into the 22nd century!). Already, in the 21st century, we have seen important changes in the UK, particularly to our population and economy. There are certain to be more big changes in the UK during your lifetime.

Who knows, by the end of the 21st century the UK may no longer exist. By then we may have split into the four smaller areas which make up the UK – England, Scotland, Wales and Northern Ireland. Alternatively, the UK may have become part of a much larger country, which could be called Europe. No one can be sure.

Your enquiry

In this chapter you will investigate the specification's two key questions.

How is the UK changing in the 21st century? You will:

- Consider how the UK's human and physical geography can affect us.
- Describe how the UK's population is changing and predict future change.
- Recognise the challenge of an ageing population.
- Explore the diversity of London's growing population.
- Understand how the UK's economy has changed.
- Locate the core economic hubs in the UK.
- Investigate the importance of London within the UK economy.
- Understand why new industries are drawn to an economic hub.

Is the UK losing its global significance? You will:

- Consider the UK's global role through international organisations.
- Investigate the UK's role in recent conflicts in the Middle East.
- Explore the UK's global influence through film and television.
- Examine how the rest of the world influences us through food.

Look at the photos opposite.

- What does each photo tell you about the UK today?
- Which photo do you think best represents the UK in the 21st century? Why?
- Are there any important images of the UK you think are missing? What are they?
- If you had to choose four photos to represent the UK, what would they be?

Human geography of the UK

On this spread you will consider how crowded the UK is and why there is a housing shortage.

A crowded country?

There are 65 million of us in the UK. Our country has one of the highest **population densities** in Europe and sometimes, living in a city, it can feel crowded. But, from a satellite in space, you get a different view of the UK (satellite image A). At night, the lights from cities show where people live.

Most of us are concentrated in a few large cities. The **land use** in the rest of the country is farmland, woodland, grassland and mountains where few people live (photos in B). It may surprise you that only 12% of the UK is urban. Even more surprising, about half the land in our cities is green open space, including parks and gardens. That means only 6% of the UK is actually built on!

A *A satellite image of the UK at night. Can you spot the cities?*

B *Land uses in the UK (3% is covered with land uses not shown here)*

UK housing shortage

With so much land that could be built on, it might seem odd that the UK has a **housing shortage**. In order to house our growing population (see spread 7.3), we need to build 240 000 homes a year. In fact, we only build about half that number. The result is that houses are too expensive for many people to buy or rent, especially in south-east England, which is the most crowded part of the country.

With so much space, why do we have a housing shortage?

- New housing needs planning permission and there is often local opposition to new homes.
- Land around many cities is protected as green belt, with strict planning controls (see spread 5.6).
- The price of land keeps rising, so landowners hold onto their land in the hope the price will rise further.

Activities

1 Look at satellite image A. How many cities can you identify? Make a list and check your ideas in an atlas.

2 Look again at image A. Write a paragraph to describe population distribution in the UK. Mention densely populated areas, such as south-east England, and sparsely populated areas like the north of Scotland.

3 Look at the photos and caption in B.

- **a** Draw a pie chart to show percentage land use in the UK (adding up to 100%).
- **b** What percentage of the land is urban? Why could this be a misleading figure?
- **c** How else could you present this data? Which is the best way? Give reasons to justify your choice.

4

- **a** Do you think more land in the UK should be used to build houses? Explain.
- **b** Where would you build houses and where would you not build them? Mention which areas of the country and which types of land you would use. Give reasons.

Your enquiry

How is the UK changing in the 21st century?

As you work through this section, make notes about anything that will help you to answer the question. You will bring your ideas together to do an activity at the end of spread 7.11.

Physical geography of the UK

On this spread you will look at the rainfall pattern in the UK and why there is water stress in some places.

UK relief

Underlying the human geography of the UK – its population distribution, location of cites and land use pattern – is the country's physical geography. A relief map of the UK shows the shape of the land (map A). Broadly, mountains are found in the north and west of the UK, particularly in Scotland and Wales. There is more lowland towards the south and east.

Relief has an important effect on weather and climate. The **prevailing wind** in the UK blows from the south-west, coming from the Atlantic Ocean. When air carrying moisture reaches land, it is forced up over mountains where it produces **relief rainfall** (diagram B).

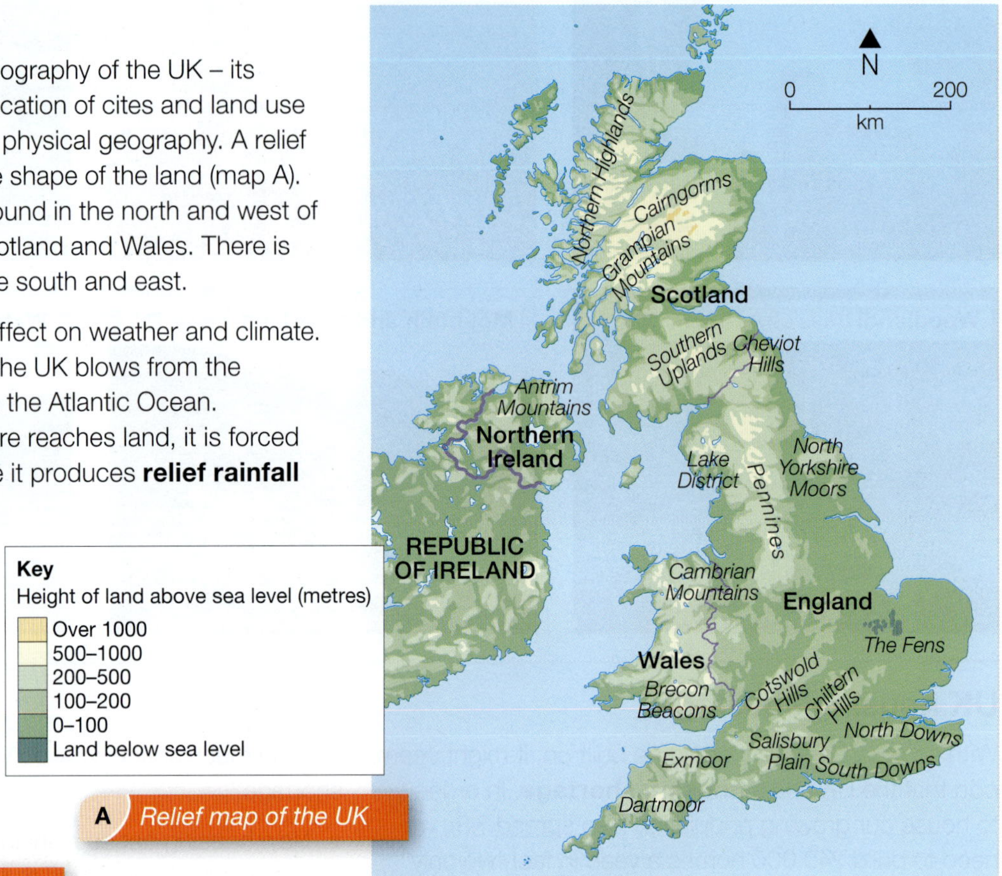

A *Relief map of the UK*

B *Relief rainfall in the UK*

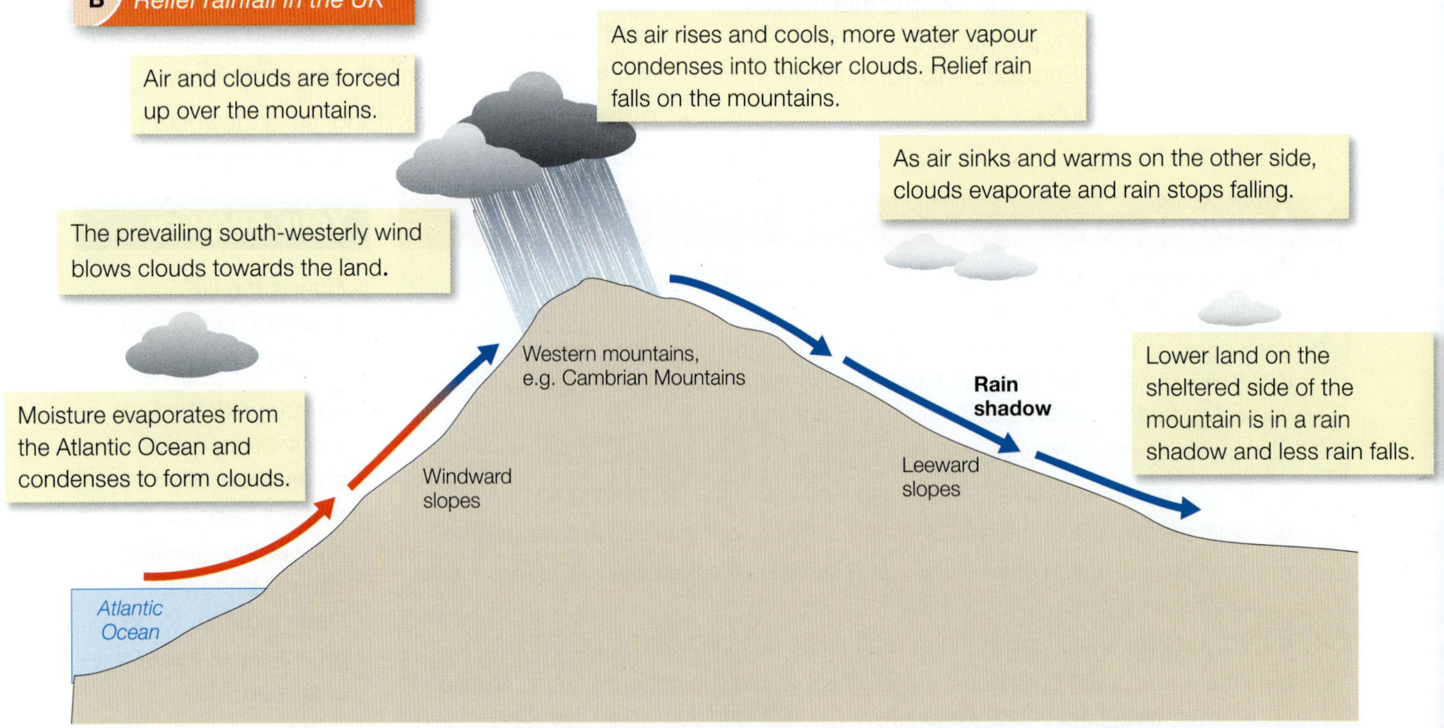

Water stress in the UK

The pattern of rainfall across the UK reflects the country's relief and the direction of the prevailing wind (map C). The highest rainfall occurs over high ground in the north and west, while less rain falls on low ground in the south and east.

The south-east is also the most crowded part of the UK (see spread 7.1), with the greatest demand for water. That leads to **water stress** in the south-east region, where the demand for water uses up a high proportion of the rainfall (map D). When water demand exceeds rainfall over a long period, it can lead to drought (see spread 1.10).

One long-term solution to the problem of water stress in the UK is to pipe water from areas with high rainfall and low demand to areas of low rainfall and high demand. Water is already transferred from the Cambrian Mountains of Wales to Birmingham. This could be done on a larger scale to transfer water from the north-west of the UK to the south-east.

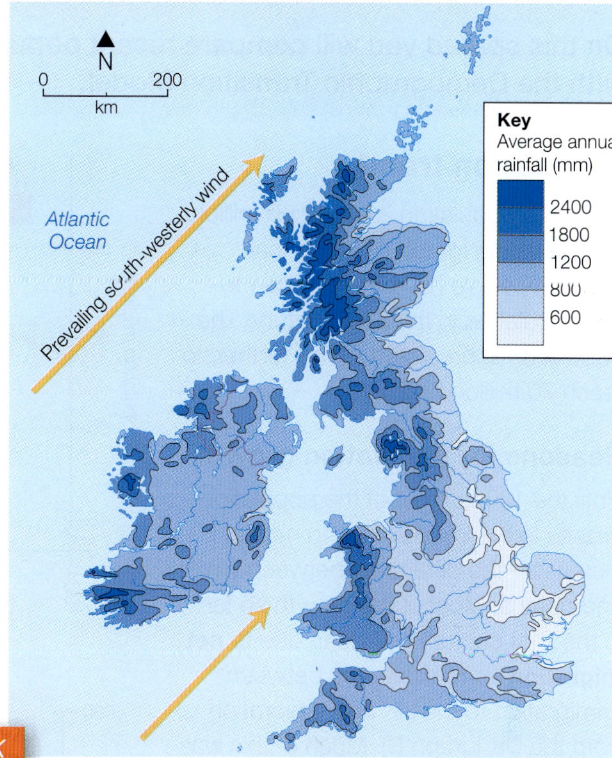

C Rainfall in the UK

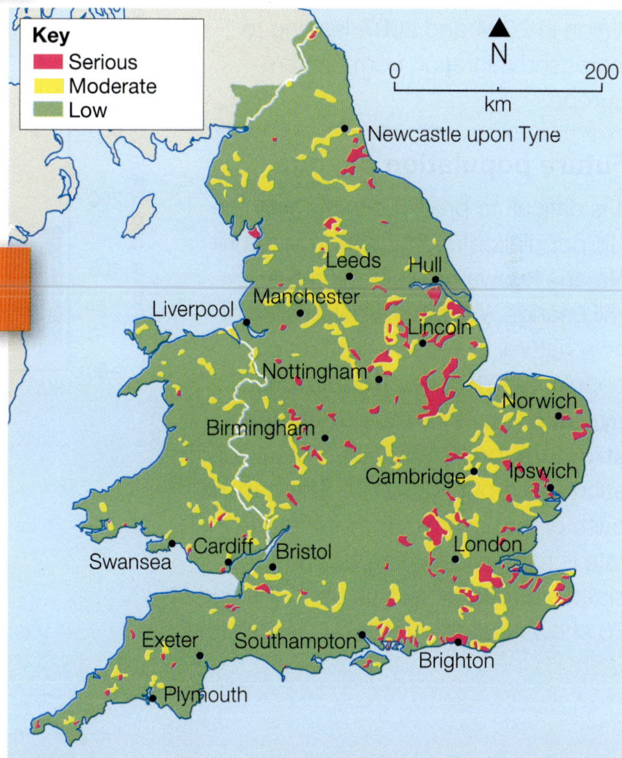

D Water stress in England and Wales

Activities

1 Look at diagram B. Explain these terms:
  - a prevailing wind
  - b relief rainfall
  - c rain shadow.

2 Look at map A and diagram B. Use them to explain the rainfall pattern in the UK on map C. Mention the following terms in your explanation: prevailing wind, relief rainfall, rain shadow.

3 Look at map D.
  - a Describe the distribution of areas with serious water stress in England and Wales.
  - b Suggest how the problem of water stress could be solved by transferring water. Use the UK rainfall pattern (map C) and population distribution (satellite image A on spread 7.1) to help you.

Practice question

Explain how population and rainfall maps of the UK can be used to predict areas of water stress. (8 marks)

7.3 The UK's changing population

On this spread you will compare recent population changes in the UK with the Demographic Transition Model.

Population trend

The UK's population is about 65 million and still rising (graph A). More than 5 million people have been added to our population in the 21st century. The population is predicted to rise further, to reach 70 million by 2030.

Reasons for population growth

Until the 1990s most of the population growth in the UK was due to natural increase – the difference between births and deaths. Most of the growth so far in the 21st century has been due to **net migration** – the difference between immigration to the UK and emigration from the UK (graph B). Much of this was a result of the expansion of the European Union in 2004 and 2007, leading to increased migration from Eastern Europe.

Future population change

It is difficult to predict exactly how the UK population might change in future. No one knows how net migration will change. We have a better idea of how natural increase will affect our population. The UK's **population pyramid** shows us the **population structure**, or the number of people in each age group (graph C). The narrow base of the pyramid shows that the birth rate is fairly low and the fact that the sides are roughly vertical up to the age of 70 shows that the death rate is also low.

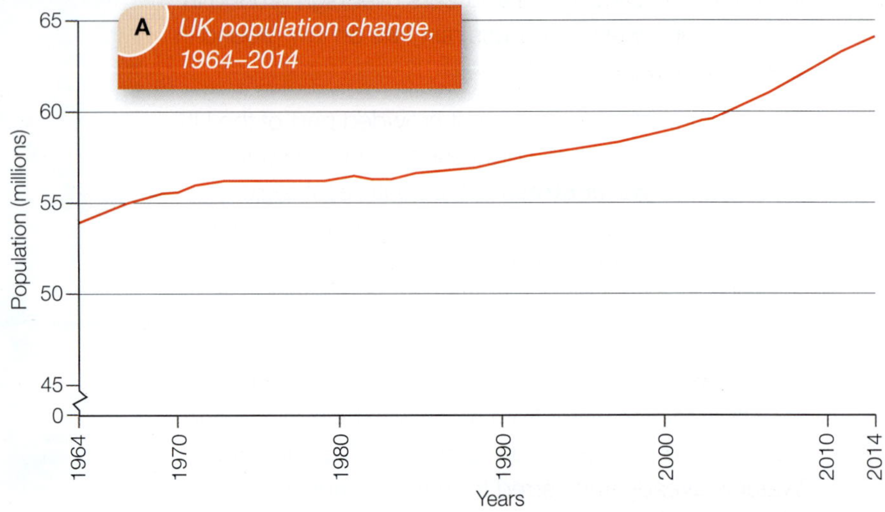

A *UK population change, 1964–2014*

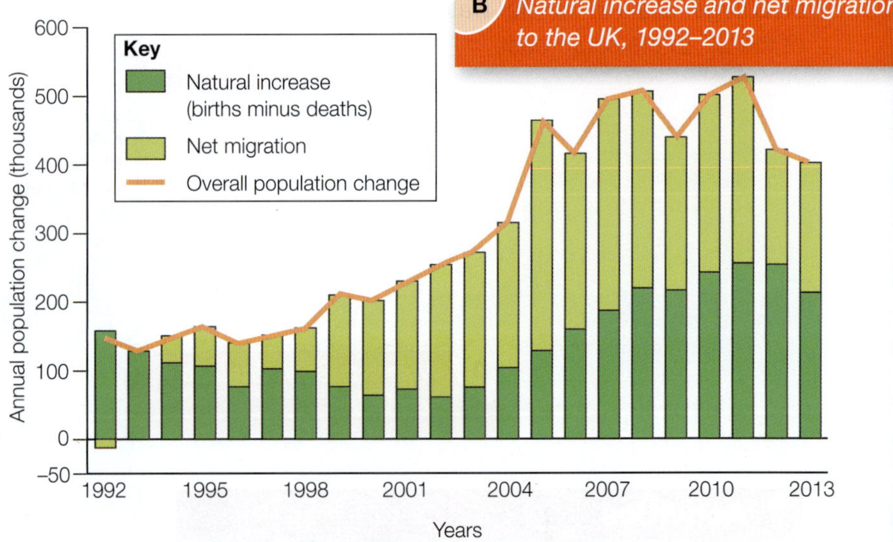

B *Natural increase and net migration to the UK, 1992–2013*

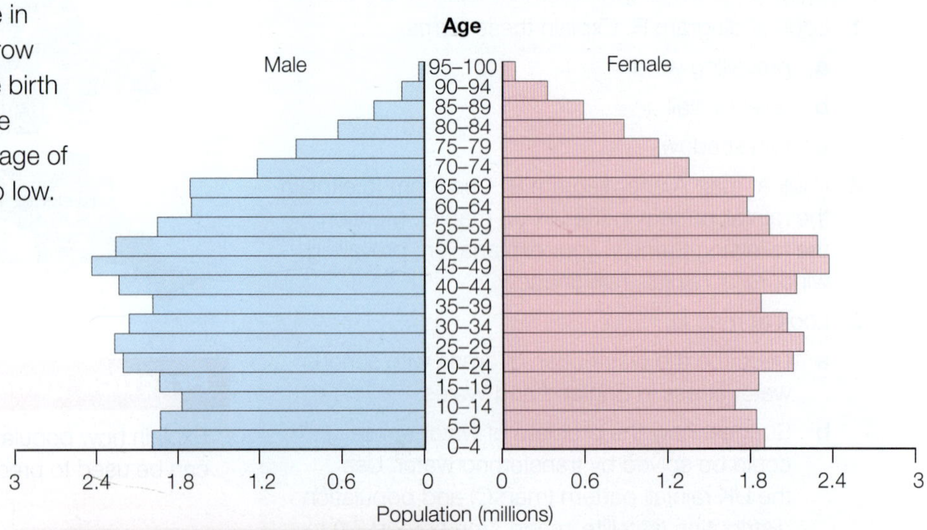

C *The UK's population structure, 2011*

Demographic transition

The study of human population is known as **demography**. Geographers describe population change over time as **demographic transition**. As countries experience economic development (see Chapter 6) they also go through stages of demographic transition. The **Demographic Transition Model** describes the stages that a country's population is likely to go through over time (graph D).

D *The Demographic Transition Model*

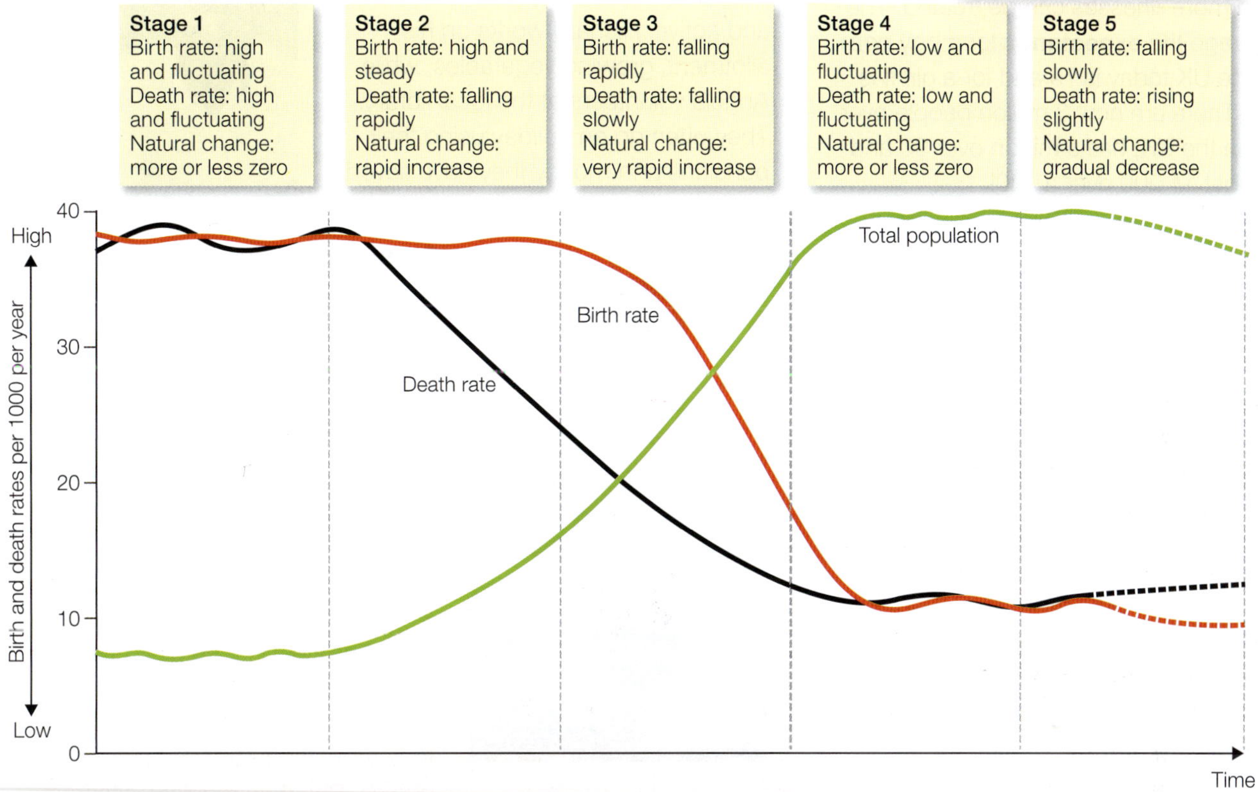

Activities

1. Look at graph A. Describe accurately how the UK's population has changed since 1964. Mention what has happened in the 21st century.
2. Look at graph B. Describe the contribution of the following factors to overall population change.
   - **a** Net migration **b** Natural increase

   Mention what has happened to both in the 21st century.
3. Look at graph C.
   - **a** The UK population pyramid is fairly narrow at the base with near vertical sides. What does this tell you about the UK's population structure and how it is likely to change?
   - **b** What does the population pyramid *not* tell you about how the UK's population might change in future?
4. Look at graph D.
   - **a** Which stage on the model best describes the UK's population now? Explain why.
   - **b** What is the problem with the model to describe the UK's population now?

Go further

Find out how the UK's population has changed in the past. In roughly which years did it pass through each stage on the Demographic Transition Model?

Practice question

Describe how the UK's population has changed so far in the 21st century and the main causes of this. (4 marks)

Our ageing population

On this spread you will investigate the stages of old age and the challenges of an ageing population in the UK.

Life expectancy in the UK

People in the UK are living longer. This is because we now enjoy better healthcare and more illnesses can be treated. The average **life expectancy** for a boy born in the UK today is 79 and for a girl it is 83. There are now 3 million people over 80 in the UK, half a million over 90 and almost 14 000 over a 100!

Of course, it is good that people are living longer. It means we are able to enjoy a longer life after we retire (panel A). However, **healthy life expectancy** – the age up to which people remain fit and active – has not increased so fast. Increasingly, it seems, there are stages of old age through which a person's health and level of independence decline.

The increase in the proportion of older people in the population puts more pressure on people of working age. They are the ones who pay taxes that pay for our health service and care for the elderly.

A *Stages of old age*

Active retirement

Richard and Angela recently retired in their sixties. They are both still fit and active. Richard works on his allotment, growing vegetables, while Angela volunteers at the local library. They often go on holiday using their pensions and money they have saved.

Semi-independence

Charlie was widowed when his wife died last year. He's got arthritis and is not quite as mobile as he used to be. He still does his own cooking, but his daughter does his shopping and pops in to clean the house. He has regular hospital appointments to check on his health.

Dependence

Mary has lived in a nursing home for the past ten years since she had a stroke. She sold her house to help to pay for her care by professional nurses. She gets regular visits from the doctor, her children and, of course, her hairdresser!

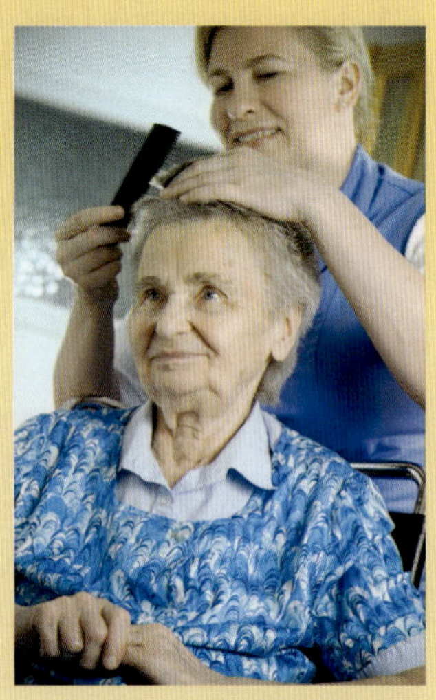

B *Predicted UK population pyramid in 2050*

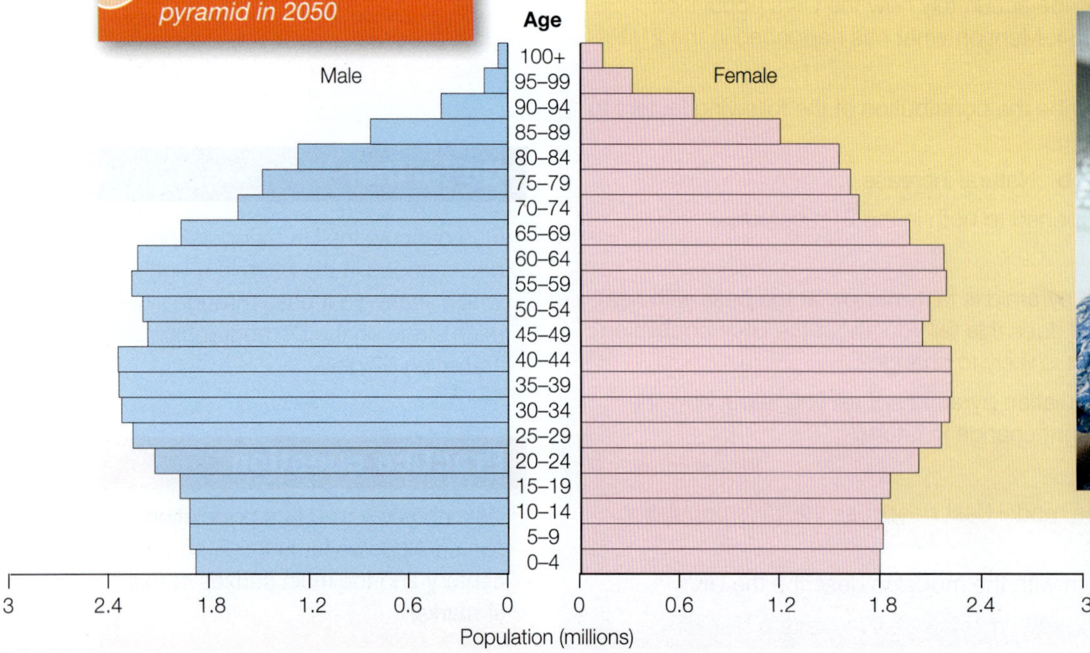

Dependency ratio

By 2050 there will be more people in the UK over 65 than under 16 (graph B). These two groups, together, make up the **dependent population** of the UK. The **dependency ratio** is the proportion of people, under 16 or over 65, who depend on the working population. Since the 1960s, the dependency ratio in the UK has not changed much but, while the number of children has fallen, the number of older people has increased (graph C).

C *Dependency in the UK*

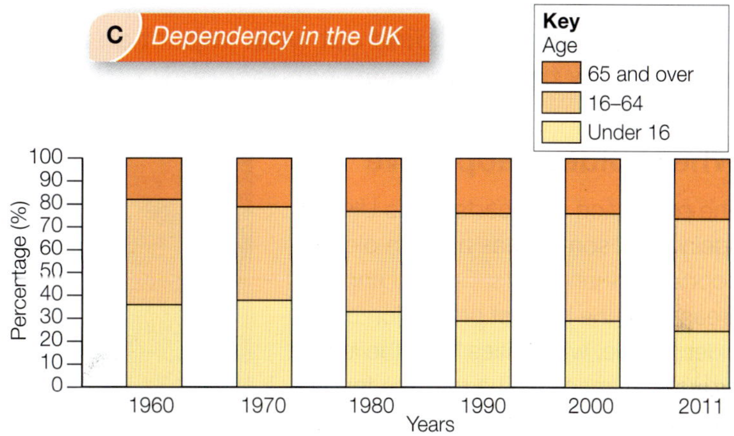

$$\text{Dependency ratio} = \frac{\text{children (under 16) + older people (65 and over)}}{\text{working population (16–64)}} \times 100$$

The cost of an ageing population

The UK's **ageing population** creates a number of economic challenges:

- More government money from the taxes we pay has to be spent on pensions. 65% of state benefits go to people over working age.
- The National Health Service needs more money. Average NHS spending on retired households is double that on working-age households.
- More money has to be spent on care for the elderly, either in nursing homes or in sheltered housing.
- People who look after their elderly relatives have less time for paid work and so earn less.
- The more money is spent on care for the elderly, the less is available for other public services like education or transport.

Activities

1 Look at the photos in panel A. Also think about older people you may know.
Think about the impact that older people have in the UK – what they contribute and what they need. Complete a large spider diagram like the one opposite. List your ideas under each heading.

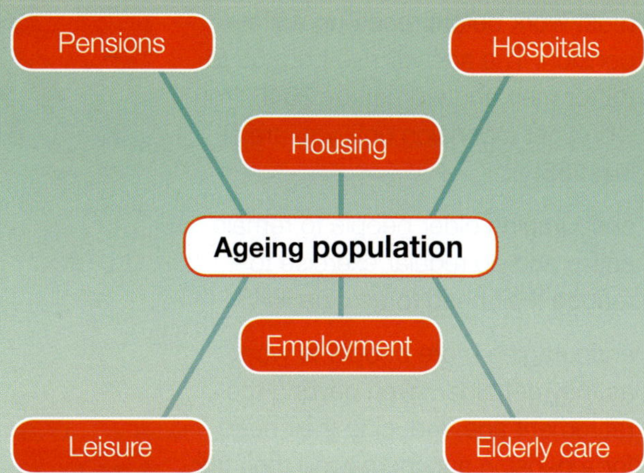

2 Compare graph B here with graph C on spread 7.3.
Describe how the UK's population structure will change from 2014 to 2050. Mention the changes in each age group:
- **a** children under 16
- **b** working age from 16 to 64
- **c** older people 65 and over.

3 Look at graph C.
- **a** Describe how the proportion of children and older people has changed since 1960.
- **b** Calculate the dependency ratio in 1960 and in 2011. How has it changed?

4 How do you think the government should prepare for an ageing population? Suggest at least three ideas.

Practice question

Describe the economic challenges faced by the UK in the 21st century due to an ageing population.
(4 marks)

7.5 Retirement dreams

On this spread you will find out what impact an ageing population has and how the government can respond.

Where older people live

The population in all parts of the UK is ageing, but some areas have an older population than others. Coastal and rural areas have the highest proportion of older people, while cities have the lowest proportion (map B). The reasons for this spatial distribuition are:

- older people often move to the seaside or countryside when they retire
- younger people often move to cities to find work
- with a younger population, cities also have a higher birth rate and so more children are born there.

A *The seaside is a retirement dream for many*

Government response to an ageing population

The UK government has begun to respond to the challenge of an ageing population. The strategies it is trying include:

- increasing the age to which people must work before receiving a state pension. The age will rise from 65 to 68 for men and women by 2046. You won't get a pension until you are at least 68!
- encouraging older people to remain active and do regular exercise to reduce their need to use the NHS
- encouraging more people to work as they get older, even beyond the retirement age, reducing their need for a pension and increasing their tax contribution
- allowing more immigration to bring more skilled labour into the UK, replacing the people who retire. At the same time, the government is also under political pressure to reduce immigration.

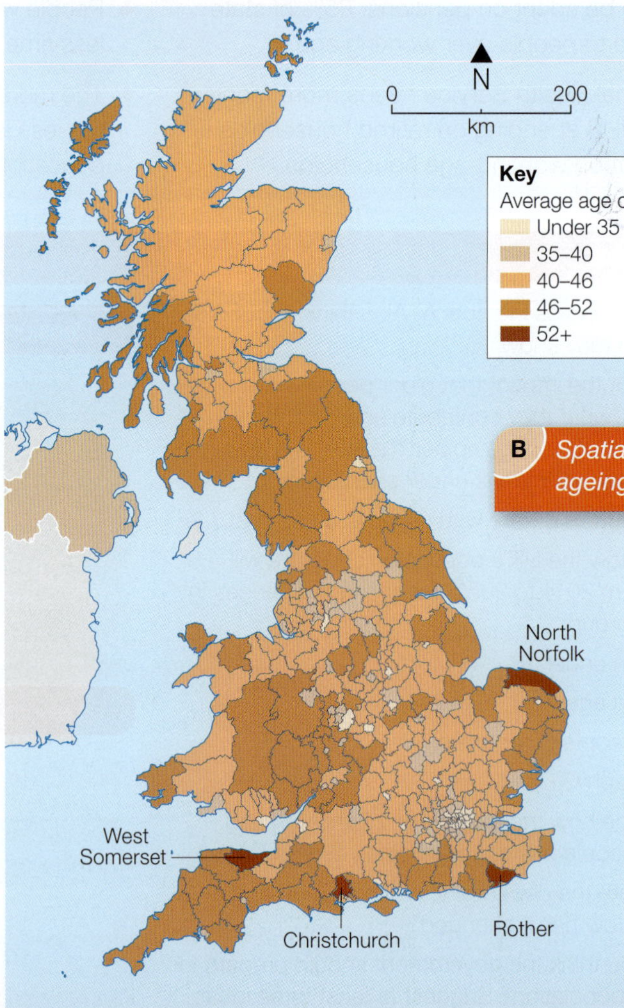

B *Spatial distribution of the UK's ageing population varies*

Seaside retirement

One area of the UK where more than half the population is over 50 is the district of Rother in East Sussex. It includes the seaside town of Bexhill (table D). The changing demography of the area creates opportunities and challenges (article C):

- Some older people earn a reasonable income from pensions and savings. They can spend their money on leisure, holidays and home improvements. This creates jobs and incomes for others.
- People have different housing requirements as they get older. They may move into a smaller home in a new location. Eventually, they may have to move to sheltered housing or a care home.
- Older people may require different levels of care, ranging from an unpaid family member to professional care in a residential home. This can also create jobs and incomes.
- Mobility is an issue for many older people. They may no longer be able to walk or drive. They require good public transport.

BEXHILL – RETIREMENT HAVEN OR BORING-ON-SEA?

Bexhill-on-Sea has been called 'God's waiting room'! So many people retire to this haven on the south coast that it's full of mobility scooters and nursing homes.

Last year's census shows that Bexhill has the highest proportion of people over the age of 100 in England and Wales. In fact, 66 of the 90 558-strong population of Bexhill's borough of Rother are centenarians.

Bexhill's reputation for long life doesn't impress everyone. Some residents think the town needs to be livelier, with more for young people. 'It's so boring here that I have to go to Hastings all the time,' says Leofwin Edwards, 32. But other young people like it in Bexhill. Emma Royston, 24, says, 'There's no crime really and you never get any bother.'

C *Adapted from news reports*

	0–14 (%)	15–29 (%)	30–44 (%)	45–64 (%)	65–74 (%)	75–84 (%)	85 + (%)
Rother, East Sussex	14.9	13.1	14.5	29.1	13.5	10.1	4.8
UK	17.3	20.2	20.8	25.3	8.6	5.6	2.2

D *Comparison of population by age in Rother District and the UK*

Activities

1. Look at photo A. Suggest why older people often want to move to the coast when they retire.
2. Look at map B.
   - **a** Describe the location of the areas with the highest average age in the UK.
   - **b** Now describe the overall pattern on the map.
3. Read article C and the rest of the information on this page.
   - **a** Do you think the writer of the article is biased? Give your reasons.
   - **b** Draw a table to list the opportunities and challenges of an ageing population for a community.
   - **c** How should a community respond to the challenges of an ageing population? Suggest at least three ways.
4. Look at table D.
   - **a** Turn the data in the table into a multiple bar graph. Draw bars for each age group, using two colours, one for Rother and the other for the UK.
   - **b** Use your graph to compare the population of Rother with the population of the rest of the UK.

Practice question

Outline how the UK government responds to the challenge of an ageing population. (6 marks)

London's booming population

On this spread you will use a range of data to find out how London's population is changing.

London's growth

London is booming. In 2015 its population reached 8.6 million, the highest it has ever been (graph B). It did reach this level once before, in 1939, just before the Second World War. However, war damage and enforced evacuation reduced the population and it continued to decline until 1991.

This time London's population is expected to keep rising and to reach 10 million by 2030, when London will become a megacity (see spread 5.2). Most of London's growth is due to the high birth rate among its young population. The city has a high proportion of young children (0–4 years old) and young adults (25–29 years old) (graph C).

Although London has large numbers of immigrants, many people leave, so net migration into the city is relatively low. It is natural increase, rather than net migration, that drives London's growth.

A *Notting Hill Carnival – a reflection of London's diverse population*

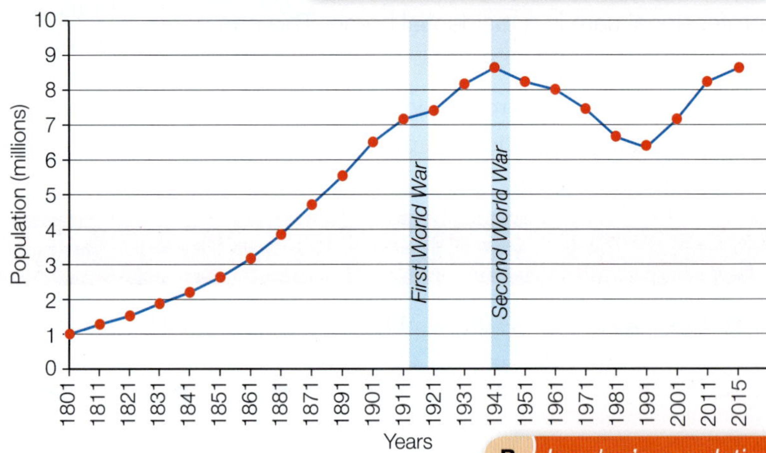

B *London's population growth, 1801–2011*

C *London's population structure, 2011*

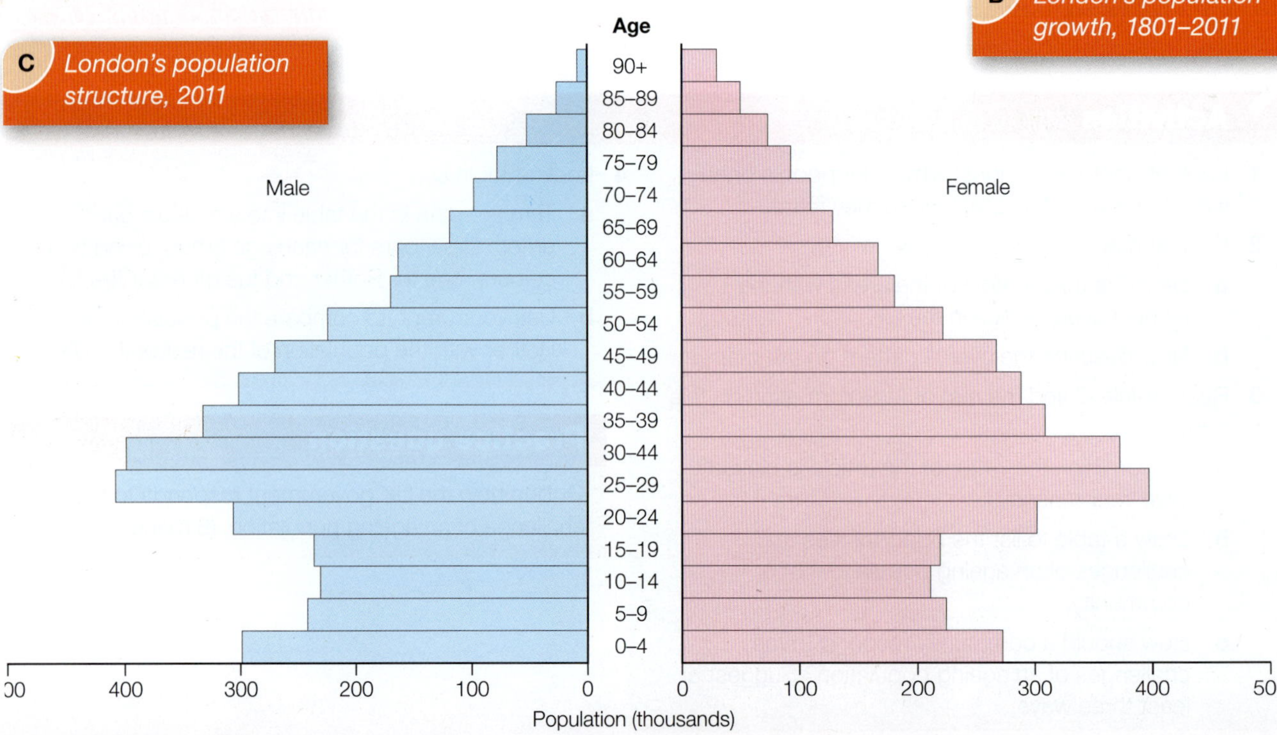

D Ethnic composition of London's population

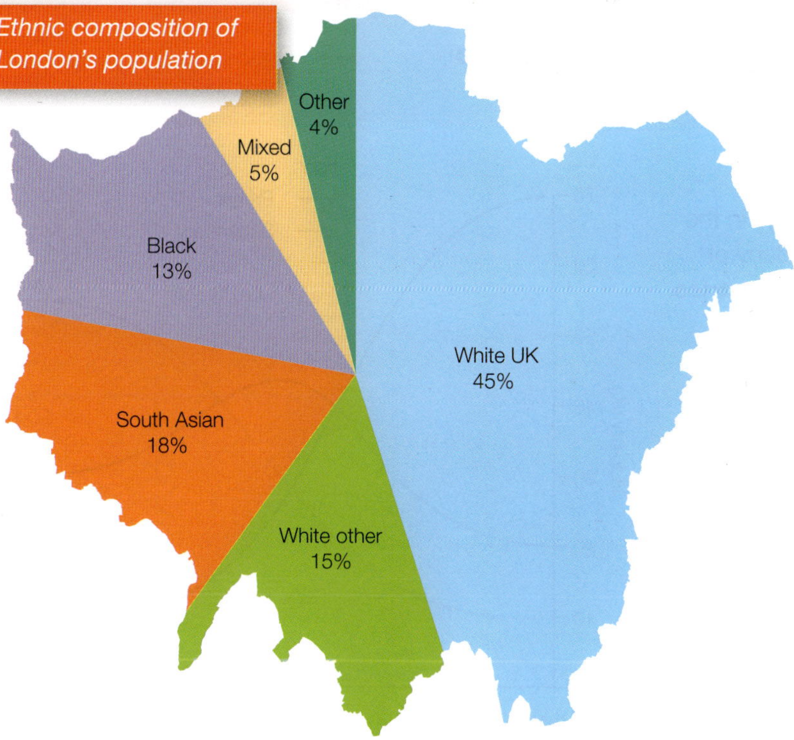

Flag	Country	London population
	UK	5 175 677
	India	262 247
	Poland	158 300
	Ireland	129 807
	Nigeria	114 718
	Pakistan	112 457
	Bangladesh	109 948
	Jamaica	87 467
	Sri Lanka	84 542
	France	66 649

E Top ten countries of birth of London residents, 2011

London's ethnic diversity

Between 2001 and 2011, London changed from a city with a majority of white British residents to a city with a majority of other ethnic groups (map D). Of course, many of this 'ethnic majority' were born in the UK with immigrant parents or grandparents.

In 2011, 38% of London's population were born abroad (table E). London is the city in the UK with the most **ethnic diversity** and the second most diverse city in the world after New York.

Activities

1 Look at graph B. Describe the growth of London's population from 1801 to 2011. Mention key figures and dates from the graph.

2 Look at graph C.
   - **a** Compare London's population structure with that of the UK in graph C on spread 7.3. Which age groups are a larger or smaller proportion of London's population?
   - **b** Try to explain the differences in population structure from what you know about the UK and London.

3 **a** Look at map D. Draw a bar graph to show the ethnic composition of London's population.
   - **b** Which do you think is a better way to present the data – as a pie chart (like map D) or as a bar graph? Give reasons for your opinion.

4 Look at table E. Complete a world map to show where London's population was born.
   - **a** Colour and label the countries in the table on the map.
   - **b** Draw arrows linking each country with London. Make the arrow width proportionate to the number of people.

Go further

Find out more about the population of your city or region in the UK. How is the population changing? What is the population structure? What is its ethnic diversity?

Practice question

For a named place in the UK [London] explain how its ethnic diversity has changed since 2001. (2 marks)

7.7 The UK's changing economy

On this spread you will compare industry in the past with the UK economy in the 21st century.

Changing industries

The past 50 years have seen a dramatic change in the UK's economy and **employment sectors**. Many of our traditional manufacturing, or **secondary industries**, have declined and been replaced by **tertiary** and **quaternary industries** (graph A). Since the 1960s, we have changed from being an industrial economy into a **post-industrial** economy.

Fifty years ago, a typical geography textbook might have listed the main industries in the UK by region (map B). Each region was associated with particular types of industry. Such a list would be harder to write now since most traditional industries have gone.

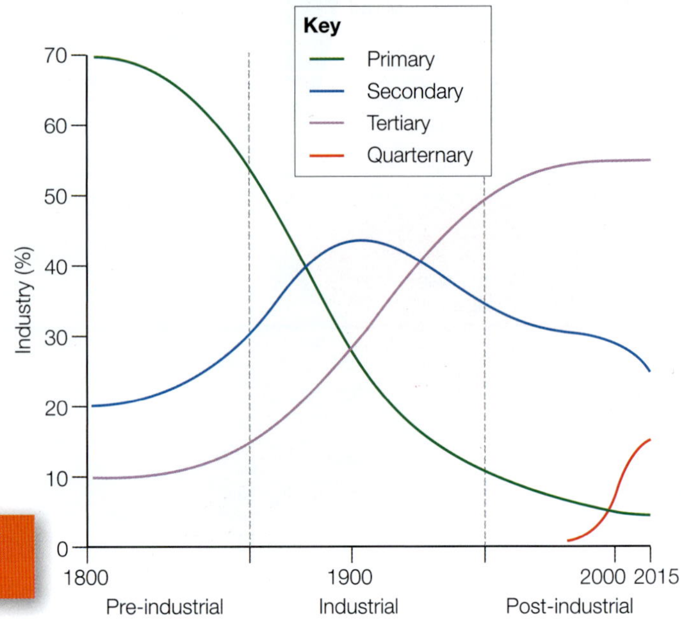

A *The changing economy of the UK*

B *Industrial regions of the UK in the 1960s*

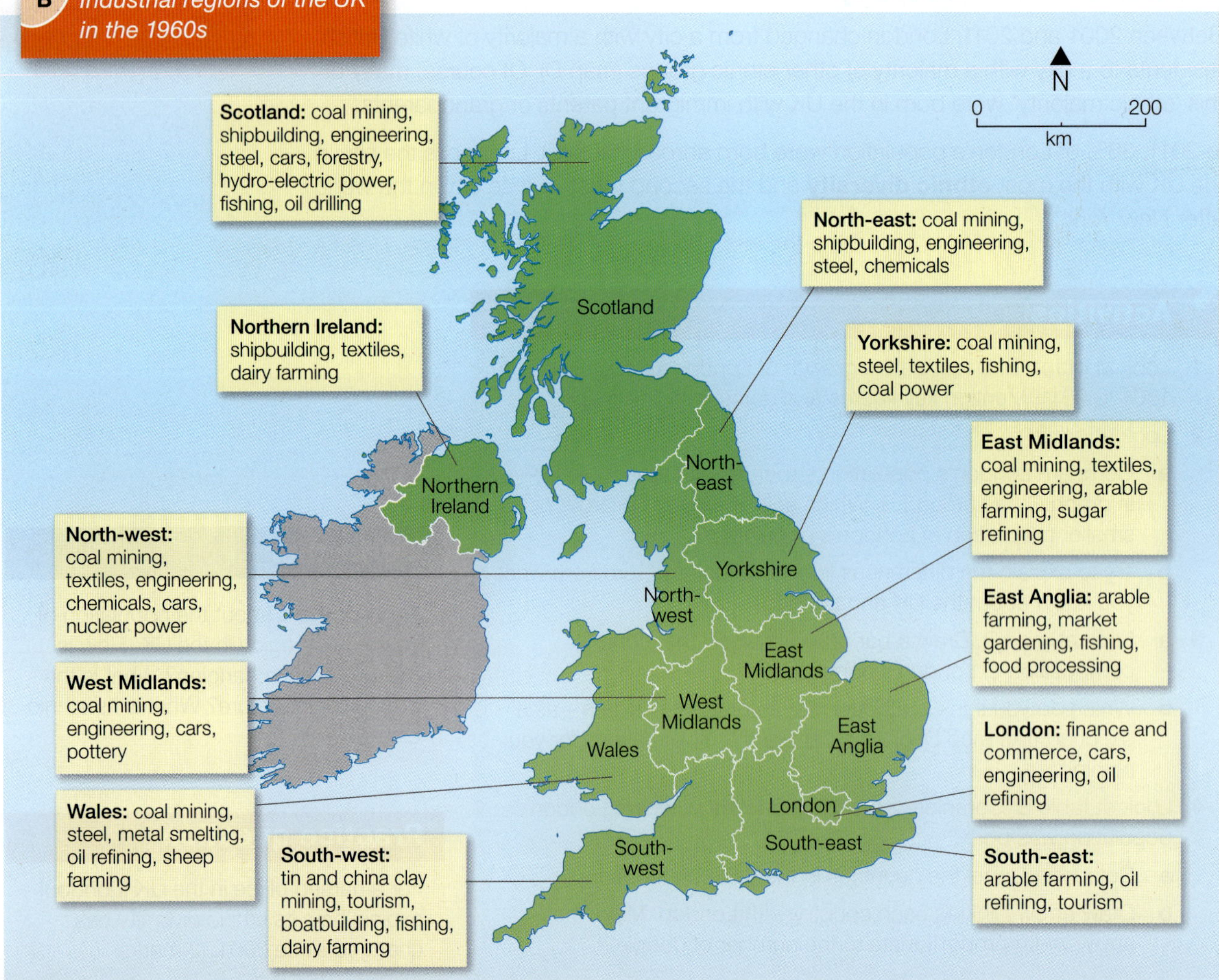

Employment and unemployment

The number of people employed in the UK is now higher than it has ever been. Of course, our population has also grown and more women work now than in the past. The types of job have also changed with the decline of manufacturing and rise of services (tertiary industry).

Despite high employment, there are still many unemployed people in the UK in the 21st century. The highest **unemployment** is found in the regions that relied most on manufacturing industry in the past (graph C). Some industries, like coal mining, shipbuilding and textile manufacturing, have almost completely disappeared.

Industrial decline in Merthyr Tydfil

One town where traditional industries have gone is Merthyr Tydfil in South Wales (photo D). Today, over a quarter of working age adults are unemployed. But, it wasn't always like that.

- In the 19th century, the ironworks in Merthyr was one of the largest in the world.
- When the ironworks closed, coal mining employed thousands of men in the 20th century.
- After coal mining declined in the 20th century, a large Hoover factory employed 5000 people.
- The factory closed in 2009, putting the last 400 employees out of work.

C *Unemployment in the UK, 2014*

D *Merthyr Tydfil*

Activities

1 Look at graph A. Describe the balance of primary, secondary, tertiary and quaternary jobs in:
  - **a** a pre-industrial economy (before 1800)
  - **b** an industrial economy (around 1900)
  - **c** a post-industrial economy (since 2000).

2 Look at map B.
  - **a** Classify the industries in the UK in the 1960s into three groups – primary, secondary and tertiary. Which was the main type?
  - **b** Why would it be difficult to draw a similar map of the UK today?

3 Look at graph C.
  - **a** Which UK regions have the highest unemployment levels now?
  - **b** How is the unemployment level today linked with the types of industry that were found in each region in the past?

Go further

a Find out what industries are found today in the region where you live.

b How have the industries in the region changed since the 1960s?

Practice question

Describe how industry has changed in one area of the UK you have studied. (4 marks)

7.8 Post-industrial UK

On this spread you will explore changes in UK employment patterns in the 21st century.

Post-industrial Merthyr

The types of work that people do in a post-industrial economy differ from work in the past. In Merthyr Tydfil, the town in South Wales that lost its only major factory in 2009, the telecommunications company EE opened a new call centre in 2013.

In a call centre, customer service operators deal with customer queries over the telephone (photo A). Many call centres have opened in areas like South Wales where traditional manufacturing jobs have gone and unemployment is high. Although low-paid, these jobs are widely welcomed as a boost to the local economy.

A *Work in a call centre*

Changing employment

Call centres are an example of new types of service employment in the post-industrial economy. There have been other changes in employment across the UK (panel B), including:

- numbers of male and female workers
- unemployment levels for men and women
- the percentage of people in self-employment
- the number of people in different types of work.

B *Changing employment patterns in the UK, 2000–2020*

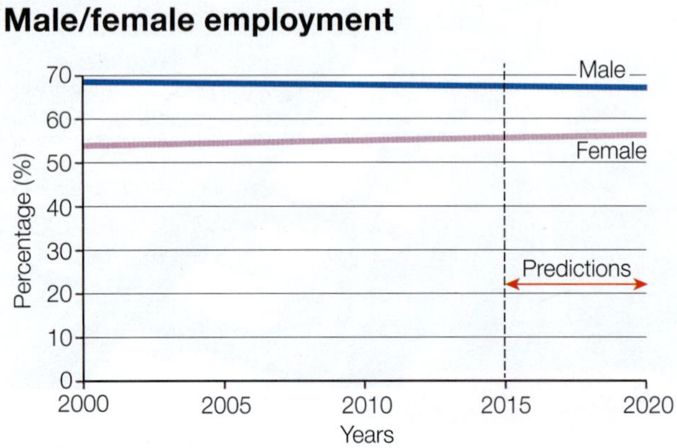

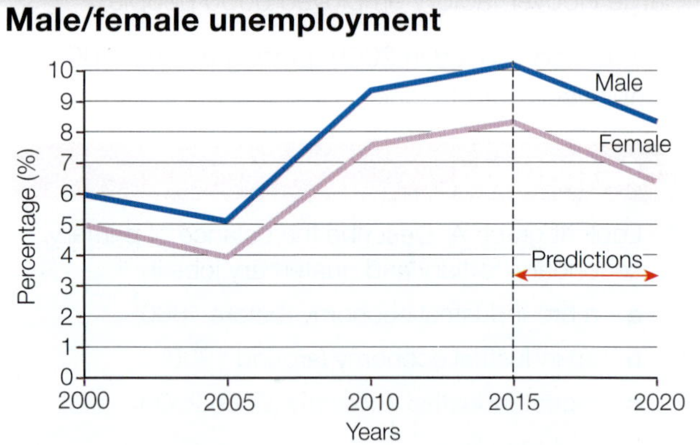

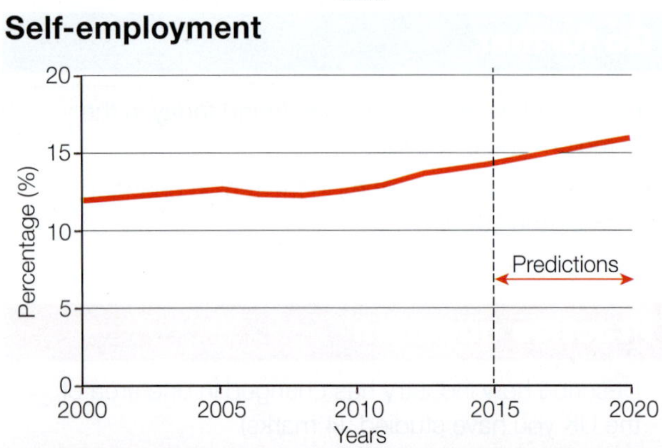

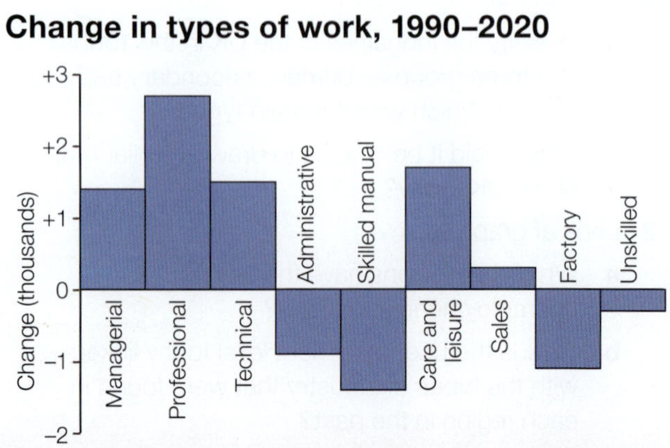

New work patterns

Years ago most people trained for one job and stayed in that job for the rest of their lives. But people in the 21st century have to work more flexibly than they did in the past. The government wants the UK to be able to compete in the global job market and that means we need to be more flexible and better qualified. New work patterns can benefit employers and employees, but there are problems too (table C).

C *Some changes in 21st century work patterns*

New work pattern	Description
Part-time work	Allows workers to fit in family responsibilities, study or do another job.
Flexi-time	Allows workers to choose when to start and finish work within set limits.
Job-sharing	Two people share the same job, which allows workers to share the load of a full-time job.
Sub-contract	An outside company or self-employed worker can be brought in to get a job done.
Zero-hours contract	An arrangement for people to be available for work, but without fixed hours and without pay when there is no work.
Teleworking/homeworking	Allows workers to spend work time away from the workplace, working via computer or telephone.

A story of self-employment

Mark Taylor is 21. He works 70 hours a week – almost twice the UK average. He works 20 hours a week for the retail company Sports Direct. The rest of the time he runs his own business providing equipment like bouncy castles and discos for a variety of events.

Sports Direct allows him to work flexibly in order to juggle his **working hours** with the demands of his business. Mark's life is typical of changing work patterns in the UK – more part-time jobs, more self-employment and more people combining two jobs.

D *Mark's business*

Activities

1 Look at photo A.
- **a** What does the term 'post-industrial' mean?
- **b** Is 'post-industrial' a good term to describe Merthyr Tydfil? Explain your answer.

2 Look at the graphs in panel B.
- **a** Write a sentence about each graph to describe the changes in employment in the UK.
- **b** Do you think these changes are good or bad? In each case, give a reason.

3 Read table C.
- **a** Complete a table like the one below to list the benefits and problems of each pattern for the employer and employee.
- **b** Overall, who do you think benefits more from changing work patterns – employers or employees?

	Employer		Employee	
	Benefit	**Problem**	**Benefit**	**Problem**
Part-time work				

Go further

Which types of work do you expect to be doing when you are 25?

- Full-time or part-time
- Employed or self-employed
- One job or more

Give reasons for your answers.

Practice question

Outline changes in UK employment in the 21st century. (6 marks)

7.9 Economic hubs in the UK

On this spread you will investigate the new pattern of core economic hubs in the UK.

The UK's economic crisis

In 2008, the UK experienced one of the worst economic crises that it has ever faced. Several large banks came close to collapse and the government had to give them billions of pounds to prevent a total disaster. The government had to borrow money to pay for other things, like schools and hospitals. Slowly, the economy is recovering as we pay off the debt.

Economic recovery is happening at different rates around the country. London and the south-east of England have recovered well and the economy is growing again (photo A). But old, former industrial areas in northern England are slower to recover.

A *Canary Wharf in London, where many large banks have their headquarters*

Economic hubs

Economists have identified cities that are the core **economic hubs** in England, where economic growth is strongest. They distinguish high growth cities, with the fastest growth since 2008 (table B), from dynamic growth cities, with the greatest potential for future growth (table C). If London were included, it would be top of both tables.

Rank	Cities	High growth score
1	Manchester	205
2	Birmingham	165
3	Milton Keynes	153
4	Bristol	150
5	Brighton and Hove	144
6	Nottingham	142
7	Leicester	137
8	Swindon	137
9	Peterborough	133
10	Cambridge	128

B *Top ten high growth cities (excluding London)*

High growth criteria include:

- **economic growth**, including the number of businesses and jobs
- **demographic growth**, including a young, educated workforce and diverse population
- **place growth**, including the number of homes and offices.

Rank	Cities	Dynamism score
1	Cambridge	176
2	Reading	146
3	Manchester	131
4	Bristol	129
5	Oxford	128
6	Brighton and Hove	127
7	Milton Keynes	123
8	Leeds	114
9	Warrington	111
10	Nottingham	107

C *Top ten dynamic growth cities (excluding London)*

Dynamic growth criteria include:

- **knowledge-intensive business** (or quaternary industry) with most potential for growth
- **high-skilled workforce** educated to degree level or above
- **new business formation** with the potential to grow into larger businesses
- **good transport connections**, including rail, road and air.

Growth corridors

The combined scores from tables B and C enabled economists to identify nine **growth corridors**, focused on cities, particularly London (map D). If the study had extended to other parts of the UK, similar corridors would have been linked to Glasgow, Edinburgh, Cardiff and Belfast.

Economic hubs in the UK have changed from the old industrial regions of the 1960s. The growth corridors extend beyond traditional city, or even regional, boundaries and are strongly influenced by major transport routes, such as the M4 from London to Bristol.

Planning for the future

The location of these economic hubs is important for future planning in the UK.

- **Cities within growth corridors** can work together with nearby cities for growth and try to avoid unnecessary competition.
- **Cities outside growth corridors** can find ways in which to make economic links with the growth corridors.
- **The UK government** can plan to promote economic growth outside the growth corridors, for example through transport links like HS2, a high-speed railway linking London to the north of England.

D *Growth corridors in England*

Activities

1 Look at photo A. Why was Canary Wharf at the heart of the following?
   a The economic crisis in 2008
   b The UK's economic growth since 2008

2 Look at tables B and C.
   a Complete a map of England to locate the top ten cities for high growth and dynamic growth. Mark and label the cities on your map. Use two colours to show the two types of city.
   b Describe the pattern on your map. Where are the high growth areas? Where are the dynamic growth areas?

3 Look at map D. Describe the pattern on the map. Mention the areas where growth corridors are located and areas without growth corridors.

4 Think about how a city outside the growth corridors could plan for economic growth. Suggest at least one thing that:
   a the city itself could do
   b the city could ask the government to do.

Go further

Investigate one of the growth corridors on map D. Find out:

- what industries are located there
- how its population is changing
- what transport connections it has.

Practice question

Describe the pattern of core economic hubs in the UK. (4 marks)

London's place in the UK economy

On this spread you will compare the changes in London's economy with changes in the rest of the UK.

London's economic growth

The economic divide between London and the rest of the UK has widened during the 21st century. While London has 13% of the UK's population, it produces 22% of the country's wealth.

Economic growth in London has been faster than in other UK regions since 1997. The gap has widened further since the economic crisis in 2008. While economic growth slowed down in other parts of the UK, it has increased in London (graph A).

Mind the gap!

London ranks higher than other UK cities for most aspects of economic performance (table B). Some people think that the gap between London and the rest of the UK is bad for the country as a whole. London attracts business that might otherwise go to other parts of the UK. Others point out that businesses and workers in London pay taxes that contribute to the UK economy. In a national survey, only 24% of people in other cities thought that London had a positive impact on their own city's economy (map C).

A *Share of economic growth in the UK regions*

B *London factfile*

	London data	UK rank (of 64 cities)
Population	8 600 000	1
Population growth (%)	1.2	5
Businesses (per 10 000 people)	463	1
New businesses (per 10 000 people)	75.5	1
Value added per worker (£)	65 800	1
High-level qualification e.g. degree (%)	46.5	4
GCSE qualifications, five A*–C (%)	62.8	5
Weekly wages (£)	634	1
Average house price (£)	461 100	1

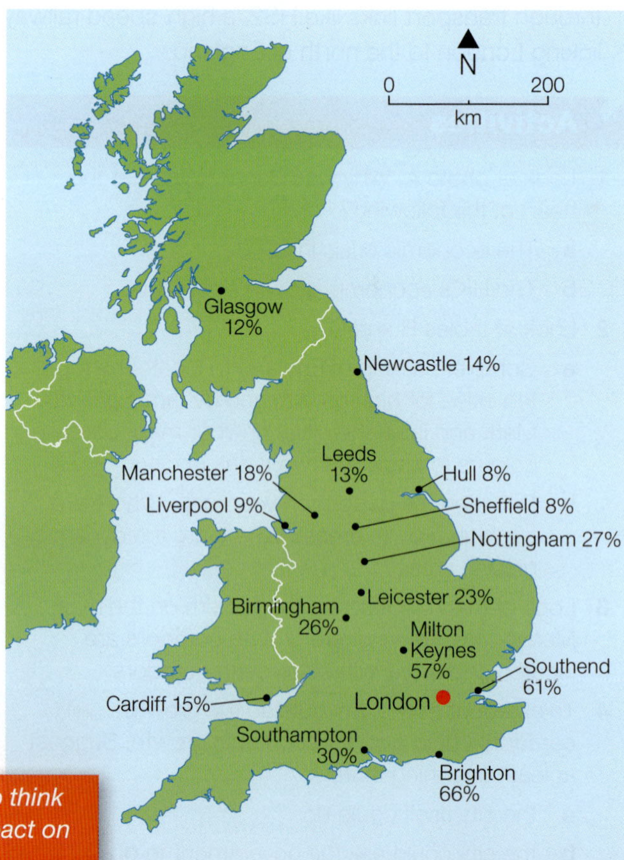

C *Percentage of people who think London has a positive impact on their own city's economy*

Silicon Roundabout

The Old Street roundabout in Shoreditch, east London (photo D) has become the focal point for hundreds of new **start-up companies** in the media and **hi-tech industries**. It has been dubbed 'Silicon Roundabout' because of all the hi-tech companies there. As well as start-ups, many well-established companies like Amazon, Facebook and Google have also invested there. Young people from around the UK, with good qualifications (story E), are attracted to London to work for these companies, taking talent away from other cities.

D *Old Street Roundabout, east London, also known as 'Silicon Roundabout'*

Decoded starts up

Kathryn Parsons is co-founder of Decoded, a company which teaches computer programming to other businesses.

'Running a start-up business is hectic, so I operate on very little sleep, relying on my alarm clock to jolt me awake at 7.00 am. I catch the Tube to our office in Old Street for 9.00 am, picking up a muffin and coffee from Shoreditch Grind on the way.

'It's a really exciting area of London to work in – there's lots of tech start-ups so it's been dubbed Silicon Roundabout, Britain's answer to Silicon Valley (in California). Our office is a huge open-plan space.

'At Decoded we teach computer programming – people learn how to code in a day. They come in having never coded anything and by the end of the day they've built their own app. Our clients tend to be really successful people from companies like Facebook or the BBC.'

E *The story of a start-up company founder*

Activities

1. Look at graph A. Compare economic growth in London with other regions of the UK. Write two sentences to compare the years before and after 2008.
2. Read factfile B. Choose two facts that are:
   - **a** a reason for London's fast economic growth
   - **b** a result of London's fast economic growth.

   In each case, explain the connection between the fact and London's economic growth.
3. Look at map C.
   - **a** Describe the pattern on the map. How do people's views of London vary with where they live in the UK?
   - **b** Explain the pattern.
   - **c** Give your view on London's impact on the economy in your city. Explain your view.

Go further

Find out more about the hi-tech industry in London. Design an advert to attract talented young people to work for a company at Silicon Roundabout. Include information about the following in your advert: type of work, wage rates, location in London, accommodation and transport, London lifestyle.

Practice question

Evaluate whether changes in London's economy in the 21st century have been good for the rest of the UK's economy or not. (6 marks)

7.11 Cambridge – an economic hub

On this spread you will explore Cambridge and consider why it has become an economic hub.

Case study

Hi-tech in Cambridge

Cambridge is an economic hub in East Anglia at the heart of the London–Cambridge growth corridor. The city has been home to one of the world's top universities for 800 years (photo A). Until recent years, Cambridge did not have many important economic activities other than education. However, in the 21st century, the city has emerged as a centre for hi-tech industry with over 1500 bio-tech and IT firms now based there.

A University students in Cambridge

B Hi-tech industry in Cambridge

Silicon Fen

The large number of new hi-tech companies clustered in and around Cambridge (map B) has given the region around the city the nickname 'Silicon Fen' after the birthplace of hi-tech industry in Silicon Valley, California. (A fen is a flat, wetland area, once typical of the area around Cambridge.)

Hi-tech industry in Cambridge began with Cambridge Science Park to the north of the city, opened in 1970 by Trinity College. It was the first of many Cambridge University colleges to make links with hi-tech industry.

C The London–Cambridge growth corridor

Abcam

Many of the hi-tech companies in Cambridge today began as small start-up businesses, formed by graduates who wanted to stay in the city when they finished at university. Some of these businesses, like the bio-tech company Abcam, have grown into successful companies (photo D). Abcam – short for 'antibodies Cambridge' – produces antibodies that are used in the treatment of diseases. The company is now worth £1 billion and employs 200 staff with PhD degrees, more than some universities!

D *Abcam headquarters at Cambridge Science Park*

Advantages of Cambridge as a business location

- There are good transport links, including the M11 to London and Stansted Airport.
- Graduates from the university provide a highly educated workforce.
- There are few traditional industries to compete for land, so rent is lower.
- The city offers a good quality of life with much open space and high-quality shops.
- There are good links between colleges and industry, helping to develop business ideas.

Disadvantages of Cambridge as a business location

- The city is overcrowded and congested – the easiest way to get around is by bicycle.
- House prices are high and still rising, making it an expensive place to live.
- Road and rail routes need to be improved to speed up connections with other cities.

Activities

1 Look at photo A. Explain the connection between the university and the growth of hi-tech industry in Cambridge.

2 Look at map B.

  a Draw an annotated sketch map of Cambridge.

  b Mark and label the features, such as the M11 motorway, that contribute to Cambridge's development as an economic hub.

3 Read the advantages and disadvantages of Cambridge as a business location. What would be the main advantages and disadvantages for the following?

  a A new graduate who wants to stay in Cambridge

  b A scientist who wants to start a new business in Cambridge

  c A large bio-tech company moving to Cambridge

Your enquiry

How is the UK changing in the 21st century?

Your parents studied geography at school in the 20th century. The UK has changed since then. Devise a crash course to bring them up to date with the geography of the UK in the 21st century. Make a list of all the key points from this section to plan your lesson. Then, at home, you can teach them what you have learned!

On this spread you will plan an economic fieldwork investigation in your area, based on the example of the Olympic Park, London.

Where to do economic fieldwork

You can do fieldwork about the economy almost anywhere in the UK, even in a rural area. However, there are more opportunities for economic fieldwork in urban areas because that's where most of the jobs are. The UK economy is changing, making it an interesting focus for fieldwork wherever you live.

The example chosen here is the Queen Elizabeth Olympic Park in east London (photo A). The 2012 Olympics has led to many new job opportunities and is changing the economy of east London.

When doing economic fieldwork, it is unlikely that you will be able to rely only on **primary data**. You will probably also need to use **secondary data** such as old maps, recent employment data or company websites.

A Here East, a new employment hub in Queen Elizabeth Olympic Park

B Economic legacy of the 2012 Olympics in east London

Hackney Wick
This old industrial area next to the Olympic Park is now home to hundreds of artists, living and working in converted factories and warehouses.

Here East
The former Media Centre for the 2012 Olympics is now transformed into a hub for creative and cultural industries. 5000 people will eventually work there.

Olympic Park
The park will attract an estimated 9 million visitors a year. This will create over 4000 new jobs in tourism, arts and leisure.

East Village
A new community now lives in what was the Athletes Village. It has a school, medical centre and shops, employing hundreds of people.

Westfield, Stratford City
This was Europe's largest new urban shopping centre when it opened in 2011 next to the Olympic Park. 10 000 people are employed there, part-time or full-time.

The International Quarter
This high-rise commercial development was built at the edge of the Olympic Park to create office space that will provide 25 000 new jobs.

Planning an economic fieldwork enquiry

1 Decide on the topic you are going to investigate. For example, you could investigate changing economic activities in your area.

2 Think of a question you could investigate or a **hypothesis** you could test. For example, the question could be '*How are economic activities in this area changing?*' or the hypothesis, '*Services are replacing manufacturing jobs in this area.*'

3 Choose suitable fieldwork activities that help you to investigate the question or hypothesis.

Economic fieldwork activities

- **Employment mapping:** Map the places where people are working.
- **Comparing workplaces:** Use old maps and photos to compare employment in the past with employment in the same area now.
- **Job classification:** Classify work into different types, such as primary, secondary, tertiary or quaternary activities.
- **Employment questionnaire:** Ask people about the work they do and how it has changed.
- **Environmental impact assessment:** Assess the positive and negative impacts of workplaces on the environment.

C *Employment questionnaire*

EMPLOYMENT QUESTIONNAIRE

Location: ______________________________

Male	Female

Age groups			
Under 25	25–40	40–55	55 +

Workplace				
Office	Shop	Factory	Public service	At home

Type of work			
Full-time employment	Self-employment	Part-time employment	Voluntary work

Job title: ______________________________

Activities

1 Think of a question or hypothesis about the UK economy that you could investigate through fieldwork in your local area.

2 Look at the suggested fieldwork activities above. Select three activities you could use to investigate your question or hypothesis. Suggest how they would help to answer the question.

3 Look at map B.
   a Classify work around the Olympic Park. You could use leisure, retail, creative, commercial (office) work in your classification.
   b Think of a way to classify work in your area of investigation. You can use different categories if you need to.

4 Look at questionnaire C.
   a Try out the questionnaire on adults you know, such as your teacher or parent.
   b How well does the questionnaire work? How could it be improved for your area? Design your own employment questionnaire.

Practice question

For an economic fieldwork investigation you have done:

a What question did you investigate? (1 mark)

b Describe one fieldwork technique you used and why you chose it. (4 marks)

c What conclusions did you come to and why? (3 marks)

Economic fieldwork – 2

On this spread you will carry out a decision-making exercise for Cambridge that you could adapt for your area.

Fieldwork

Planning a decision-making exercise

Economic fieldwork could take the form of a decision-making exercise. For example, you could put yourself in the position of a person or a company thinking of starting a new business or moving to a new location (see right). To choose the best location you would have to consider a range of factors, including:

- land availability
- suitable buildings
- the cost of land or rent
- transport for workers, customers and/or goods
- employment of workers
- the market for selling your goods or services
- competition or co-operation with other businesses
- the surrounding environment.

Fieldwork activities

To help you to make your decision, you can use a range of fieldwork activities:

- **Interviews** with business owners to find out which factors would be important in choosing a location
- **Transport surveys** to find out train, bus and cycle routes and what car parking is available
- **Land use surveys** to identify vacant land or buildings and other similar businesses that may be in competition
- **Pedestrian counts** to find out about the number of potential customers (if this is relevant).

A decision-making exercise for a business in Cambridge

What would be the best location for each of these businesses in Cambridge?

- A new café
- A web design company hoping to expand
- A large pharmaceutical company moving to the city

Interview

What does your company do?
Web design. We design websites for other organisations.
Why did you choose this location?
It was cheap office accommodation when we started the business.
How many people do you employ?
15 people, but we expect to double over the next two years. Most of our employees are graduates with experience in computing or graphics.
Are you planning to move? If so, where to?
We need more office space. We're looking for a new office with good broadband connections in a fairly central location.
What is the most important factor for your location?
Good transport. We need to be able to get to work easily. Also, although we communicate mostly online, it is good to meet clients face to face.

A *Interview transcript with the owner of a web design company*

Transport survey

Location: Station Road, Cambridge
Distance to station: 200 metres
Train destinations: London, Norwich, Ely, Peterborough, Bishop Stortford
Distance to bus stop: 50 metres
Bus destinations: City centre, Cherry Hinton, Fulborn, Addenbrookes
Distance to cycle route: 200 metres
Bike parking provision: 50 cycles (shared)
Car parking provision: 200 cars (shared)

B *Transport survey*

C *Business locations in Cambridge; OS map, scale 1:25 000*

Activities

1 Look at the map and photos in C.

- **a** Give six-figure references for the three sites on the map.
- **b** Measure the distance from the city centre (448 587) for each site.
- **c** Describe the three sites from the photos and map evidence. Write one more sentence for each site.

2 Read interview transcript A.

- **a** List the location requirements for this expanding company.
- **b** Make similar lists for:
  - a new café
  - a large pharmaceutical company.

3 Choose the best locations on map C for the three businesses in the photos. In each case, give reasons for your decision.

Go further

Plan an economic fieldwork investigation you could do in your area.

- **a** Think of an enquiry question to investigate or plan a decision-making exercise.
- **b** Choose one fieldwork technique to use. Explain why you chose it.
- **c** Think of one or more locations to carry out your investigation or decision-making exercise. Explain why you chose those locations.

The UK's global role

On this spread you will explore the UK's political role around the world as a member of international organisations.

The British Empire

There was a time when the **British Empire** spanned the globe. At its peak, Britain ruled about one-third of the world's land surface (map A). There was a saying, 'The Empire on which the Sun never sets', because it was always daytime somewhere in the British Empire.

The benefits to Britain from its Empire were great, through the vast resources it obtained. But the Empire didn't come without a cost, mainly to the people colonised by Britain (see Chapter 6). Some people's land was taken from them by force. Others were taken into slavery or recruited to fight in the British Army. This led to resistance and rebellion against the British in many corners of the globe.

Eventually, the British Empire disintegrated during the 20th century as, one by one, countries gained their independence.

A *All areas that were ever part of the British Empire*

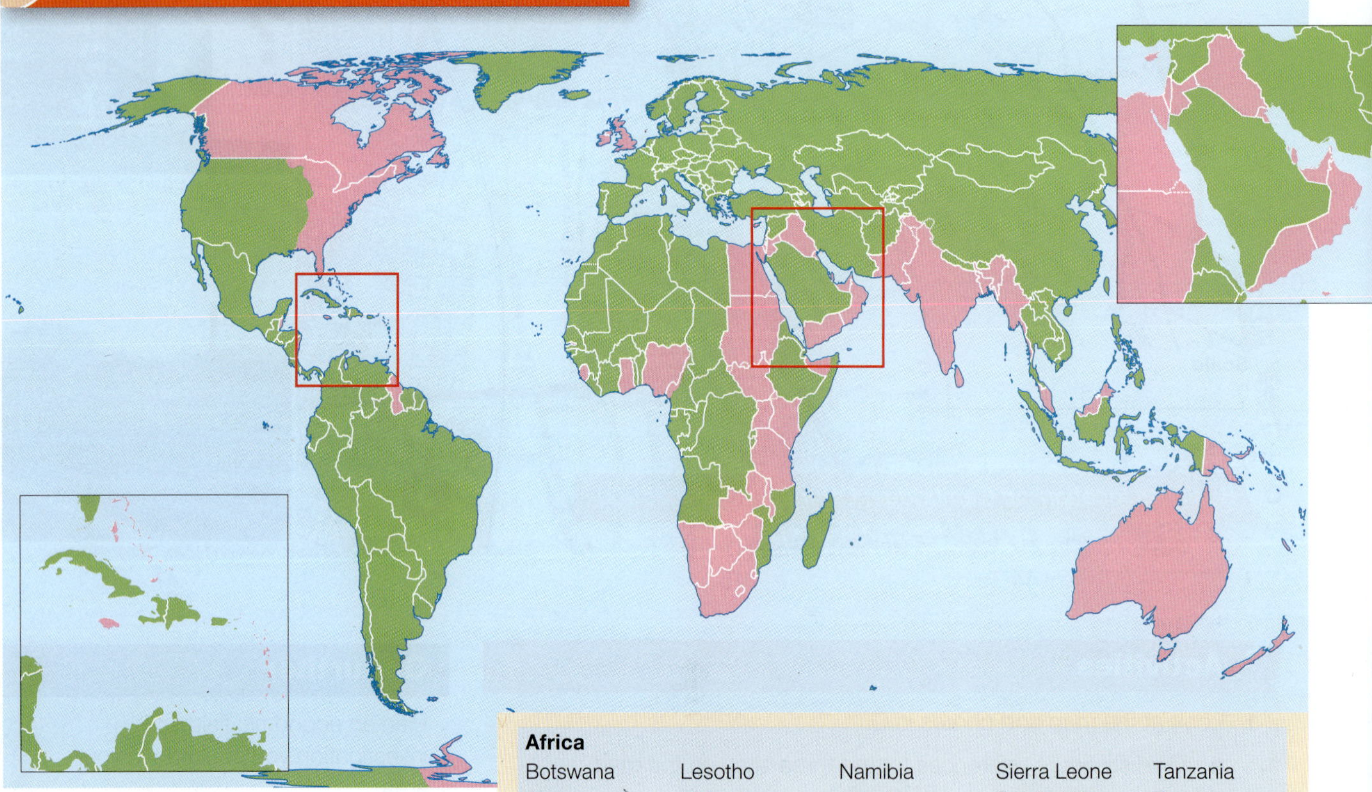

The Commonwealth

Many countries that were once part of the British Empire are now members of the **Commonwealth** (panel B), a voluntary association of 53 countries. Unlike the Empire, which Britain ruled, all Commonwealth countries have equal status. They share common political values, including the promotion of democracy, human rights and trade.

Africa
Botswana
Cameroon
Ghana
Kenya
Lesotho
Malawi
Mauritius
Mozambique
Namibia
Nigeria
Rwanda
Seychelles
Sierra Leone
South Africa
Swaziland
Uganda
Tanzania
Zambia

Asia
Bangladesh
Brunei
India
Malaysia
Maldives
Pakistan
Singapore
Sri Lanka

Caribbean and America
Antigua & Barbuda
Bahamas
Barbados
Belize
Canada
Dominica
Grenada
Guyana
Jamaica
St Lucia
St Kitts & Nevis
St Vincent & the Grenadines
Trinidad & Tobago

Europe
Cyprus
Malta
United Kingdom

Pacific
Australia
Fiji
Kiribati
Nauru
New Zealand
Papua New Guinea
Samoa
Solomon Islands
Tonga
Tuvalu
Vanuatu

B *Commonwealth member countries*

The UK's influence

The UK now maintains its global influence not by having an empire, but through its membership of international organisations (panel D). They have helped to shape the UK's policies in areas such as defence, trade and international relations.

The UK is one of the five members of the United Nations (UN) Security Council, which gives us an important say over any UN military action. The other members are Russia, China, France and the USA – all among the world's biggest military spenders (graph C).

C *The world's top military spending countries*

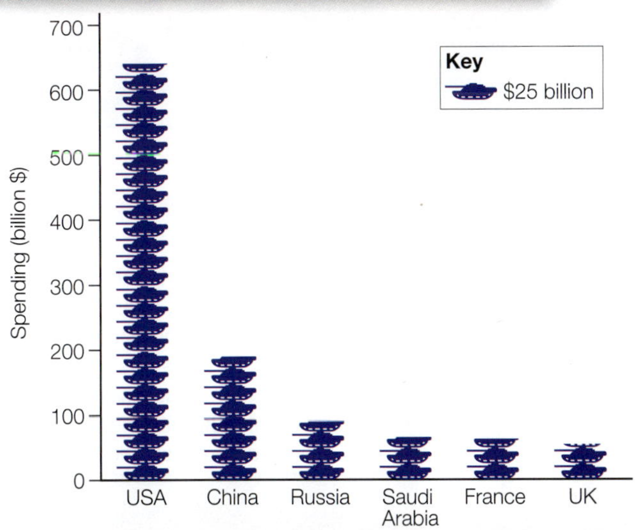

The United Nations (UN)

The UN is an inter-government organisation established to promote international co-operation. It was set up in 1945 at the end of the Second World War and now has 193 members – almost all the countries in the world. It aims to maintain international peace and security, promote human rights and economic development, protect the environment and provide aid after disasters.

North Atlantic Treaty Organisation (NATO)

NATO is a military alliance of 28 countries in North America and Europe. The Treaty states that member countries should defend each other if any of them is attacked by another country. This happened for the first time after the terrorist attacks on the USA on 11 September 2001 and led to the invasion of Afghanistan and Iraq.

European Union (EU)

The EU is a political and economic union of 28 countries in Europe. It has a **single market** for trade and an agreement to allow the free movement of people, goods, services and money between member countries. There is also **monetary union** between 19 countries (not including the UK), which all use the Euro as their currency.

D *International organisations of which the UK has been a member in the 21st century*

Activities

1 Look at map A.

- **a** Describe the distribution of the British Empire around the globe. Mention the names of continents and oceans.
- **b** 'The Empire on which the Sun never sets'. Why was this a good description of the British Empire?

2 Look at panel B.

- **a** Identify the Commonwealth countries on map A. Use an atlas to check.
- **b** Which countries that were in the British Empire are not in the Commonwealth? (Not many!)

3 Read about the international organisations in panel D. Do you think the UK should be a member of the following organisations?

a The UN **b** NATO **c** The EU

In each case, give reasons for your answer.

4 Look at graph C.

- **a** Is military spending a good indication of the global influence a country has? Why, or why not?
- **b** What other factors, apart from military spending, should be used when choosing members of the UN Security Council? Think of at least three.

Your enquiry

Is the UK losing its global significance?

Through this section, make notes about anything that will help you to answer the question. You will bring your ideas together to do an activity at the end of spread 7.17.

The UK in the Middle East

On this spread you will investigate the UK's role in the Middle East, a region of conflict.

Case study

Conflict in the Middle East

The UK once had a dominant **political role** in the Middle East. Until the mid-20th century, parts of the region were still in the British Empire. As the countries became independent and the UK withdrew, the USA took its place as the most influential power in the Middle East.

Over the past 50 years, the Middle East has seen many of the world's worst conflicts (photo A). Although the UK no longer has any control, it still has a strong political interest in what happens in the region, as:

- the UK's trade with countries in the Middle East is growing
- Islamic terrorism, connected with the Middle East, threatens security in the UK.

A *The bombing of Iraq in 2003 by NATO forces, including the UK*

B *Major conflicts in the Middle East in the 20th and 21st centuries*

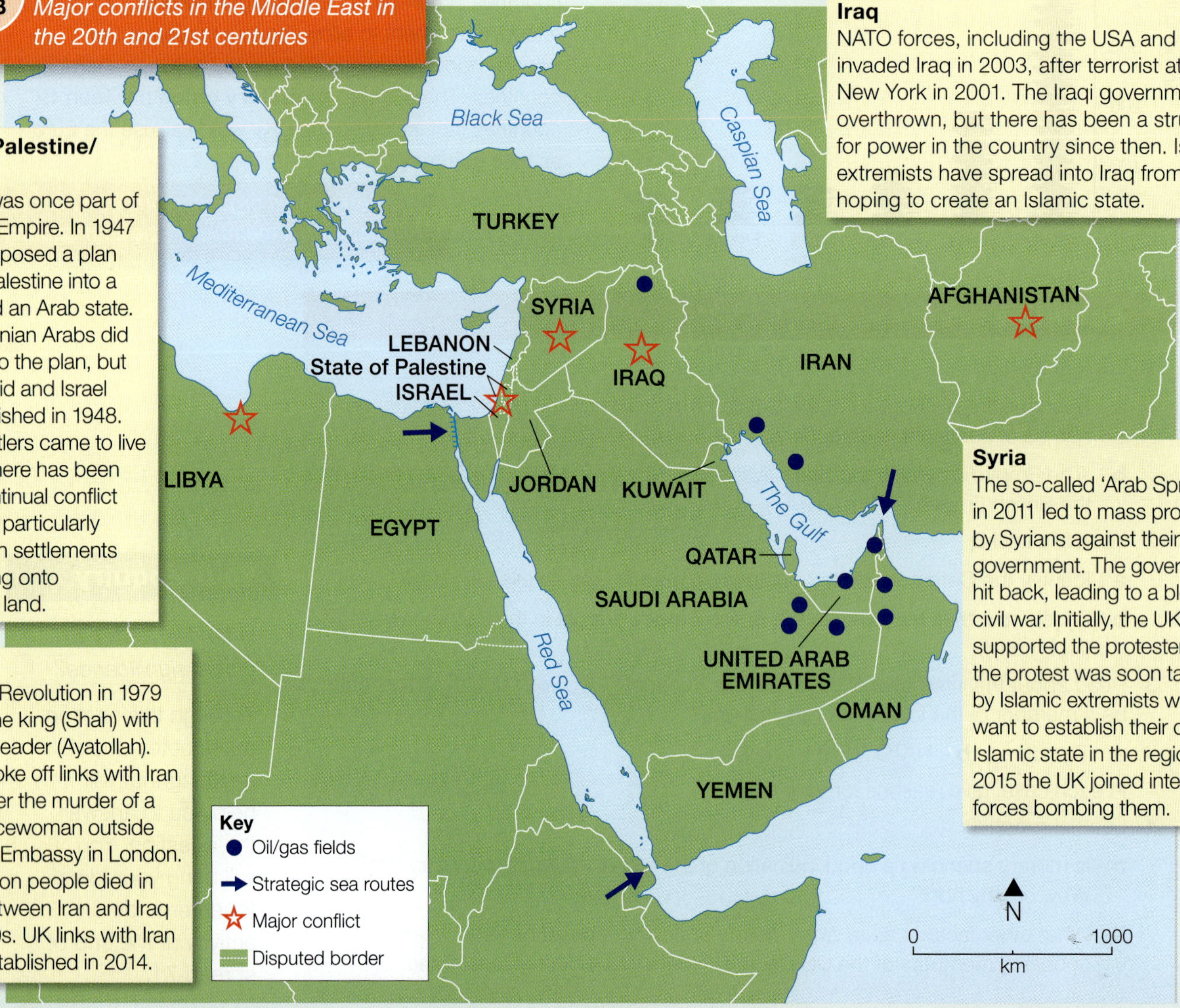

Iraq
NATO forces, including the USA and UK, invaded Iraq in 2003, after terrorist attacks in New York in 2001. The Iraqi government was overthrown, but there has been a struggle for power in the country since then. Islamic extremists have spread into Iraq from Syria, hoping to create an Islamic state.

State of Palestine/ Israel
Palestine was once part of the British Empire. In 1947 the UN proposed a plan to divide Palestine into a Jewish and an Arab state. The Palestinian Arabs did not agree to the plan, but the Jews did and Israel was established in 1948. Jewish settlers came to live in Israel. There has been almost continual conflict ever since, particularly over Jewish settlements encroaching onto Palestinian land.

Syria
The so-called 'Arab Spring' in 2011 led to mass protests by Syrians against their government. The government hit back, leading to a bloody civil war. Initially, the UK supported the protesters, but the protest was soon taken over by Islamic extremists who want to establish their own Islamic state in the region. In 2015 the UK joined international forces bombing them.

Iran
An Islamic Revolution in 1979 replaced the king (Shah) with an Islamic leader (Ayatollah). The UK broke off links with Iran in 1984 after the murder of a British policewoman outside the Iranian Embassy in London. Over a million people died in the war between Iran and Iraq in the 1980s. UK links with Iran were re-established in 2014.

Trade with the Middle East

Trade with the Middle East is the key factor that affects the UK's policy in the region. The UK exports many goods and commodities, particularly military exports like aircraft and weapons (graph D). At the same time it imports oil and gas (graph E), of which the Middle East is the world's largest supplier (photo C). As the UK's own supply of oil and gas in the North Sea begins to run out, imports from the Middle East will grow.

While many UK companies, particularly oil companies, invest in the Middle East, companies from the Middle East also invest here. The Dubai company, DP World, has opened the largest new port in the UK on the River Thames. Companies from Qatar have invested in new buildings, such as the Shard in London.

The UK's priority in the Middle East is to maintain friendly relations in order to increase trade. Previous attempts at military intervention in Iraq and support for rebel groups in Syria have failed and led to the rise of Islamic extremist groups like ISIS.

C *An oil well in Qatar, where much of the UK's gas comes from*

D *UK exports to Middle East countries, 2013*

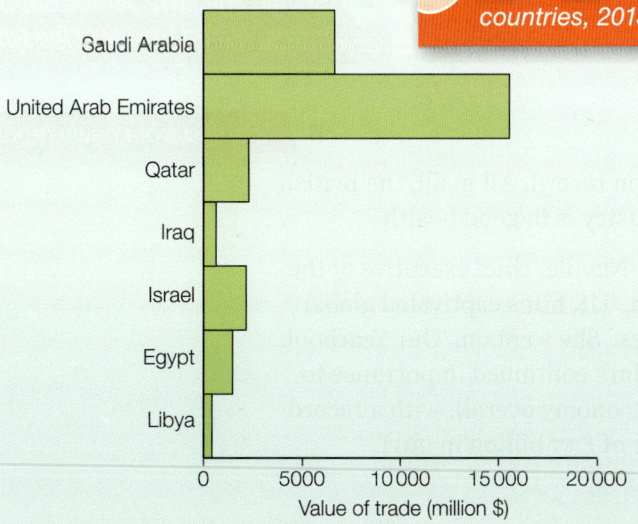

E *UK imports from Middle Eastern countries, 2013*

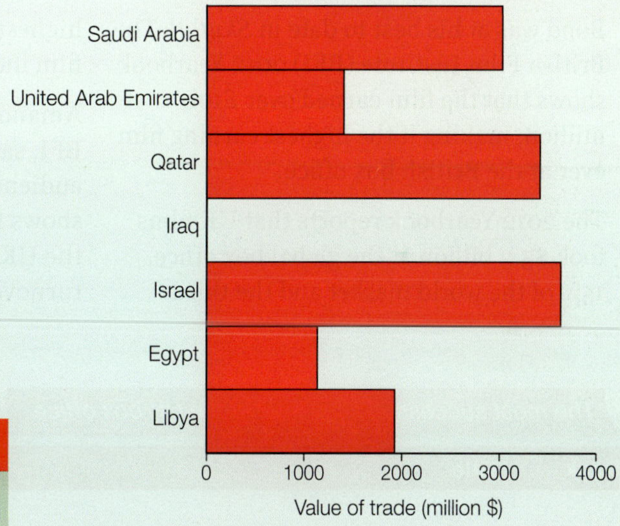

Activities

1 Look at photo A and map B.

- **a** Why did NATO forces invade Iraq in 2003?
- **b** Was the invasion a success or failure? Give reasons for your answer.

2 Look at map B.

- **a** Draw a timeline from 1948 to the present day. Label events that have happened in the Middle East at the correct point on the line.
- **b** Outline the UK's role in any of the events.

3 Look at photo C and graphs D and E.

- **a** Describe UK trade with the Middle East. Include exports and imports. Which countries are the UK's main trade partners?
- **b** How could this influence the UK's relationship with Middle Eastern countries?

Go further

Events in the Middle East are often in the news. Find out what events have happened recently.

- How do any recent events connect with what you have read here?
- How is the UK involved in any recent events?

Practice question

Outline the UK's role in one global conflict [in the Middle East] through its involvement in international organisations. (4 marks)

7.16 The influence of UK media

On this spread you will consider the global influence of UK media and their impact on the UK economy.

The UK media industry

The UK has one of the most critically acclaimed **media** industries in the world. Film-making and TV production were well-established industries in the UK during the 20th century. One of world's best-loved British film characters – James Bond – dates back more than 50 years.

The film and TV industry has risen to new heights in the 21st century. In 2012, the Bond movie *Skyfall* became the most successful film ever at the UK box office (photo A). Many other box office successes since 2000 have been filmed here, with British production teams, even if they were funded by US film companies (table C).

A *Skyfall, the 2012 James Bond movie, filmed mainly in the UK*

B *Adapted from news reports*

SKYFALL ACHIEVES NEW HEIGHTS

Bond was at his best to date in *Skyfall*. The British Film Institute (BFI) 2012 Yearbook shows that the film earned over £100 million, making it the highest earning film ever at the British box office.

The 2012 Yearbook reports that UK films took $5.3 billion at the global box office, 15% of the world market and the third highest on record. All in all, the British film industry is in good health.

Amanda Neville, chief executive of the BFI, said, 'UK films captivated global audiences.' She went on, 'Our Yearbook shows film's continued importance to the UK economy overall, with a record turnover of £7.7 billion in 2011.'

C *The most successful films ever at the UK box office*

Rank	Title	Country of origin	UK box office (£ million)	Year
1	*Skyfall*	UK/USA	102.9	2012
2	*Avatar*	USA	94.6	2009
3	*Titanic*	USA	80.1	1997
4	*Toy Story 3*	USA	74.0	2010
5	*Harry Potter and the Deathly Hallows Part 2*	UK/USA	73.1	2011
6	*Mama Mia!*	UK/USA	68.5	2008
7	*Harry Potter and the Philosopher's Stone*	UK/USA	66.1	2001
8	*The Lord of the Rings – the Fellowship of the Ring*	USA/New Zealand	63.0	2001
9	*The Lord of the Rings – the Return of the King*	USA/New Zealand	61.1	2003
10	*The Lord of the Rings – the Two Towers*	USA/New Zealand	57.6	2002
11	*Star Wars Episode 1*	USA	56.4	1999
12	*The Dark Knight Rises*	UK/USA	56.3	2012
13	*Casino Royale*	UK/USA	55.6	2006
14	*Harry Potter and the Chamber of Secrets*	UK/USA	54.8	2002
15	*The Pirates of the Caribbean – Dead Man's Chest*	USA	52.5	2006

UK media exports

The UK does not just make films and TV programmes for fun. There is also plenty of money to be made. Creative industries, including film and TV, are now worth over £70 billion a year to the UK economy, creating 1.7 million jobs.

As well as films, UK TV programmes are also exported around the world (panel D). Some are sold as ready-made programmes to other countries, dubbed in the local language if necessary. The formats for other programmes, more specific to UK culture, are sold to be adapted for the local culture in other countries. One advantage for the UK film and TV industry is that English is widely spoken around the world, including the USA, one of the main markets for UK media.

D *Some UK TV programmes successful around the world*

***Downton Abbey*:** a period drama set in the UK in the early 20th century. Shown in 150 countries with a global audience of 160 million.

***Come Dine with Me*:** a reality show where strangers compete to be the best dinner party host. Shown in 36 countries.

***Who Wants to be a Millionaire?*:** a quiz show with a £1 million prize. Shown in 107 countries, it inspired a successful British film, *Slumdog Millionaire*.

***Top Gear*:** a car show that test-drives cars and performs stunts. Shown in 170 countries with a global audience of 350 million.

***The Great British Bake Off*:** a baking show where people compete for the title of Britain's best amateur baker. The format has been licensed to over 20 countries worldwide.

***Peppa Pig*:** a children's cartoon show featuring a pig family. Shown in 170 countries. It was created by three college students who have become millionaires.

Activities

1. Look at photo A. How do films like *Skyfall* contribute to:
   - **a** the UK's influence around the world?
   - **b** the economy in the UK?
2. Read article B. Which of the factors listed below do you think are important in the global success of the British film industry? In each case, give a reason.
   - **a** English is a widely spoken language around the world.
   - **b** The UK has a film-making tradition and there are many film studios.
   - **c** The UK has a well-educated workforce.
   - **d** The UK has beautiful landscapes and many historic buildings.
3. Look at table C. Choose one film that you have seen.
   - **a** Why do you think that it was successful around the world?
   - **b** How would it influence the image of the UK around the world?
4. Read the list of TV programmes in panel D.
   - **a** Why do you think these programmes have been successful in other countries? Give a reason for each programme you have seen.
   - **b** Think of another UK TV programme you have seen. Suggest whether it would be successful in other countries or not.

Practice question

Describe how the UK's media industry has a global influence. (6 marks)

7.17 # The great British takeaway

Case study

On this spread you will explore the contribution of different ethnic groups in the UK to the food we eat.

Takeaway meals

Our eating habits in the UK are changing. Not long ago, a takeaway meal used to be an occasional treat. According to a survey in 2013, the average person in the UK now spends £110 per month on takeaway meals – that works out at about 12 takeaway meals a month (graph A).

There are a number of reasons for our change in eating habits:

- Average incomes have increased more than the cost of takeaway meals.
- We lead busier lives and have less time to cook.
- We travel more and experience a greater variety of food from around the world.
- More people from other parts of the world live in the UK and some of them open takeaway restaurants.

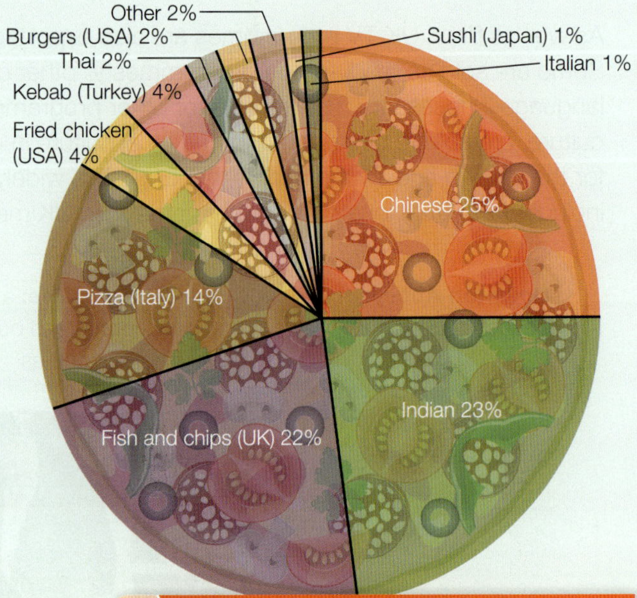

A The UK's most popular takeaway meals

Chinese takeaways

Our most popular type of takeaway food is Chinese. Chinese food arrived in the UK in the 19th century when Chinese sailors settled here and opened the first restaurants. During the 1960s, more people came from Hong Kong, opening takeaway restaurants all around the country. Every town in the UK now has at least one Chinese takeaway. In cities like London and Manchester, Chinese restaurants have grouped together in an area called Chinatown (photo B).

B Chinatown in Manchester

British food

Over the years, British cuisine has not had the best reputation. Perhaps that is why only one of the UK's top ten takeaway meals in 2013 – fish and chips – is British in origin. Other well-known British dishes include:

- full English breakfast
- Sunday roast
- shepherd's pie
- bangers and mash
- regional specialities like Cornish pasties, Yorkshire pudding and Scottish haggis.

One reason for the previous lack of variety in British food could have been the periods of wartime rationing during the 20th century when many types of food were not available in the UK. In the 21st century, a wider range of good British food is appreciated, making new restaurants and TV cookery programmes very popular.

The mystery of chicken tikka masala

Chicken tikka masala recently replaced fish and chips as the UK's favourite takeaway meal. It's interesting that chicken tikka masala is not a traditional dish found in India. So where does it come from?

It is alleged to have been invented in the UK. The story goes that a customer at an Indian restaurant in Glasgow complained that his curry was too dry. The chef added tomato soup to the curry. It proved popular and so chicken tikka masala was born.

Whether it is true or not, the story illustrates the way in which food from other cultures can be absorbed into and helps to enrich British **cultural life**. There is good evidence of this in TV cooking programmes like *Masterchef* or in the variety of restaurants that can be found on every high street.

C *Chicken tikka masala – the UK's favourite takeaway meal*

Activities

1 **a** Carry out a survey to find out what the favourite takeaway is among your class. Draw a pie chart like chart A to show your results.

b Think of a different way to present the results of your survey, other than a pie chart. Which method do you prefer, and why?

2 Look at chart A.

a Compare the results of your class survey with the survey of the UK's favourite takeaway meals. How can you explain any differences? (Clue: think about the population of your area and the types of takeaway available.)

b Carry out a similar survey at home with your parents. Find out what their favourite takeaway was when they were young and what it is now. Use the results to show how food culture in the UK has changed.

3 Which of these two statements do you think is more true?

- We are losing our traditional British cuisine to food from other cultures.
- Food from other cultures is being absorbed into British cuisine.

Give reasons for your opinion.

Your enquiry

Is the UK losing its global significance?

a Draw up a balance sheet to show how the UK has lost or gained global significance. Include all the things you have learned about the UK in this section and any other ideas of your own. Two examples are done for you.

The UK is losing global significance	The UK is gaining global significance
It has lost the British Empire.	It is part of the Commonwealth.

b Rank all the ideas in your table in order of importance, with 1 for the least important. Add the numbers in the two columns to draw up the balance sheet.

c Overall, is the UK losing or gaining global significance?

8 Resource Reliance

What is this topic about?

In order to live healthy lives humans need basic resources like food, water and energy. Some people and countries do not have enough resources, while others use more than their fair share. As the world's population increases, there is more demand on Earth's ability to supply all the resources that people need. In order to increase the supply of resources, humans are damaging the environment in ways that might threaten the future of the planet.

Food is an essential resource. A key challenge for the future is producing enough food to feed the growing numbers of people. There are different approaches to providing the nutritious food that people need. These range from cheap small-scale attempts to expensive hi-tech projects. There are advantages and disadvantages to each approach, but humans must find sustainable ways of meeting their future needs.

Your enquiry

In this chapter you will investigate the specification's two key questions.

Will we run out of natural resources? You will:

- Understand the reasons why resources are in short supply.
- Explore different approaches to providing more food, water and energy.
- Evaluate the impact on the environment of trying to provide more resources.

Can we feed nine billion people by 2050? You will:

- Understand what is meant by food security and how it can be measured.
- Identify which parts of the world do not have food security.
- Consider two different perspectives on the link between population growth and food supply.
- Examine a range of statistics about food security in Tanzania.
- Investigate how Tanzania has attempted to achieve food security in the past and in the present.
- Explore a range of different approaches to achieving food security.
- Evaluate the sustainability of different approaches to achieving food security.

Look at the photo opposite.

- Where do you think this photo was taken?
- Do you think these children have all the food they need?
- How might the food that you eat be different to the food these children eat?
- Who is responsible for providing these children with food?

More resources please

On this spread you will learn why demand for Earth's resources is increasing.

Tortilla riots!

In February 2007, 70 000 people in Mexico City marched in protest against the rising price of tortillas (photo A). The price increased because Mexico had a poor corn harvest so there was less **supply** – less corn to make tortillas. At the same time **demand** for corn increased because it can also be used to make **biofuels** for cars.

Since 2007 food riots have taken place in many other countries but they are part of a bigger problem. As long ago as 1972 a report entitled *The Limits to Growth* warned that Earth had a finite (limited) amount of resources. The report showed that demand for essential resources like **food**, **water** and **energy** was growing more quickly than the planet can supply them.

A *Huge crowds in Mexico City protested against the rising price of tortillas*

Demand outstripping supply

Demand for resources like food, energy and water is rising so quickly that supply cannot always keep up. Three main factors lead to this situation:

- population growth
- economic development
- finite and threatened supply.

This spread outlines the global situation, but access to resources can vary dramatically in different locations around the world.

Population growth

There are currently 7.3 billion people living on Earth and the population is expected to increase to 9 billion by 2050. The world's population has risen rapidly (graph B) in the last century because of the following two factors:

- death rates have fallen because of advances in medical knowledge and vaccination programmes, better diets, and improved sanitation and living conditions
- birth rates in some countries have remained high.

With more and more people added to the world's population, demand for resources will increase.

B *World population growth*

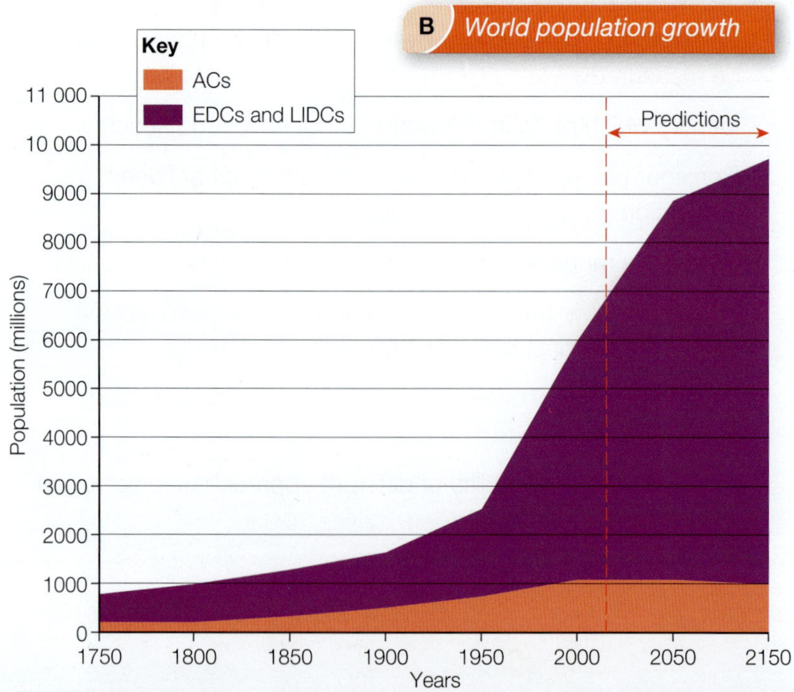

Economic development

As **low-income developing countries (LIDCs)** and **emerging and developing countries (EDCs)** develop they need more energy for industry. People also tend to want a lifestyle similar to that in **advanced countries (ACs)**, so they use more energy in their homes and want to own more goods like televisions, cars and clothes. Therefore global demand for energy is rising (graph C).

Demand for water is also increasing. Development means people have water piped into their homes and they have more appliances that use it, but water is also used for industry and agriculture (graph D).

Development also means people consume more food. They tend to want a similar diet to that in ACs, which means eating more meat. This is less efficient to produce — it takes 7 kg of grain to make 1 kg of beef.

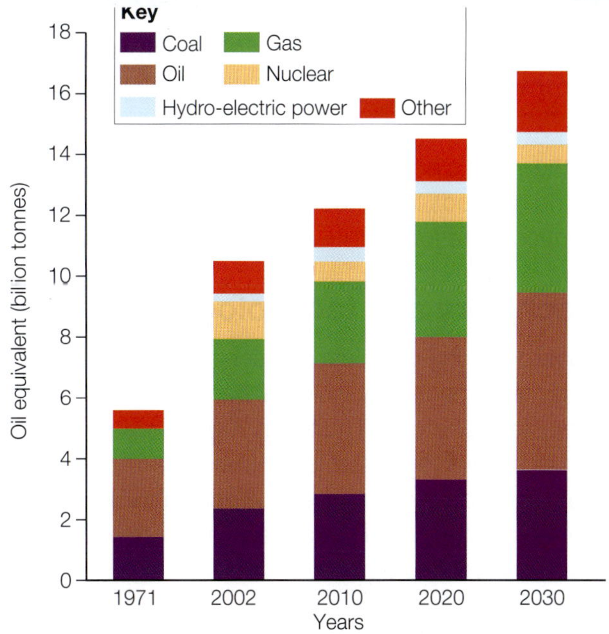

C Increasing demand for energy

Finite and threatened supply

At the moment food supply is keeping up with demand, but in some years various factors mean there is not enough food to meet demand.

- Bad harvests are caused by climate change (see spread 2.8).
- Important crops are used for biofuels rather than food.
- Increased meat consumption means grain is grown to feed animals rather than humans.

The world currently relies on fossil fuels for energy and their availability is limited. Many scientists predict that **'peak oil'** (when oil production reaches its maximum level before declining) will be reached by 2020. As supplies of fossil fuels decline, alternative fuels like solar energy will become increasingly important.

There is a limited amount of water on Earth. The availability of fresh water is reducing because of climate change. At the same time more water is being used in homes, agriculture and industry. 2.7 billion people suffer from water scarcity for at least one month a year.

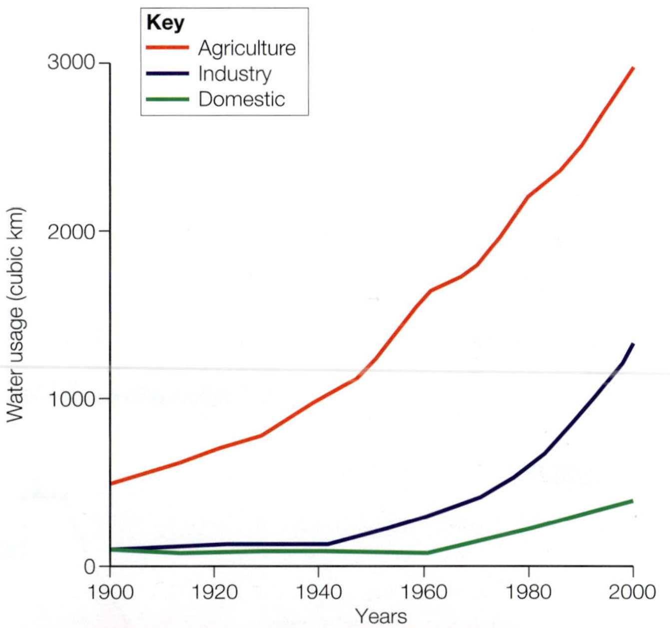

D Global use of water in different sectors

Activities

1. Look at photo A. Write a script for a 30-second news bulletin about the tortilla riots in Mexico.
2. Look at graph B. Explain why the world's population has increased so rapidly since 1900.
3. **a** Look at graph C. Describe the changes in world demand for energy.
   b Look at graph D. Describe the changes in the world's use of water.
   c Why do countries use more energy and water as they develop?
4. Explain what might happen to food supply if everybody was a vegan (someone who doesn't eat any animal products like meat or cheese).

Your enquiry

Will we run out of natural resources?

As you work through this section, make notes about anything that will help you to answer the question. You will bring your ideas together to do an activity at the end of spread 8.4.

8.2 Farming and fishing for food

On this spread you will discover how mechanised farming and commercial fishing affect the environment and ecosystems.

As the world's population has grown, farming and fishing methods have had to develop in order to produce more food.

Machinery and technology in farming

Modern farming uses lots of machinery and technology. Today farm machinery includes tractors and combine-harvesters (photo B). The technology includes GPS (global positioning systems) on tractors and computer-programmed milking parlours. **Mechanisation** replaces humans and makes farming more efficient. This increases the amount of food that is grown and harvested.

Commercial farming is usually highly mechanised, especially in ACs. Commercial farms produce crops to sell for a profit. This is different to **subsistence farming** where farmers grow food and animals to feed their families (photo A). Most of the world's farmers are still small-scale subsistence farmers.

A *Subsistence farming*

B *How mechanised commercial farming affects the environment*

Commercial fishing

Globally, fish consumption has increased from 9.9 kg per person a year in the 1960s to 19.2 kg in 2012. Fishing has developed to meet demand and commercial boats catch as many fish as they can. This affects ocean ecosystems. Over-fishing and **by-catch** have negative impacts but some **commercial fishing** uses even more destructive methods (panel C).

Fish farms

Just under half the world's fish supply is grown in the indoor pools or ocean cages of fish farms. In theory these are a solution to the problems caused by ocean fishing. However, fish farms often breed disease. Their waste is flushed into the ocean or fish escape and diseases spread to surrounding ecosystems.

C How commercial fishing affects the environment

Over-fishing: For decades more fish have been caught than are replaced naturally. Many fish species such as Bluefin tuna are nearly extinct. This can upset the balance within an ecosystem (diagram D). Fishing quotas (limits) are often put in place to try to stop over-fishing.

By-catch: Commercial fishing uses enormous nets and very long fishing lines. Fish and marine animals are caught by accident (by-catch), especially if a quota gives permission to catch only one species. Everything else is thrown back into the sea, often dead or dying.

Bottom trawling: Huge nets with metal weights drag along the ocean floor. They catch and destroy everything in their path. This can permanently destroy delicate ecosystems. By-catch can be 90% of a total catch and might include centuries-old pieces of coral.

Poison and explosives: Some fishermen spray cyanide into the ocean to stun fish and make them easier to catch. This kills everything in the ecosystem. For every fish caught using this technique a square metre of coral reef is destroyed. Explosives can also be used to kill fish and are just as destructive.

D The ocean ecosystem

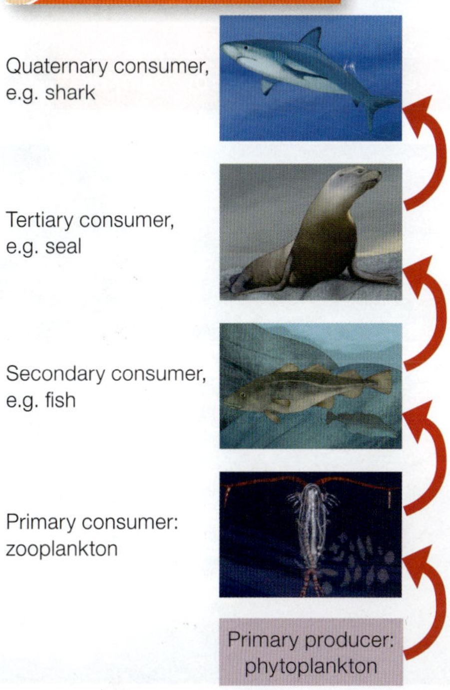

Activities

1. Look at photos A and B. List three differences between commercial farming and subsistence farming.
2. List the impacts that mechanised farming has for non-living parts (soil, air, water) and living parts (plants, insects, animals) of farm ecosystems.
3. Study photo C.
   - **a** Do you think the fishermen planned to catch this shark?
   - **b** Make a leaflet about the problems caused by commercial fishing.
4. Look at diagram D. Explain the impacts on the ocean ecosystem if:
   - **a** lots of secondary consumers are caught
   - **b** lots of quaternary consumers are caught as by-catch.

Practice question

Describe two impacts of commercial fishing on ocean ecosystems. (4 marks)

8.3 Using fossil fuels for energy

On this spread you will learn why obtaining and using fossil fuels damages the environment and ecosystems.

Taking Earth's energy

Most of the world's energy comes from fossil fuels like coal, oil and gas. They are extracted by mining, drilling and **fracking** (see spread 10.4). These processes cause **deforestation**, water shortages and pollution, which have many further impacts on the environment and **ecosystems**.

Using fossil fuels releases **greenhouse gases** into the atmosphere and causes climate change, which also impacts on ecosystems (see spreads 2.7–2.9).

Deforestation and energy supplies

Obtaining energy resources causes deforestation – natural vegetation is removed, and plant and animal habitats are destroyed. This can devastate ecosystems and threaten animal species with extinction (photo A).

- **Gathering wood:** More than 2 billion people depend on wood for cooking and heating. Most is collected when it is already dead, but demand for charcoal (which is made from wood) is causing deforestation, especially in Africa and Asia.
- **Mining, drilling and fracking:** Removing energy resources from the ground requires large-scale deforestation (see spread 4.6). With open-cast mining, the land is completely stripped to get at coal under the surface (photo B). Vegetation is also removed to create underground mining sites and build infrastructure, such as roads, railways and pipelines.

A *Orangutans are threatened in Malaysia as more rainforests are used to grow palm oil*

Biofuels instead of fossil fuels

The oils in plants like sugar, corn and palm oil can also be used as fuel. Demand for these has increased because they release fewer greenhouse gases than fossil fuels. Clearing land to grow these biofuels has also increased deforestation.

Water shortages

Huge quantities of water are used to get oil and gas out of the ground, to clean coal and manage the dust created. Fracking, for example, involves blasting huge amounts of water and chemicals into the ground to force gas and oil deposits to the surface. Water is taken from groundwater supplies, starving agriculture and ecosystems of water.

B *An open-cast coal mine in an Indonesian rainforest*

Pollution

- **Soil pollution:** Mining or drilling into the ground releases toxic chemicals contained in the rocks. These can be washed onto the surrounding land, polluting the soil. This means vegetation doesn't grow, altering the diversity of plants, insects and animals.
- **River pollution:** Toxic chemicals from mines or drilling can also be washed into local rivers (photo C). In addition, some processes leave huge spoil tips of waste material, also polluted with chemicals, above ground. Rain washes loose material from these tips into rivers. This type of pollution kills fish and plant species.
- **Groundwater pollution:** Chemicals released in mining can leach into and pollute groundwater supplies, affecting plant growth. In the USA, people living near fracking sites can set fire to water running out of their taps because it contains large amounts of methane (photo D).
- **Ocean pollution:** Oil leaks and spills can be very destructive. For example, when the Deep Water Horizon oil well exploded in the Atlantic Ocean near Florida in 2010 about 4.9 million barrels worth of oil leaked into the ocean. This affected the ocean ecosystem, killing dolphins, turtles and fish, and contaminating the food chain.
- **Air pollution:** Mining creates lots of dust containing dangerous chemicals like sulphur and arsenic. This pollutes the air as well as the soil and water. Burning fossil fuels releases sulphur dioxide and nitrogen oxides, which dissolve in rainwater to create acid rain. This kills plant life and pollutes water courses. In addition, when coal is burnt, it releases mercury, which gets into the food chain.

C *A polluted river near a coal mine*

D *Water on fire*

Activities

1. a. Explain why deforestation takes place to obtain energy supplies.
   b. Look at photos A and B. Explain how plants and animals are affected by deforestation.
2. Compile a table with these headings: Water pollution, Soil contamination, Air pollution. Under each heading list two ways that ecosystems are changed by obtaining energy supplies.
3. a. Arrange these statements in order to show how mercury moves up the food chain (see spread 4.1).
   - Small fish eat plankton.
   - Mercury pollutes the ocean when it rains.
   - Burning coal releases mercury into the atmosphere.
   - Humans catch and eat large fish.
   - Tiny plankton in the ocean absorb mercury.
   - Large fish eat small fish.

   b. Draw a diagram to illustrate this process.

Go further

Research the environmental impacts of the Deep Water Horizon oil well explosion. How do you think the ocean ecosystem would have been affected? (Clue: look at diagram D on spread 8.2.)

Practice question

Describe two ways in which ecosystems are altered by obtaining energy supplies. (4 marks)

8.4 Providing enough water

On this spread you will learn how systems to secure water supplies alter the environment and ecosystems.

A shortage of fresh water

About 71% of Earth's surface is covered with water. 96.5% of this water is saltwater (table A). Demand for freshwater is increasing and many parts of the world suffer water shortages. A key challenge is to provide water where people need it. This is done by building reservoirs to store water and artificial channels to move it from one place to another.

Water source	Percentage of total water on Earth	Percentage of freshwater this provides
Oceans and seas	96.54	0.0
Ice caps, glaciers and permanent snow	1.74	68.6
Groundwater	1.69	30.1
Other water, including rivers, lakes, swamps, permafrost and the atmosphere	0.03	1.3

A *The world's water stores*

Reservoirs

Reservoirs provide a store from which water can be taken when it is needed. Natural lakes are used, but reservoirs are often specially built artificial lakes. They can be created when a dam is built across a river or filled with water diverted from other rivers.

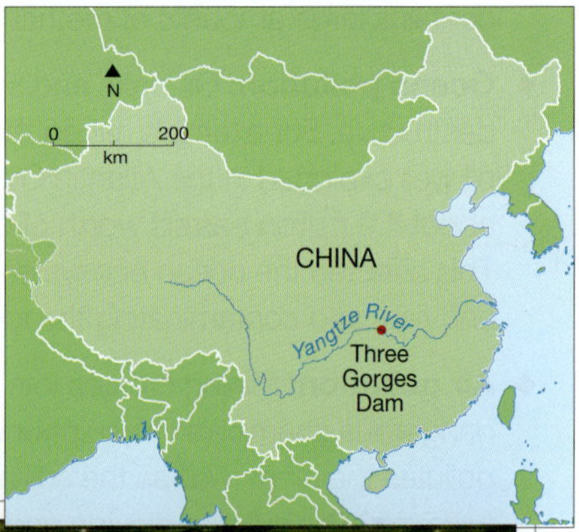

Flooding land to create a reservoir can be very destructive to the surrounding environment. The Three Gorges Dam in China is one of the world's largest dams (photo B). Its reservoir is nearly 650 km long (the distance from London to Edinburgh). Thousands of towns and villages were submerged and 1.3 million people were forced to move to make way for it. The flooding destroyed many natural habitats on land and in the river.

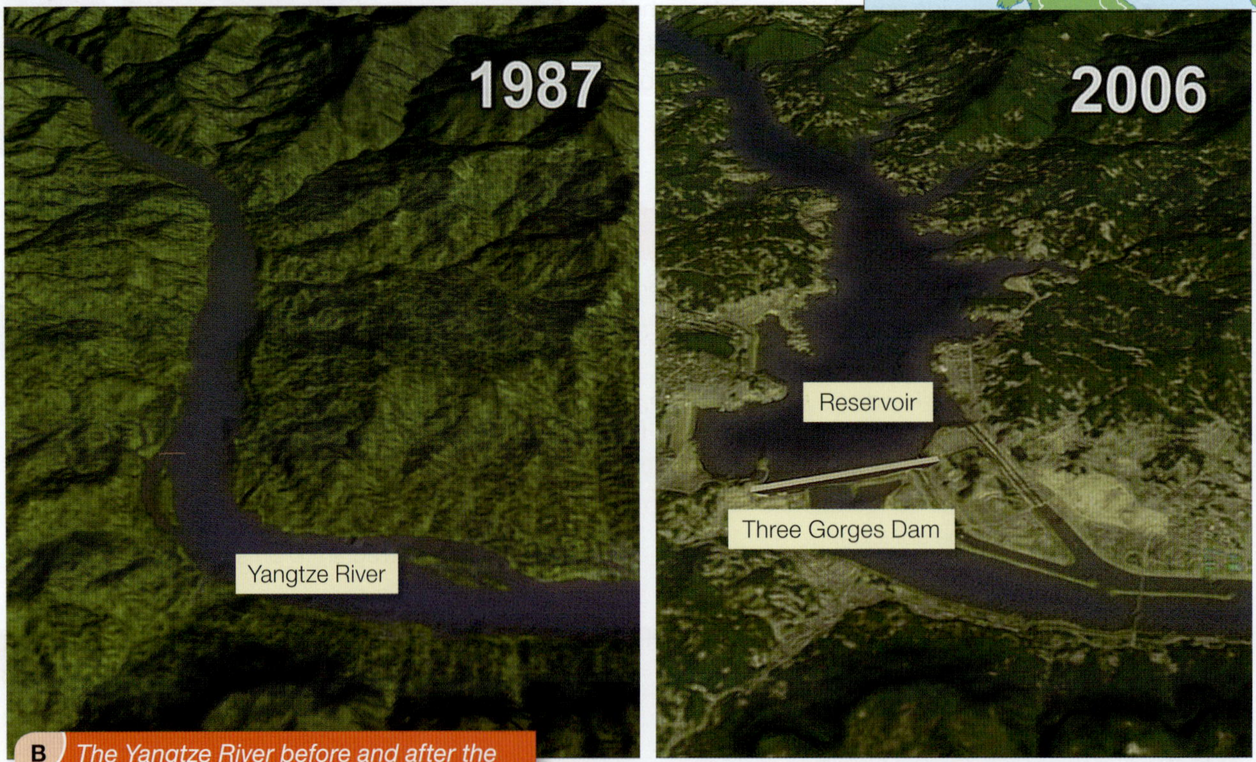

B *The Yangtze River before and after the Three Gorges Dam was built*

Damming impacts on ecosystems

Building a dam can have many effects on ecosystems upstream and downstream of the dam (photo C).

Upstream

- Land here is flooded, destroying plant and animal habitats.
- Floodwater covers towns and industry, polluting the river.
- Water temperatures become warmer on the surface and colder deep down, threatening species with extinction.
- Deep water can become a 'dead zone' with not enough oxygen for species to survive.
- The dam traps sediment, reducing water quality and covering sites where fish lay their eggs.
- The dam blocks fish migration routes. Some dams include 'fish ladders' so that fish like salmon can get upriver to breed.

Downstream

- Water temperatures might be higher because river levels are reduced.
- The dam releases oxygen-starved water, killing fish downstream.
- With less sediment, the river has more energy, increasing erosion.
- Diversity of fish and other river species is reduced.
- Lower groundwater tables affect plants and wildlife.
- Flooding is reduced so fewer nutrients are deposited, changing floodplain ecosystems.
- Less sediment means deltas and estuaries are eroded and seawater extends further inland.

C *The effects of building a dam on ecosystems upstream and downstream*

Water transfer schemes

Water transfer schemes are artificial channels for moving water from areas with lots of water to areas of shortage (see spread 12.4). They often take stored water from a reservoir but can involve diverting or splitting a river.

Moving water in this way can have environmental impacts for ecosystems at both ends of the channel. Many are similar to those created by dams. Species might also be transferred to an ecosystem where they do not belong.

Activities

1. Look at table A. Create one or two graphs to display the information. Which type of graph do you think is the most suitable? Why?
2. Look at the photos in panel B. List three differences you can see between the two views.
3. Look at photo C.
   - **a** List three ways that building a dam changes ecosystems upstream and three changes downstream. Include at least one living and one non-living factor in both lists.
   - **b** Explain why these changes are happening.

Your enquiry

Will we run out of natural resources?

Obtaining enough food, water and energy resources for the world's population is a key challenge for the future.

Decide whether food, water or energy is the most challenging issue. Produce a poster or presentation, outlining:

- why you think food, water or energy presents the most challenging issue
- why there is a shortage of this resource
- the potential environmental impacts of trying to provide enough supplies of this resource.

Food security

On this spread you will learn about the physical and human factors that influence food security.

About 800 million people do not have enough food. The World Health Organisation defines **food security** as 'when all people at all times have access to sufficient, safe, nutritious food to maintain a healthy and active life'. There are three aspects to food security (chart A), which can be influenced by physical and human factors.

Physical factors influencing food security

Food security is affected by a range of physical factors, which are things to do with the natural environment (photo B). These influence whether land is suitable for growing food and how successful food production might be.

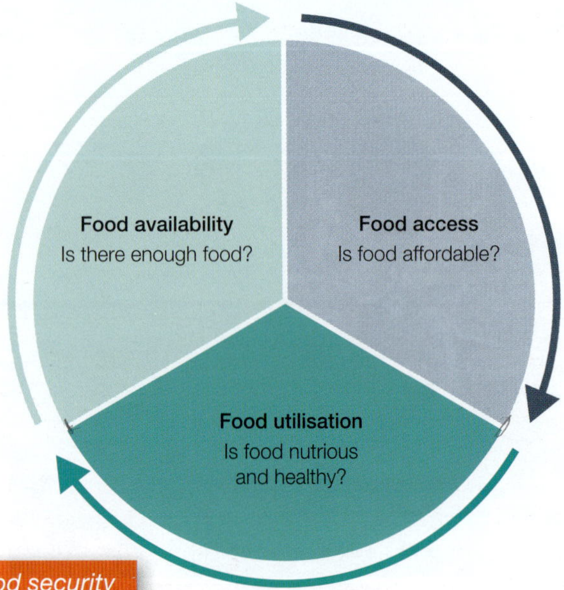

A *Aspects of food security*

B *A number of physical factors affect food production*

Human factors influencing food security

The world grows enough food to feed 10 billion people. Food insecurity is not just a question of growing enough food. It is also about whether people are able to access the food that is grown.

- **Poverty:** People suffer from food insecurity when they cannot afford to buy food. Farmers may not be able to afford seeds or tools.
- **Distribution and infrastructure:** Food must be distributed to those who need it. Poor transport infrastructure makes it difficult to get food to where it is needed.
- **War and conflict:** Conflicts disrupt farming and make distributing food more difficult. War can force millions to flee from their homes to places where there is not enough food, leading to food emergencies.
- **Land ownership:** Farmers often do not own the land they farm. Land might be bought by TNCs and used to grow export crops, so there is less land for growing food for local people.
- **Waste:** Lots of food is wasted. For example, in Vietnam, as much as 25% of rice goes to waste because of poor transport, storage and processing. In the UK, 4.4 million apples are thrown away every day.
- **Climate change:** Climate change is affecting rainfall patterns and temperatures, which threatens food production (see spread 2.8).

C *A lot of food produce is wasted in ACs*

D *Food aid is given to refugees in Somalia*

Famine in Somalia

260 000 people died in a famine in Somalia between 2011 and 2012. A two-year drought had killed livestock and reduced harvests by 60%. Food prices rapidly increased and, despite warnings of a crisis, the international community was very slow to provide aid. When aid finally arrived, conflict made it difficult to distribute.

Activities

1 a Write out the definition of food security.
  b Look at chart A. Draw a picture or cartoon to illustrate the three aspects of food security.

2 Draw a mind-map showing the factors influencing food security. Draw cartoons on your mind-map to represent each factor.

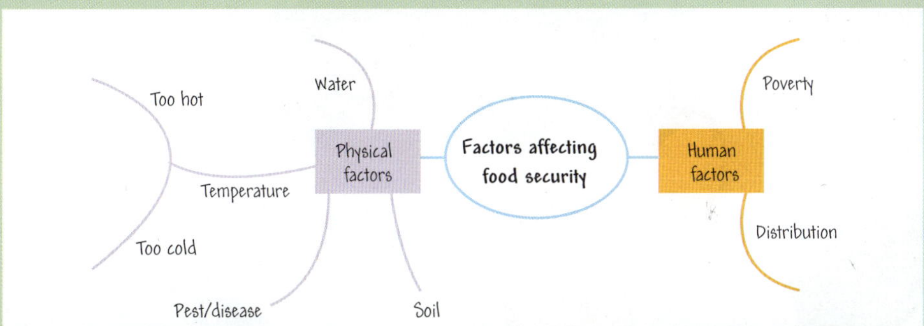

3 Do you think climate change should be listed as a human factor influencing food security? Explain and justify your ideas. (Clue: look at spreads 2.5–2.6.)

4 Were physical or human factors to blame for the famine in Somalia? Give reasons for your answer.

Your enquiry

Can we feed nine billion people by 2050?

As you work through this section, make notes about anything that will help you to answer the question. You will bring your ideas together to do an activity at the end of spread 8.15.

Measuring food security

On this spread you will explore and evaluate different measures of food security.

Food security varies across the world. Some people and places are more food secure than others. But how is food security measured?

Wealth and health measurements

Food security could be measured through wealth indicators like:

- GNI/capita (average wealth per person) at **purchasing power parity (PPP)** (see panel below table A)
- percentage of population living on less than a dollar a day
- percentage of household income spent on food.

Health and dietary indicators can also be used, such as:

- life expectancy
- mortality (death) rate of children under five
- **average daily calorie consumption** (table A).

Composite measures

Composite measures combine a number of different indicators.

The Global Hunger Index

This index (map B) ranges from 0 to 100, with 0 indicating zero hunger and 100 the most hunger. The index is calculated using four indicators:

- **Undernourishment:** the percentage of the population who are undernourished.
- **Child wasting:** the proportion of children under five who have low weight for their height.
- **Child stunting:** the proportion of children under five who are short for their age.
- **Child mortality:** the death rate of children under five.

Country	Average calories per day per capita	GNI/capita at PPP ($)
Austria	3800	43 840
Belgium	3721	40 280
USA	3688	53 960
Germany	3549	44 540
France	3531	37 580
UK	3432	35 760
Canada	3399	42 610
Cuba	3258	18 520
Brazil	3173	14 750
China	3036	11 850
Chile	2908	21 030
Japan	2723	37 630
Vietnam	2690	5 030
Indonesia	2646	9 260
Niger	2489	910
Nepal	2443	2 260
Rwanda	2188	1 430
Ethiopia	2097	1 350
Kenya	2092	2 250
Eritrea	1640	1 180

A *Average daily calorie consumption per capita and GNI/capita at PPP for selected countries*

PPP means purchasing power parity, a way of saying that the cost of living in each country has been taken into account. That's because a US dollar buys more in some countries than in others.

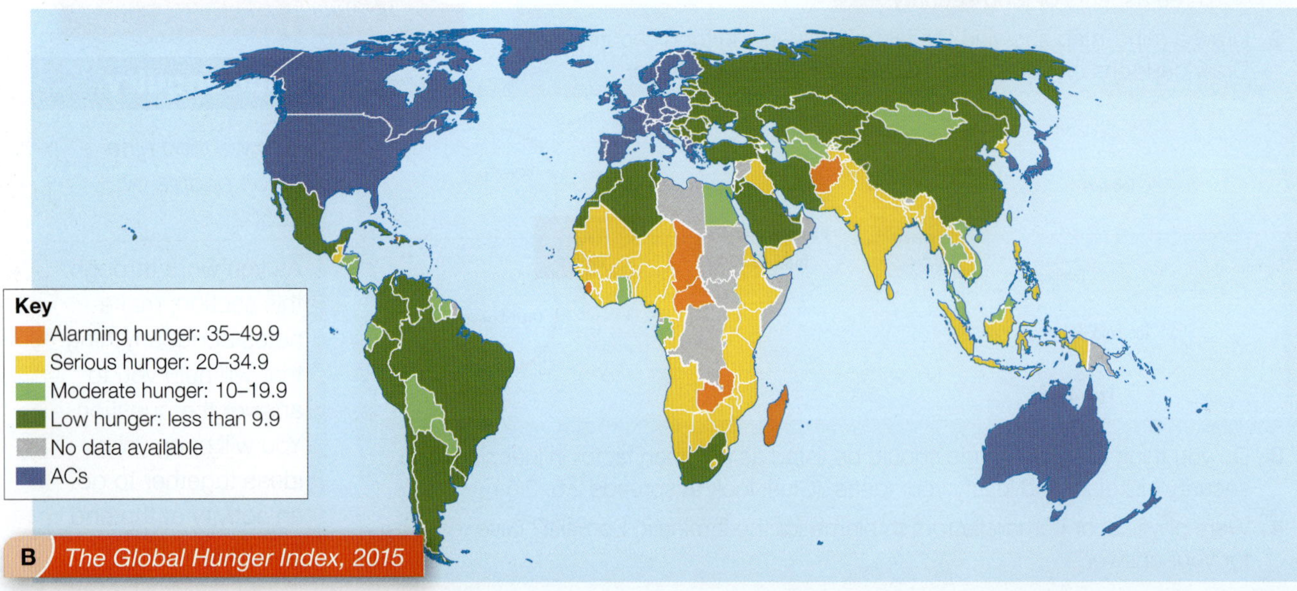

B *The Global Hunger Index, 2015*

The Global Food Security Index

This index combines 28 different indicators, looking at the core issues of food availability and affordability, quality and safety to categorise countries as shown on map C.

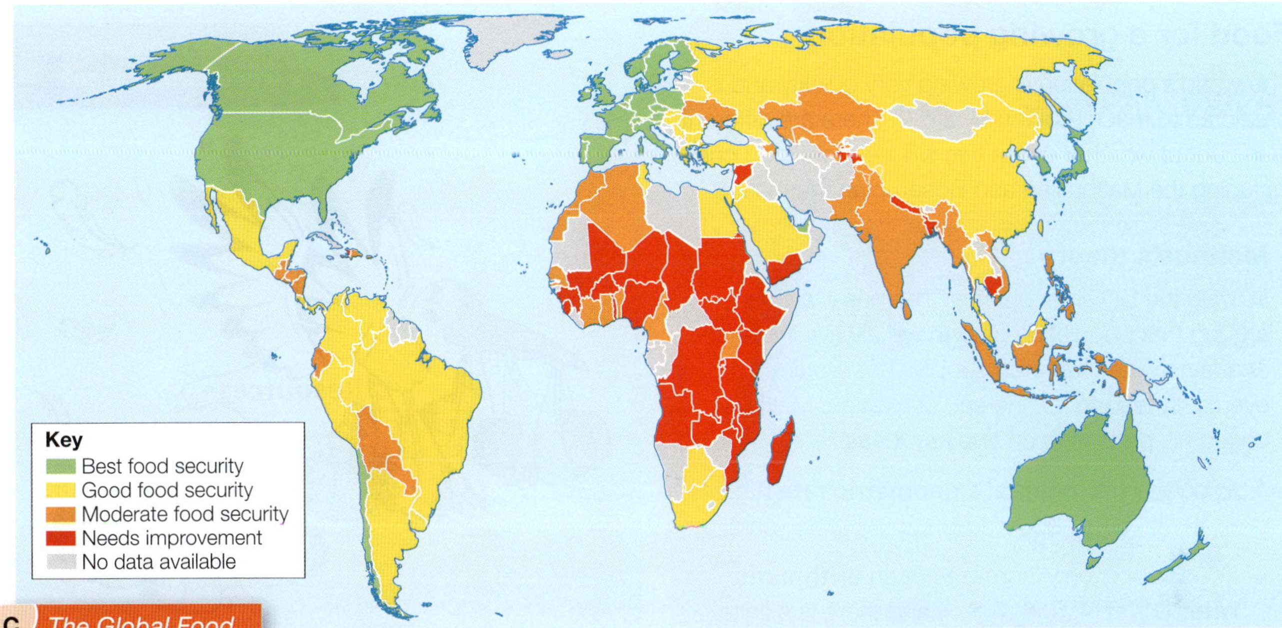

C *The Global Food Security Index*

Activities

1 a Look at the wealth and health measurement indicators. List these indicators in order of how effective you think they are in measuring food insecurity.

b Explain your reasons for deciding which indicators are most and least effective.

2 a Look at table A. Draw a scatter graph to look at the relationship between average daily calorie consumption and GNI/capita at PPP.

b Draw a line of best fit and describe what your graph shows. Mention any countries that don't fit the general pattern.

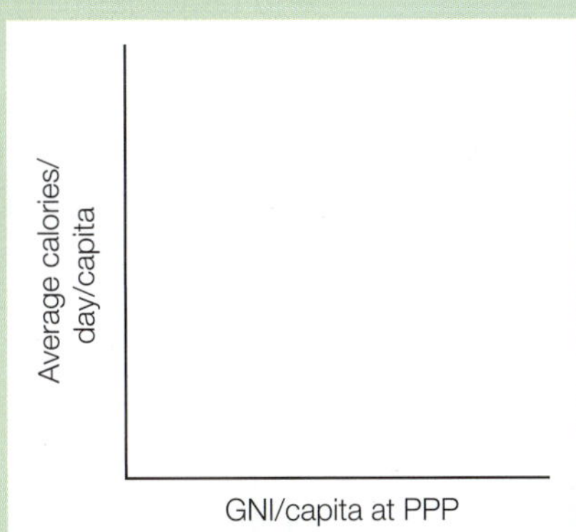

3 Why do you think the Global Hunger Index uses three measures of child health rather than adult health?

4 Look at map B. Describe the global pattern of hunger.

5 a Look at maps B and C. Name any countries that are given the same rating on map B but different ratings on map C.

b Why do you think there are differences between how the maps rank different countries?

6 Look at photo D and maps B and C. Do you think the two maps are misleading in any way? Explain your ideas.

Food banks in the UK

Food security has recently become a big political issue in the UK. Food banks are established by charities and give three days' worth of food for people and families who cannot feed themselves. In 2014, 1.1 million people used food banks in the UK.

D *A food bank in the UK*

Practice question

Look at map C. Describe the location of countries with moderate food security. (3 marks)

On this spread you will analyse two different theories of the relationship between population growth and food supply.

Food for a growing population

The world's population is growing very quickly and is predicted to reach 9 billion by 2050. There are different ideas about whether or not this will lead to a food crisis, including the Malthusian and Boserupian theories.

Malthus's theory

In 1798 the British economist Thomas Malthus argued that population growth will always increase at a faster rate than food supply. This would eventually lead to famine and starvation. Malthus based his prediction on two key ideas:

- Population increases at a **geometric rate** (1, 2, 4, 8, 16, 32, 64…), doubling in size every 25 years.
- Food supply only increases at an **arithmetic rate** (1, 2, 3, 4, 5, 6, 7, 8…) and there is a finite amount of land on which food can be grown.

Malthus believed that the different rates of increase would lead to a crisis (graph B), resulting in positive and preventative checks on population growth.

- **Positive checks:** These are not good, but include disease and famine, which increase the death rate and reduce the total population size.
- **Preventative checks:** These are things people can do to reduce population growth and prevent a crisis. They include delaying marriage and having smaller families. Malthus was very religious and did not support contraception, but modern-day neo-Malthusians, who follow his ideas, support large-scale contraception programmes, especially in LIDCs.

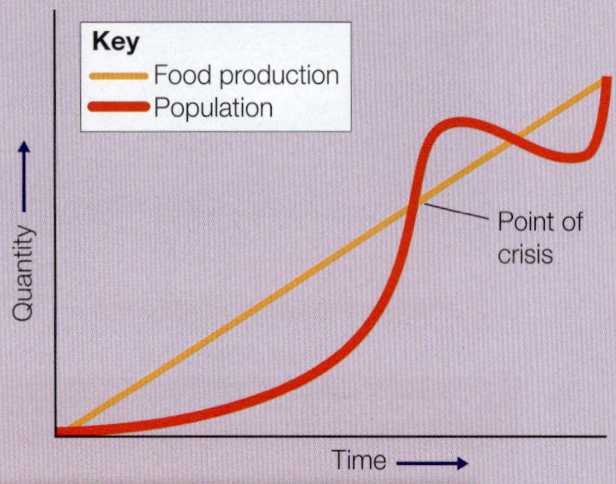

B Malthus's theory of population and food supply

A *Can we provide for the world's population?*

Boserup's theory

In 1965 the Danish economist Esther Boserup outlined a different relationship between population growth and food supply. She argued that when population growth begins to approach the limits of food production (Malthus's crisis point), this creates an incentive for people to invent new technology and better farming techniques. Examples include irrigation to divert water onto previously unused land, chemical fertilisers and higher yielding crops to grow more food on the same amount of land (see spread 8.13). These improvements allow more food to be produced and population growth can continue.

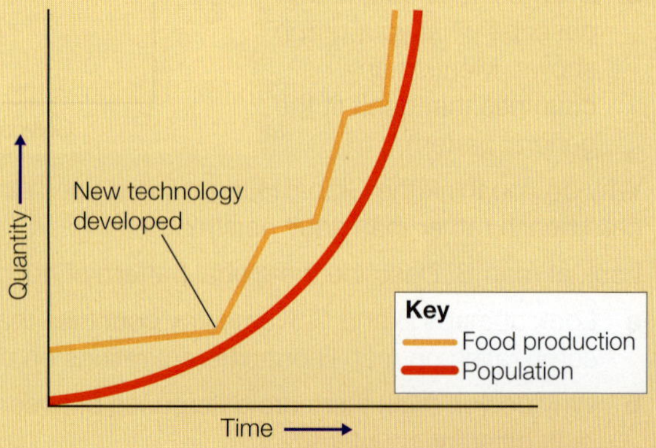

C *Boserup's theory of population and food supply*

D *Global food production and population growth since 1992*

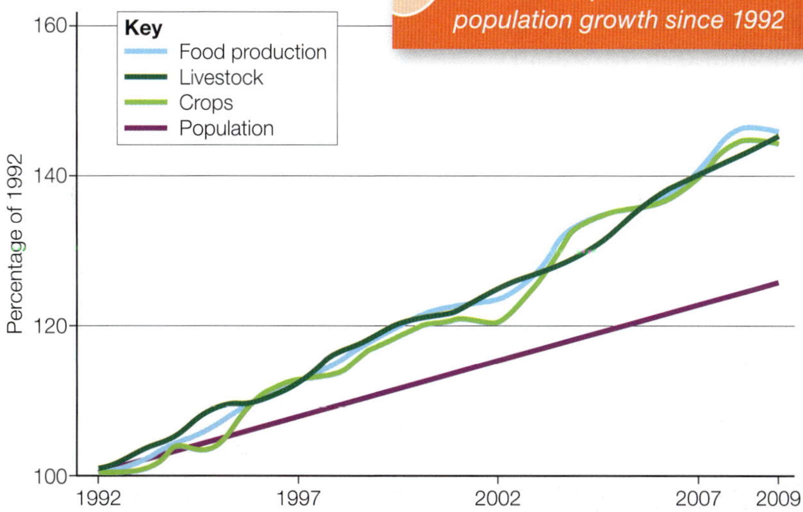

E *Boserup argued that technology increases farm production*

Activities

1. Look at cartoon A. Explain what you think its message is.
2. Copy graph B. Add these labels at appropriate places on your graph:
   - Food production increasing arithmetically (1, 2, 3, 4…)
   - Population increasing geometrically (1, 2, 4, 8…)
   - Widespread famine – a positive check
   - Preventative checks needed
3. **a** Write a definition of 'positive check', including one example.
   b Write a similar definition of 'preventative check'.
4. Copy graph C. Add these labels to your graph at places where they might have happened:
   - Population size approaching crisis point
   - New technology like ploughs are invented, increasing food production
   - Better seeds are developed, further increasing food production
5. **a** Sort these statements into the correct order to show Boserup's theory of the relationship between population and food supply.
   - Population increases.
   - Farming technology improves.
   - Population continues to increase.
   - More food is needed.
   - More food is produced.

   b Draw a flow diagram using similar statements to illustrate Malthus's theory of the relationship between population and food supply.
6. Copy the statements below. Decide which is most likely to be said by:
   a someone who supports Malthus's theory
   b someone who supports Boserup's theory.
   - People should be encouraged to marry later.
   - I am optimistic about the future.
   - People are a valuable resource.
   - Famine is inevitable.
   - We will always develop technology to solve our problems.
7. Look at graph D. Does it suggest Malthus's or Boserup's ideas are correct?
8. Some scientists think that events like the 2007 food riots (see spread 8.1), wars in the Middle East and the ebola outbreak that started in 2014 show that we are approaching a crisis like Malthus predicted. Do you think they are correct? Give reasons for your answer.

Go further

Use the internet to find out about a bet in 1980 between Julian Simon (a Boserup supporter) and Paul Ehrlich (a Malthus supporter). It cost one of them more than $500!

Practice question

Compare Malthus's and Boserup's theories of the relationship between population growth and food supply. (6 marks)

8.8 Food security in Tanzania

On this spread you will study a range of statistics to find out how hunger and food security have changed in Tanzania.

Introducing Tanzania

Tanzania (map A) is one of the poorest countries in the world and has a population of 51 million. Most people live in rural areas (photo B) but 5 million live in the largest city – Dar es Salaam (photo C).

A *Tanzania is a neighbour of Zambia (see Chapter 6), in East Africa*

B *A rural village in Tanzania*

C *Dar es Salaam: Tanzania's largest city*

Food and hunger indexes

Tanzania is ranked 98th out of 109 countries in the Global Food Security Index (see spread 8.6) with a score of 33.7 out of 100 (the lower the score, the less food security).

In the Global Hunger Index (GHI) (see spread 8.6) Tanzania is ranked 89th out of 116 countries with a score of 28.7 (the higher the score, 0–100, the more hunger) and a rating of 'serious'. Graph D shows Tanzania's GHI with one of the four indicators used to calculate the GHI.

Undernourishment

Over the last couple of decades the change in the percentage of people in Tanzania who do not have enough to eat is quite different from the trend in Africa as a whole and in the rest of the world (graph E).

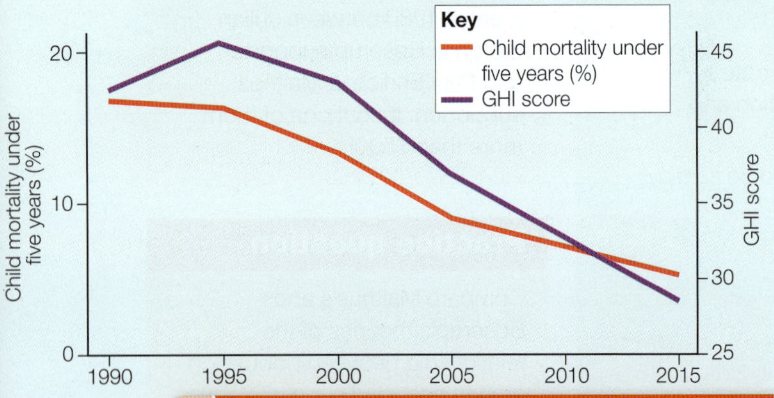

D *GHI in Tanzania, with the percentage of children under five who are underweight, 1990–2014*

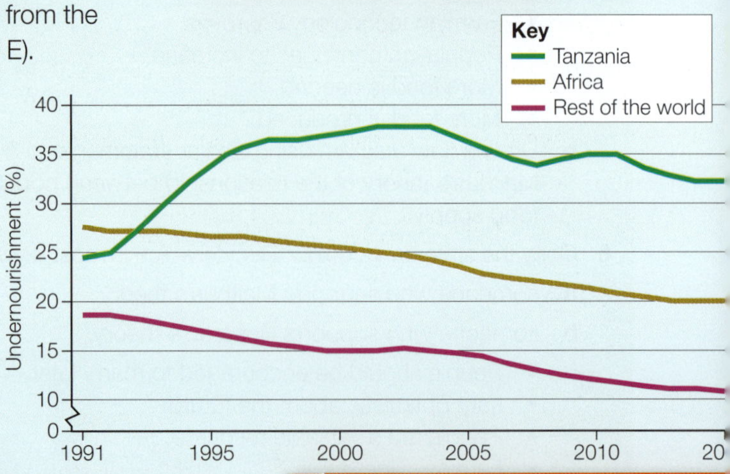

E *Change in the percentage of people who are undernourished*

Distribution of food

Graph F shows how well food is distributed in Tanzania, Africa and the rest of the world. It does this by measuring whether there is sufficient food available. If the percentage is over 100, there is more than enough food for everyone to have the minimum daily amount of food needed. Any hunger would be because the food is not shared out equally between the people who need it. If the percentage is below 100, there is not enough food for everyone.

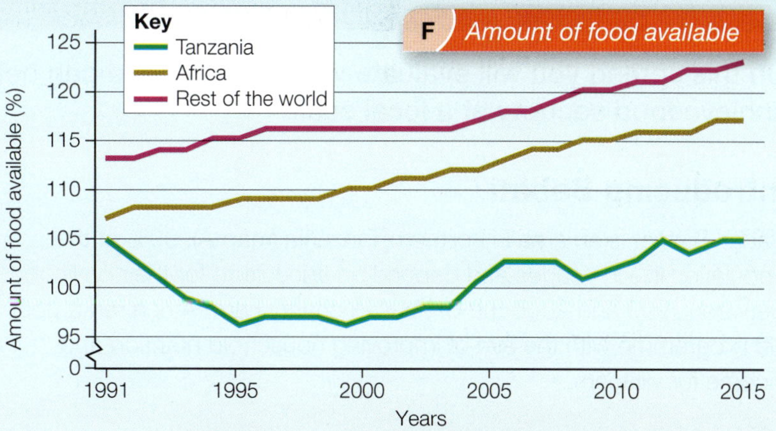

F *Amount of food available*

Calorie intake

Tanzania has a similar sized population to Uganda and Kenya. The average number of calories per person per day in Tanzania has gradually been rising compared to these neighbours (table G), but it is still below Uganda.

Year	Tanzania	Kenya	Uganda
1964	1696	2203	2262
1969	1704	2347	2357
1974	1897	2327	2349
1979	2228	2326	2126
1984	2248	2044	2167
1989	2283	2079	2303
1994	2019	2022	2217
1999	2056	2105	2251
2004	2169	2018	2347
2009	2137	2092	2260

G *Average calorie supply per person per day*

Poverty as a measure of food security

In countries like Tanzania many people live on less than \$1.25 a day (table H). They might be able to grow what they need to eat, but poverty means they have very little money to spend on food or seeds.

Year	Percentage of population
1992	72.0
2000	84.2
2007	67.9
2012	43.5

H *Percentage of the population in Tanzania living on less than \$1.25 a day*

Activities

1. Look at photos B and C. Where do you think food insecurity is more likely? Give your reasons.
2. Look at graph D. Describe the changes in Tanzania's GHI score.
3. Look at graph E.
   - **a** Has the percentage of undernourished people in Tanzania increased or decreased since 1991? Write a few sentences describing this change, including data in your answer.
   - **b** Describe how the percentage of undernourished people in Tanzania has changed compared with Africa as a whole.
4. Look at graph F.
   - **a** In which years was there not enough food to feed people in Tanzania?
   - **b** Write a paragraph comparing changes in the amount of food available in Tanzania, Africa and the world between 1991 and 2015.
5. Look at table G.
   - **a** Draw a graph to compare changes in the average calorie supply per person per day in each country between 1964 and 2009.
   - **b** Calculate the mean and median daily calorie intake for each country over the same period.
6. Look at table H. Draw a line graph to show how the percentage of people living on less than \$1.25 each day is changing. Be careful because the years in the table are not equally spaced.
7. Look at sources D–H again. Which do you think is the best indicator of food insecurity in Tanzania? Give reasons for your answer.

Practice question

For a country you have studied [Tanzania], describe how one piece of data shows food security changing over time. (4 marks)

Goat Aid in Babati, Tanzania

On this spread you will evaluate whether Goat Aid can help achieve food security at a local scale.

Case study

Introducing Babati

Babati District is an area in northern Tanzania (map A). 90% of the population live in villages and depend on agriculture for their livelihood. Between 1999 and 2006 the UK-based charity Farm Africa ran a goat aid programme with the aim of improving household nutrition and income for villagers.

Goat Aid

The goat aid programme is an example of a **bottom-up aid** project (see spread 6.17 for another in Zambia). After talking to villagers about their needs, Farm Africa imported Toggenburg goats at a cost of £400 each. In total £200 000 was invested. Toggenburg goats were chosen because they produce up to 3 litres of milk a day. Before the goats were imported, the villagers were trained in how to keep and care for the goats, including how to treat simple diseases.

The goats were given to small groups of villagers 'on credit'. This means that the cost had to be repaid. Farm Africa believes that if goats are given for free, people will not value and care for them. Villagers in Babati soon valued the goats because of all the benefits they bring (table D and photos B and E).

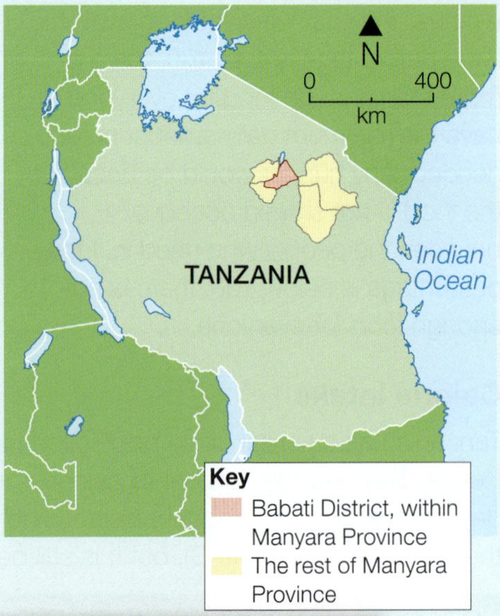

A Babati District in northern Tanzania

B Toggenburg goats

I was trained by Farm Africa on dairy goat management and breeding. Every year I get an average of 16 young goats. I sell some and have used the money to buy an ox cart and two bulls for tilling my land, iron sheet and bricks for my house, banana seedlings and equipment for spraying my animals against ticks. The bulls have reduced the time for tilling my 1.4 hectare farm from three days to two hours. I regularly have meat for my family that I could not afford before.

The milk from my goats gives my family a more nutritious diet and the goats' manure has improved my crops. The money from selling spare milk, butter and cheese, and my extra crops, has helped me to build a modern house and sheds for my animals. It also means I can afford to send my children to school so they get a good education. I beg my fellow farmers to form more dairy goat groups so they can improve their families' nutrition and income too.

C Villagers explain how Goat Aid has changed their lives

Spending power

Use	Percentage of income spent
Buying food	11.3
Paying medical bills	19.1
Paying school fees	25.2
Building and repairing houses	12.2
Buying land	0.9
Buying consumer goods	20.0
Improving crop farming	5.2
Increasing flock size	5.2
Goat house repair	0.9

D *How villagers spent the extra income gained from owning goats*

E *Villagers improve their farming methods with their extra income*

Assessing success

One way of judging how successful Goat Aid has been is to compare the incomes of farmers in the scheme with farmers who are not in the scheme. Table F shows the profit farmers made from farming and other business activities in Tanzanian schillings. One US dollar is worth about 2000 schillings.

Activity	Profit: farmers in the Goat Aid scheme	Profit: farmers not in the Goat Aid scheme
Crop production	998 496	444 580
Animals	384 692	190 950
Other business activities	820 666	128 350
Total	2 203 854	763 880

F *Income for dairy and non-dairy goat farmers*

Criticisms

Despite many benefits, there are some criticisms of small-scale projects like this:

- Goats require lots of water, which is a scarce resource.
- Goats' hooves and grazing can damage the land and lead to desertification.
- Veterinary bills can be expensive if goats get sick.
- Small-scale projects are not ambitious enough and have little impact on a country's food security.
- The number of hoofed animals in Africa has increased from 275 million in 1961 to 655 million in 2005. During that time poverty has increased not decreased!

Activities

1. **a** Why is it sensible to develop aid projects through talking to villagers?
   b Do you agree with Farm Africa that aid like this should not be given for free? Give reasons.
2. Look at the two speech bubbles. List three benefits that having goats brought to each villager.
3. Look at table D. Draw a bar graph or pie chart showing how villagers spent their extra income from the goat aid programme.
4. Look at table F. Why do you think farmers in the goat scheme made more profit from their crops compared to farmers not in the scheme?
5. Look at sources C–F. Draw a mind-map of the social, economic and environmental advantages and disadvantages of Goat Aid in Babati.

Go further

Use a search engine to find out what types of places Farm Africa believes Goat Aid is not suitable for.

Practice question

For a country you have studied [Tanzania], describe the key features of an attempt to achieve food security at a local scale. (6 marks)

Tanzania's bid to grow all its own wheat

On this spread you will evaluate the effectiveness of a past attempt to achieve national food security in Tanzania.

Case study

Grow your own

In 1967 the Tanzanian president argued that the country should grow all its own food. The need for food security became even more important after a severe drought in 1973 and 1974. By this time, 90% of Tanzania's maize and 80% of its wheat was imported. In 1975 Tanzania relied on emergency food aid shipments for the first time in its history.

Tanzania–Canada Wheat Programme

Tanzania's president asked Canada for help in growing wheat because Canada had expertise in growing large amounts of wheat using modern technology. Between 1968 and 1993 Canada provided $95 million of aid.

The wheat programme covered 26 400 hectares (graph B) in Hanang District in northern Tanzania (photo A). That is the same as 34 320 Wembley football pitches! Canada helped to develop suitable seeds and provided expertise, training, chemical fertilisers and machinery, like tractors. At first the chemicals and equipment were free, but eventually Tanzania had to pay for them.

However, the Hanang Plains were previously used by the Barabaig people, who are a nomadic tribe, moving from season to season. Forcing the Barabaig off their land caused conflict and has been controversial.

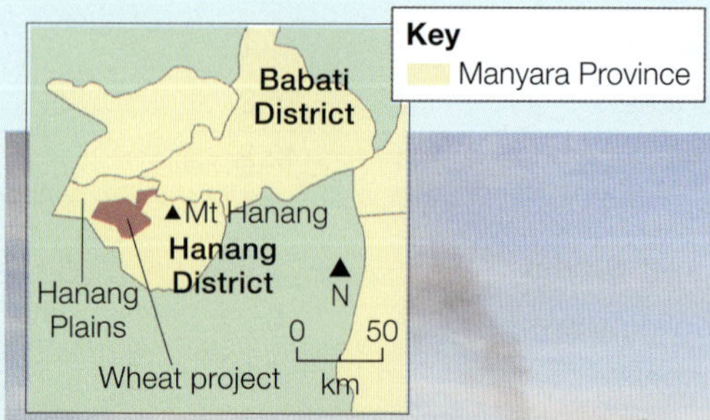

A *The Hanang Plains now look like the Canadian wheat prairies*

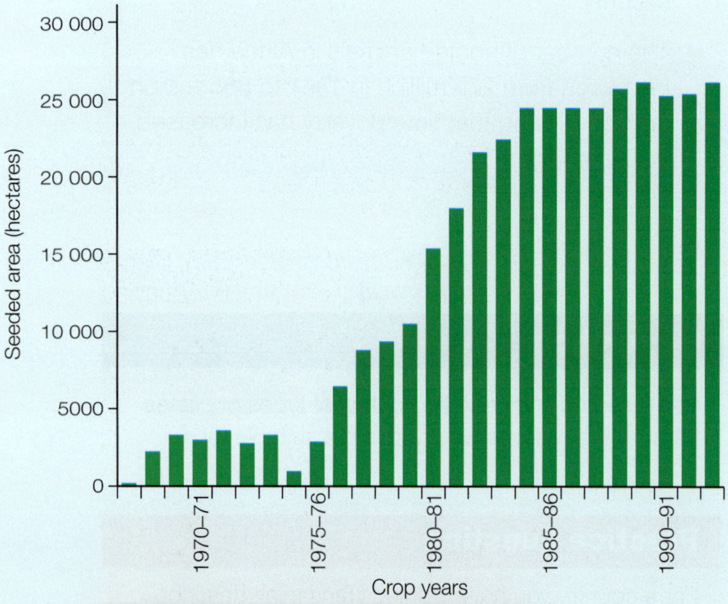

B *How the area under wheat production increased, 1967–1993*

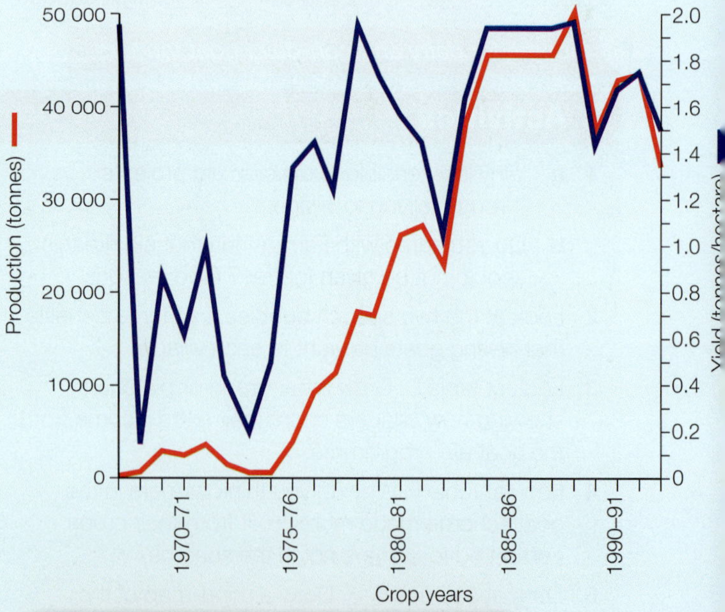

C *Wheat production and yield at the Hanang Wheat Complex, 1967–1993*

Impact on the Barabaigs

The Barabaigs used to graze their cattle on the Hanang Plains. The lakes there provided one of the few sources of water. People say Barabaig villages were burnt down and Barabaigs were raped, beaten, fined and imprisoned as they tried to carry on their traditional way of life. Critics say the wheat programme threatened the livelihoods of 40 000 people.

Some say the project was a success

- It provided about 60% of all Tanzania's wheat (graph C).
- 121 Tanzanians received training in wheat production.
- 150 mechanics gained skills in maintaining farm machinery.
- Up to 400 people worked on the farms.
- Road, rail and electricity connections were improved.
- In the 1992 drought, Tanzania was the only southern African country not to rely on food aid.

Others say the project was a failure

- The yield was mostly low (graph C). It would have been cheaper to import wheat.
- Only one crop was grown, so biodiversity and soil fertility decreased.
- After harvesting, rain washed the topsoil away.
- Most Tanzanians eat maize and cannot afford bread made from wheat. Low-technology maize production might have been better.
- Tanzania could not afford spare parts and fuel for the tractors and combine-harvesters. Many were used for spare parts or just rusted unused.
- Tanzania had to buy spare parts from Canada. The biggest winners were the Canadian manufacturers.
- Only a few jobs were created.
- The food security of the Barabaigs got worse.

Activities

1. Write a speech bubble for the president, explaining why Tanzania needed to grow its own wheat.
2. Create a factfile showing the key details of the wheat programme. Include details about: the dates, costs, size, aid from Canada, conditions on the aid, jobs created, amount of wheat produced.
3. Look at graphs B and C.
   - **a** Describe how wheat production changed over time.
   - **b** Explain why the amount of wheat produced sometimes increased, even when yields decreased.
4. **a** Copy the statements below. Is each one a fact or an opinion?
   - The wheat programme was the best way of improving Tanzania's food security.
   - 60% of Tanzania's wheat was grown in Hanang District.
   - Tanzania received modern farm technology through the project.
   - The project cost Canada $95 million.
   - Improving maize production would have been more appropriate.

   b Explain whether you agree or disagree with each opinion.
5. Imagine you are interviewing a Barabaig. Write five questions you would ask them about the impact of the wheat programme on their lives.

Go further

Research further the impact of the wheat programme on the Barabaigs. Try to find answers to your questions in activity 5.

Practice question

Describe one past attempt to provide food security at a national scale. How effective was this? (8 marks)

Tanzania's current bid for food security

On this spread you will learn about a large project aimed at improving national food security in Tanzania.

Case study

Southern Agricultural Growth Corridor of Tanzania

The Southern Agricultural Growth Corridor of Tanzania (SAGCOT) is a large project started in 2010. It aims to improve farming in Tanzania in a '**growth corridor**'. This strip of land stretches from east to west (map A). The land is very fertile and connected to the port at Dar es Salaam by main road and the TAZARA railway (see spread 6.13).

The Tanzanian government, foreign governments, aid agencies and TNCs are investing millions of dollars to improve infrastructure like roads, railways and irrigation. They want to develop a modern agricultural economy, creating six key cluster areas with better connections to national and world markets. This will help food security by producing more food. It will also earn money by selling food to other countries. It is claimed that by 2030 this project will create thousands of jobs and lift millions out of poverty (map A).

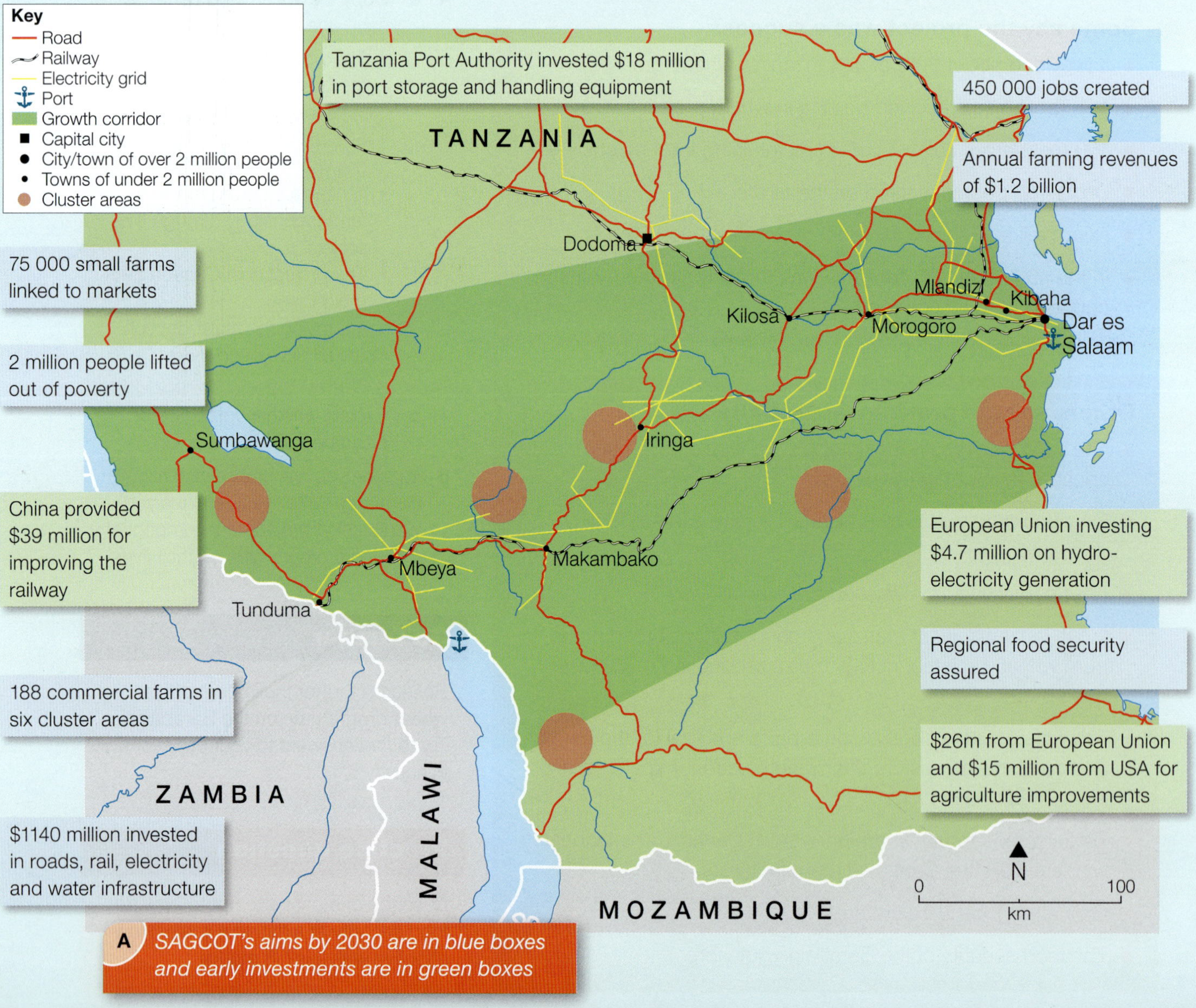

A *SAGCOT's aims by 2030 are in blue boxes and early investments are in green boxes*

Hub and out-grower model

Food security will be improved through a hub and out-grower model. The idea is that improvements to the infrastructure will attract big commercial farms to the six cluster areas. Each farm will act as a hub and improve things like irrigation, seeds, fertiliser, local transport and storage facilities. Small farmers (out-growers) working on their own land nearby will also be able to use these. They will also be helped by expertise and training from the commercial farms (diagram B).

B *Hub and out-grower model*

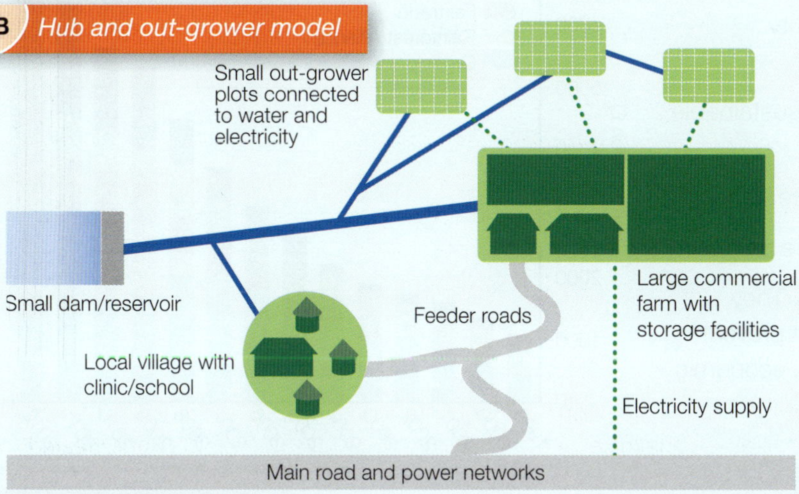

Early successes

One early success is the Kilombero Plantation. A commercial farm has invested millions of dollars in tractors, irrigation, a rice mill and storage. The central farm doubled its rice yield. In addition, 7300 other rice growers in 11 villages are now connected to better facilities and have increased their rice production. Some farms produced eight times more rice! As well as increasing the amount of rice grown, they have also been able to get a better price by having improved access to markets.

Four developments near Morogoro (map A) have also had success. Two tobacco-processing factories have been built, offering advice to local farmers and a market for their tobacco. A sunflower-processing plant has increased farm incomes and a factory making farm equipment has increased its output.

Criticisms

The SAGCOT project has faced some criticisms, including:

- Most of the money invested benefits large commercial farms.
- A lot of promised investment has not actually been given.
- Small landowners have not been involved in decision-making.
- Nomadic tribes have lost access to water for their animals.
- Some small landowners have lost their land to make way for big plantations.

Activities

1. Look at map A. Give two reasons why the shaded green area was chosen as a growth corridor.
2. Create a poster that the Tanzanian government could use to attract a TNC to invest in a commercial farm in the SAGCOT area. It should illustrate the benefits for the TNC.
3. Study diagram B. Order the statements below to create a flow diagram of the hub and out-grower model.
   - Local farms connect to the irrigation and electricity on commercial farms.
   - Government improves the main roads and electricity supplies.
   - Commercial farms are attracted to the area because it has better infrastructure.
   - Small farmers sell spare food to national markets.
   - Commercial farms sell food to national and international markets.
   - Commercial farms improve the local infrastructure, such as irrigation.
   - Better infrastructure and access to fertiliser mean small farmers grow more food.
4. Explain how SAGCOT affects food security for:
   - **a** people who live in urban areas like Dar es Salaam
   - **b** small farmers
   - **c** nomadic tribes.
5. Do you think SAGCOT is making any of the same mistakes as the Tanzania–Canada Wheat Programme (see spread 8.10)? Give reasons for your answer.

Practice question

For a current national food security scheme in a country you have studied [Tanzania], explain two ways food security will be improved. (4 marks)

8.12 Attempts to achieve food security – 1

On this spread you will learn how choosing Fairtrade products and eliminating food waste can make food production more sustainable.

You can make a difference

Ethical consumerism means buying products that have positive social, economic or environmental impacts. It can also help achieve food security. Examples might include:

- Rainforest Alliance products, promoting sustainable agriculture and protecting biodiversity
- Freedom Foods, which ensure animal welfare standards
- **Fairtrade** products, ensuring farmers are paid fairly.

Ethical products are often more sustainable. They minimise negative impacts today and do not prevent future generations from meeting their social, economic and environmental needs.

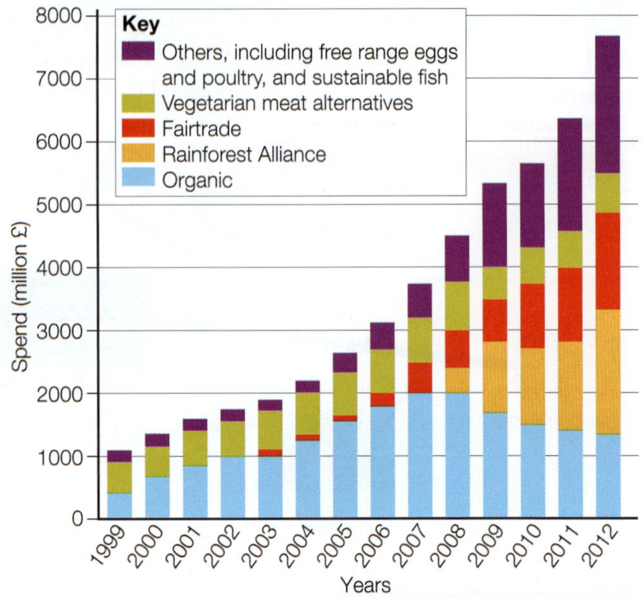

A *Sales of ethical products in the UK*

Making trade fairer

Fairtrade is a global movement that gives farmers a fairer price for their products. The FAIRTRADE Mark was developed in 1994 and shows that the product has been certified to meet standards set by Fairtrade International. Today there are over 4500 Fairtrade products on sale, including bananas, chocolate, tea and coffee (graph A).

In order to achieve Fairtrade certification for their products, farmers organise into a **cooperative** (a group that works together). If a product has the FAIRTRADE Mark, it means that:

- Farmers have received the Fairtrade minimum price, which should cover the cost of sustainable production. This means farmers can plan for the future knowing what their income might be.
- Farmers have received the Fairtrade Premium, an additional payment on top of the minimum price. This is paid to the co-operative, which decides how to use it to benefit the community. It can be used on school buildings, medical facilities, projects to improve farming practices or anything else that is important to the community (poster B).
- Farmers are trained in better farming methods and, in order to achieve the FAIRTRADE Mark, they must stop using the worst chemical fertilisers and pesticides.
- Workers' rights are protected and their working environment is safe.
- Farmers are able to connect to international markets, allowing them to sell more produce.

Fairtrade can be successful at improving incomes for local farmers, but it is also important for national economies. In Caribbean countries like the Windward Islands, Fairtrade bananas make an important contribution to GNI and provide jobs in rural areas.

B *Poster promoting Fairtrade tea*

Food waste

About one-third of all food grown for humans gets lost or is wasted. This is 1.3 billion tonnes of **food waste** a year. Food is wasted at a number of stages in the food supply chain (diagram C).

In LIDCs most wastage happens in the early stages of the food supply chain, but in ACs most waste takes place at the retail and consumer stages (see spread 11.4).

Ethical choices to reduce food waste

Buying local: Locally grown food is less likely to be wasted during transportation.

Eating ugly: Many people in ACs only like to eat perfect-looking fruit and vegetables, so lots of food is wasted. Most is thrown away during harvesting, but food can get to supermarkets before being rejected. The Ugly Food movement campaigns for supermarkets to sell imperfect food.

- When the French supermarket Intermarché started selling 'inglorious' fruit in 2014, it received 500 000 Facebook likes in three days (advert D).
- In Britain, Asda started selling 'wonky' vegetables in some stores in 2015 in a campaign supported by a famous TV chef.

Benefits of reducing food waste

Reducing food waste means that farmers receive more income because more of their produce is sold. With extra income, farmers in LIDCs and EDCs could afford a better and healthier diet. Reducing food waste is also better for the environment because growing food that is not consumed wastes energy and creates extra CO_2 emissions.

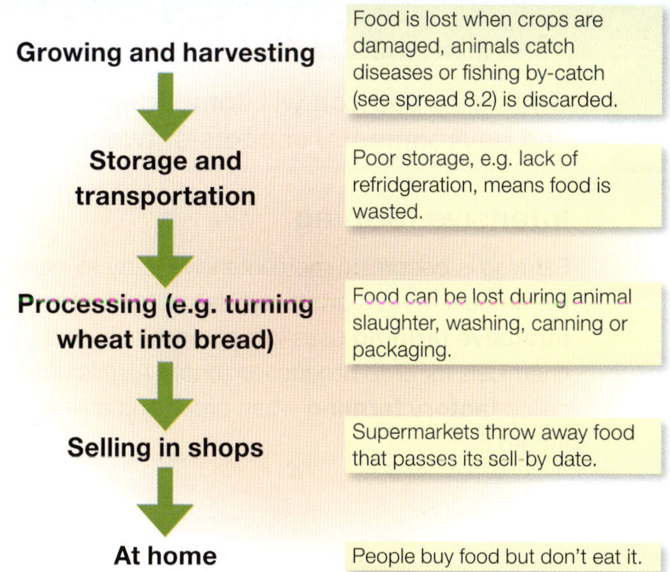

C *Waste occurs at all points in the food supply chain*

D *An advert for ugly apples*

Activities

1 Write your own definition of the term 'ethical consumerism'.

2 a Look at graph A. Describe the change in spending on ethical products since 1999. Include statistics and refer to specific types of product.

  b Which ethical product type was most money spent on in 2012?

3 Study poster B and the section on fairtrade.

  a List one social, one economic and one environmental benefit of fairtrade.

  b Explain how fairtrade helps farmers now and in the future.

  c Fairtrade products cost a little more than others. Write a script for a radio advert persuading people to buy Fairtrade products.

4 Study diagram C. Is it more important to reduce waste at the start or end of the food supply chain? Explain your ideas.

5 Look at poster D and the section on ethical choices.

  a Explain how selling ugly food improves incomes for farmers.

  b In what ways is buying ugly food more environmentally sustainable?

Practice question

Using an example, explain how ethical consumerism can improve food security. (6 marks)

8.13 Attempts to achieve food security – 2

On this spread you will compare the social, economic and environmental impacts of two farming methods.

Intensive farming

Farming is becoming more intensive to try to increase the amount of food produced (see spread 8.2). **Intensive farming** uses lots of machinery and chemicals for crop production (photo A). It is often called **factory farming** when producing animals (photo B).

A *A cropduster sprays a banana plantation with pesticide*

Advantages of intensive crop farming

- Huge amounts of food are produced, helping to achieve food security.
- It makes the most of the land, giving very high yields.
- Fruit, vegetables and wheat are very cheap to produce.
- The food is sold cheaply in ACs, making a healthy, balanced diet more affordable there.
- Many countries have laws controlling the amount of pesticides used.

Disadvantages of intensive crop farming

- Chemical fertilisers, pesticides and herbicides pollute the environment (see spread 8.2).
- Chemical sprays are bad for the health of farm workers.
- People eat the chemicals if food is not washed properly. This may cause an increase in cancer.
- Single crops reduce the biodiversity of plants, animals and insects.
- Fertilisers are a major source of greenhouse gases (see spread 2.5).

The rise of agribusiness

Intensive food production is being taken over by **agribusiness**, mainly in LIDCs, with big companies buying huge amounts of land. More food is grown, but the main motive may be profit, not food security. The food is often exported or sold at prices local people cannot afford. Big companies leave less land for small farmers to grow food to feed their communities. This can lead to local food shortages.

B *Caged pigs in a factory farm*

Factory farming animals

- Animals are kept indoors in cramped conditions (photo B). Sometimes they spend their entire lives in a cage that is not big enough to turn around in, a situation now banned in the UK and the European Union.
- To stop disease spreading, animals are fed antibiotics. Some scientists think this is making antibiotics for humans less effective because bacteria are developing resistance to the antibiotics.
- Factory farms produce huge amounts of waste that can pollute the land and water supplies.
- The costs of production are reduced, so the products are cheaper and more people can afford them.

Organic farming

Organic farming bans the use of chemicals. To keep the soil fertile, farmers rotate a wide variety of crops so the same crop is not always grown in the same field. Compost from dead plants and animal manure is used as fertiliser. Animals are raised in more natural conditions (photo C).

There are disadvantages, but in the long run organic farming might be better because it doesn't have any of the negative effects of intensive farming.

C *Organic pigs that are free to roam*

Activities

1. a. Explain the disadvantages for people of using chemical sprays on crops.
   b. Explain why organic farming is better for the environment.
2. Look at graph A on spread 8.12. Why do you think the amount of money spent on organic food has declined since 2007?
3. Look at photos B and C, and the information about different types of farming.
   a. Create a table listing the advantages and disadvantages of organic farming.
   b. Colour code these as social, economic or environmental.
   c. Create and colour code a table for intensive farming in the same way.
   d. Look at your two tables. Explain which approach you think is more sustainable. Think about social, economic and environmental sustainability in your answer.
4. Assess the extent to which organic farming meets the three pillars of food security: food availability, food access and food utilisation (see spread 8.5).
5. Intensive farming produces a lot more food more cheaply than organic farming. Does the need for food security outweigh the disadvantages of intensive farming? Explain your ideas.

Practice question

Explain how environmentally sustainable one method of improving food production is. (4 marks)

Attempts to achieve food security – 3

On this spread you will develop an opinion about genetically modified crops.

Laboratory food

Genetically modified (GM) crops are an example of using technology to achieve food security. Genetic modification takes DNA from one species and inserts it into another species. For example, scientists are trying to develop frost-resistant strawberries by inserting DNA from the Arctic flounder fish (which can survive at very low temperatures) into strawberries. The aim is to produce a strawberry that can be grown at low temperatures.

A *GM crops are developed in laboratories*

Advantages of genetic modification

Genetic modification has many potential benefits.

- GM crops engineered to resist drought and frost will grow in places currently not suitable for food production.
- Crops can be modified with DNA harmful to pests and insects, which reduces the need for pesticides.
- Crops can be modified to be resistant to certain herbicides. So farmers can spray their crops with the herbicide to kill the weeds but not the crops. This increases the amount of food produced.
- Food with additional health benefits can also be engineered (panel D).

Lessons from the past: Green Revolution

Some concerns about GM crops come from the **Green Revolution**. In the 1960s scientists developed rice and wheat seeds that produced very high yields. They did this by cross-breeding rather than genetic modification and some results were very positive. In India rice yields increased by 30%. However, the new strains of rice and wheat were only suited to intensive farming. They needed lots of fertilisers, pesticides and water. Many poor farmers got into debt when one bad crop meant they could not afford the next year's seeds and fertiliser. Biodiversity also decreased as many large farms only grew the new types of rice and wheat.

Disadvantages of genetic modification

GM crops are not natural and some people believe there are many potential problems.

- GM crops might not be safe to eat because they have been artificially engineered.
- Their pollen might spread and contaminate other plants, which will permanently change natural plant species.
- There are concerns of 'superweeds' and 'superbugs' developing that are resistant to herbicides and pesticides.
- There are also concerns that biodiversity will decrease. Farmers might use GM seeds instead of a wider variety of traditional seeds. This will lead to less diverse ecosystems.
- GM seeds are made by a small number of TNCs that might control global food supply. Their main motivation is likely to be profit not food security.

B *Protesters destroying a field of GM crops*

GM seeds will provide far better productivity, better drought tolerance, and if the safety is proven, then African countries will be among the biggest beneficiaries. *(Bill Gates)*

GM foods currently available are not likely to present risks for human health. *(World Health Organisation)*

Genetically modified organisms can spread through nature and interbreed with natural organisms, thereby contaminating non-GM environments and future generations in an unforeseeable and uncontrollable way. *(Greenpeace)*

GM crops are taking us down a dangerous road, creating the classic conditions for hunger, poverty and even famine. Ownership and control concentrated in too few hands and a food supply based on too few varieties of crops are the worst option for food security. *(Christian Aid)*

With appropriate choices GM technology can offer one approach to sustainable farming by reducing the need for chemical inputs. This benefits the consumer, the farmer and the environment. (*Anne Glover, Professor of Biology, Aberdeen University*)

C *Experts can't agree*

Golden rice

500 000 children lose their sight every year because of a lack of vitamin A. This could be prevented through golden rice, which is genetically modified with added vitamin A. This shows the potential of genetic engineering to produce more nutritious food.

However, it might be better to focus on providing people with a balanced diet of fruit and vegetables. Golden rice encourages people to eat one food type, which will never be a healthy diet.

D *Golden rice promises health benefits*

Activities

1 Write your own definition of the term 'genetically modified crops'.

2 Make a copy of the opinion line below, making the arrow longer. Place each opinion in source C somewhere on the line. Judge how strong each opinion is – the stronger the opinion, the closer to the end of the line.

I totally agree with GM crops. ⟷ I totally disagree with GM crops.

3 Draw a mind-map of the possible social, economic and environmental advantages and disadvantages of GM crops.

4 Do you think GM crops are environmentally sustainable? Explain your ideas.

5 Place yourself somewhere on your opinion line. Explain why you have put yourself there.

Go further

There are other technological approaches to food security. **Hydroponics** involves growing plants indoors in water. Research the advantages and disadvantages of this approach.

Practice question

Using an example, evaluate the sustainability of a technological attempt to achieve food security. (6 marks)

Small-scale approaches to food security

On this spread you will learn how bottom-up approaches like urban gardens and permaculture can help to achieve food security.

Urban gardens

About 15% of the world's food is now grown in urban areas. **Urban gardens** are an example of a **bottom-up approach** to food security. Millions of extra small gardens could be just as effective at providing food security as large-scale intensive agriculture and expensive technological approaches. This is especially the case for the world's poor.

In poorer countries urban agriculture in small community gardens provides lots of healthy food that would be too expensive to buy (photo A). Some urban gardens grow up on rubbish dumps, improving the local environment. Urban gardens can also include livestock such as chickens.

However, there can be problems with urban gardens. Livestock can transmit diseases and dirty waste water is sometimes used to water the crops.

A *A community urban garden in a Nairobi slum*

A factory with a difference

Chongqing has a population of about 9 million people and is one of the fastest growing cities in China. Workers at one of its many factories have created a farm on its roof (photo B). The farm has an area of 10 000 square metres and factory staff work there in their breaks. The vegetables they grow are taken home for free. The farm has chickens, goats and geese, and is so big that a tractor is used to help with the farming.

Chongqing is not the only city where rooftops are being used to grow food. They are becoming more and more common across the world, e.g. in New York and Tokyo.

B *Factory workers exercise on the rooftop farm*

Permaculture – learning from nature

Think about a natural area, such as a forest, that you know. There are many different plant species living together. The forest and its plants survive and thrive without humans adding fertiliser to help plants grow or pesticides to kill weeds and bugs.

Permaculture is an approach to growing food that tries to achieve food security by copying nature. It brings two words together – 'permanent' and 'agriculture'.

C *A garden using the principles of permaculture*

Key ideas in permaculture

Diversity: Instead of growing a single crop, permaculture grows food by mixing many species together (photo C). Different plants use different nutrients so the soil does not become exhausted. They also produce food at different times of the year, giving a year-round supply of food.

Nature instead of chemicals: Different plants attract different insects. A variety of insects helps keep harmful bugs away without the need for chemical sprays. If a garden had a slug problem, permaculture would introduce a pond with frogs rather than use slug pellets. The variety of plants means diseases are less likely because diseases thrive when there is a single species to attack. Plants dying at different times of year put nutrients back into the ground without the need for chemical fertilisers.

Stacking: As in a rainforest, permaculture grows more food in the same amount of space by 'stacking' (diagram D). This means growing tall trees, small trees, shrubs and vegetables together to maximise the amount of food on the same area of land.

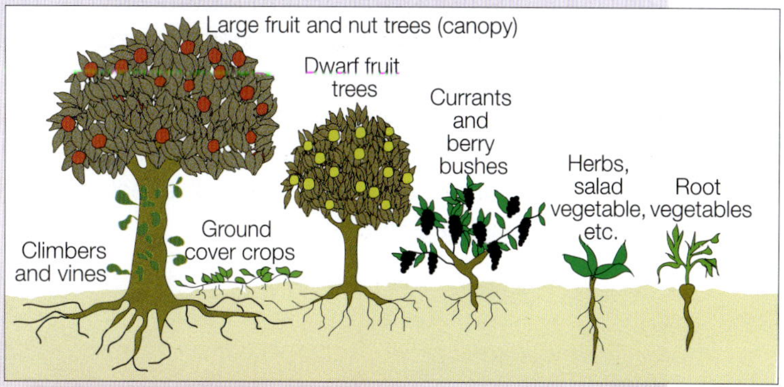

D *Stacking can grow more food in the same space*

Activities

1 Look at photo A. List the places on your journey to school that could be used as urban gardens.

2 a Draw a Venn diagram like the one below. Place each statement below about the impacts of urban gardens in the correct place.

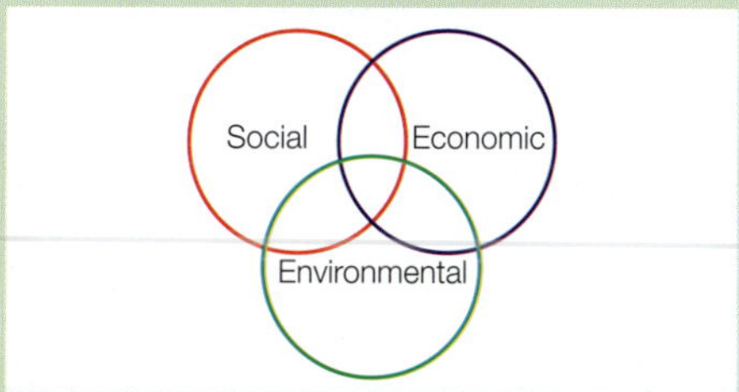

- The food grown is healthy and nutritious.
- Growing food does not cost much.
- Urban gardens create a safe place for children to play.
- Diseases can spread in urban livestock.
- Urban gardens create a community space.
- Extra food can be sold for profit.
- Urban waste is used as fertiliser.
- Urban gardens remove CO_2 from the atmosphere.
- Dirty waste water is used to water crops.
- Rubbish tips are turned into green spaces.
- Rooftop farms insulate buildings and lower heating costs.

b Highlight the positive and negative impacts in different colours.

c Using your diagram, explain whether you think urban gardens are a sustainable way to achieve food security.

3 a Write a definition of permaculture.

b Explain why permaculture does not need chemical fertilisers and pesticides.

c How does permaculture produce food in a sustainable way?

Your enquiry

Can we feed nine billion people by 2050?

You have been invited to speak at a meeting about food security. Write a two-minute speech, arguing one of the following points:

- Large-scale technological approaches to growing more food are the best solutions for feeding 9 billion people by 2050.
- Small-scale bottom-up approaches are the best way of feeding 9 billion people by 2050.
- A mixture of large and small projects are needed to feed 9 billion people by 2050.

In your speech try to:

- use some data to introduce the problems with achieving food security
- outline different theories on the issue
- refer to examples to add detail to your argument.

Hazards in the USA

On this spread you will investigate the location, frequency and impacts of natural hazards in the USA.

Links

This decision-making exercise links to other topics:

- Global Hazards (Chapter 1)
- Urban Futures (Chapter 5)
- Dynamic Development (Chapter 6)
- Changing Climate (Chapter 2).

USA – a hazardous country

The USA is one of the world's most hazardous countries. It can suffer from a range of natural weather hazards and tectonic hazards that are, all too frequently, deadly for people and costly to the economy. Sometimes these hazards are dramatic, like the tornadoes that regularly hit the mid-west of the USA (photo A). Other hazards are less dramatic but just as deadly, like the heatwaves that affect many parts of the USA in summer. While some hazards affect the whole country, others affect particular areas (map B).

A *A tornado hits the mid-west of the USA*

B *Where hazards happen in the USA*

Frequency and impacts of hazards

The frequency and impacts of hazards vary. Although 11% of the world's active volcanoes are in the USA, eruptions do not happen very often. The most recent, deadly, eruption in the USA was Mt St Helens. It erupted explosively in 1980, killing 57 people (photo C).

Other hazards occur much more frequently. Partly because of its size, the USA suffers more from tornadoes than any other country. On average, there are over 1000 tornadoes in the USA each year. Some of these can be deadly and cause huge amounts of damage (charts D–F).

C *Mt St Helens erupted in 1980*

Key
- Drought
- Earthquakes
- Extreme temperatures (heatwave and snow)
- Floods
- Storms (hurricanes and tornadoes)
- Wildfires
- Other

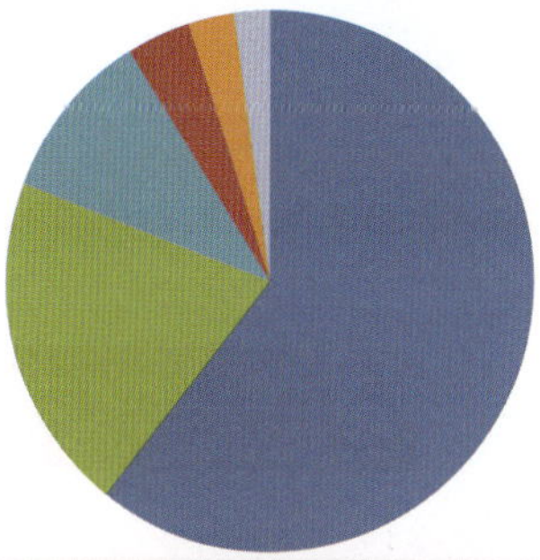

D *Frequency of hazards in the USA, 2005–2014*

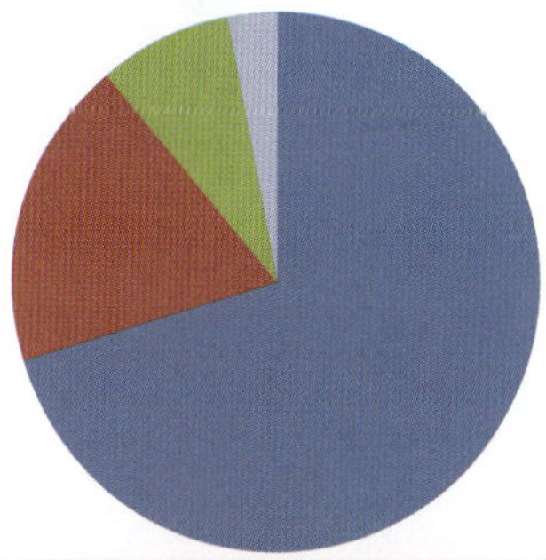

E *Deaths caused by hazards in the USA, 2005–2014*

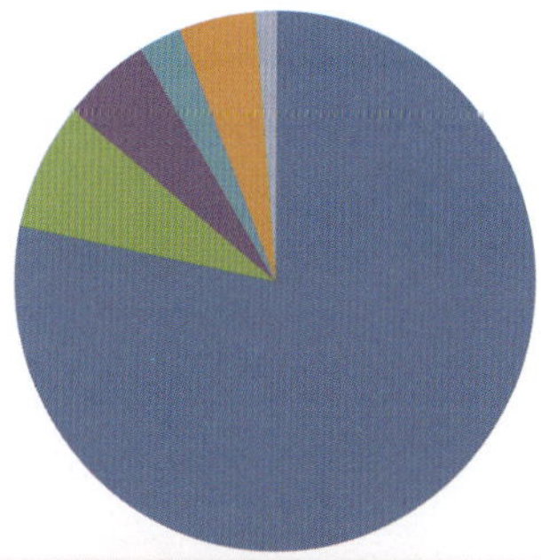

F *Economic costs of hazards in the USA, 2005–2014*

Activities

1. Sort all the hazards mentioned on this spread into two groups – weather hazards and tectonic hazards.
2. Look at photos A and C. Using all the information on this spread, compare tornadoes and volcanic eruptions in terms of:
   - **a** type of hazard
   - **b** location
   - **c** frequency
   - **d** impacts.

   You could complete a table to help you.
3. Look at map B.
   - **a** Describe the distribution of six hazards shown on the map. Write a sentence about each hazard.
   - **b** Try to explain their distribution, e.g. *Wildfires happen mainly in the west where there are more mountains and forests.*
4. Compare graphs D–F.
   - **a** Which is the most serious hazard in the USA? Give three reasons.
   - **b** Name a hazard that happens quite frequently, but with low economic cost and no deaths. Suggest why this is.
   - **c** Name a hazard that happens less frequently, but causes more deaths. Suggest why.
   - **d** Name a hazard that happens less frequently, but has more economic costs. Suggest why.

Go further

How hazardous is the UK compared to the USA? Find out more about similar hazards in the UK to answer this question.

Your decision-making enquiry

How should the USA prepare for the next natural disaster?

As you work through this chapter, make notes about anything that will help you to answer the question. You will bring your ideas together to do an activity at the end of the chapter.

From hazard to disaster

On this spread you will find out how hazards turn into disasters, using Hurricane Sandy as an example.

Hurricane Sandy, 2012

Hazards turn into disasters when they affect people. When Hurricane Sandy hit the US coastline in 2012, 73 people died and it caused $65 billion worth of damage. The hurricane made a direct hit on New York, the USA's largest city, with a population of over 8 million (photo A). If the hurricane had hit the coast further north or south, the death toll and damage could have been far less.

The fact that Hurricane Sandy was so costly could be seen as bad luck. But building a city there was bound to be risky because the east coast of the USA is known to be vulnerable to hurricanes. People can choose where they live and large numbers of Americans choose to live in places that are vulnerable to natural hazards (map B).

A Flooding in New York after Hurricane Sandy

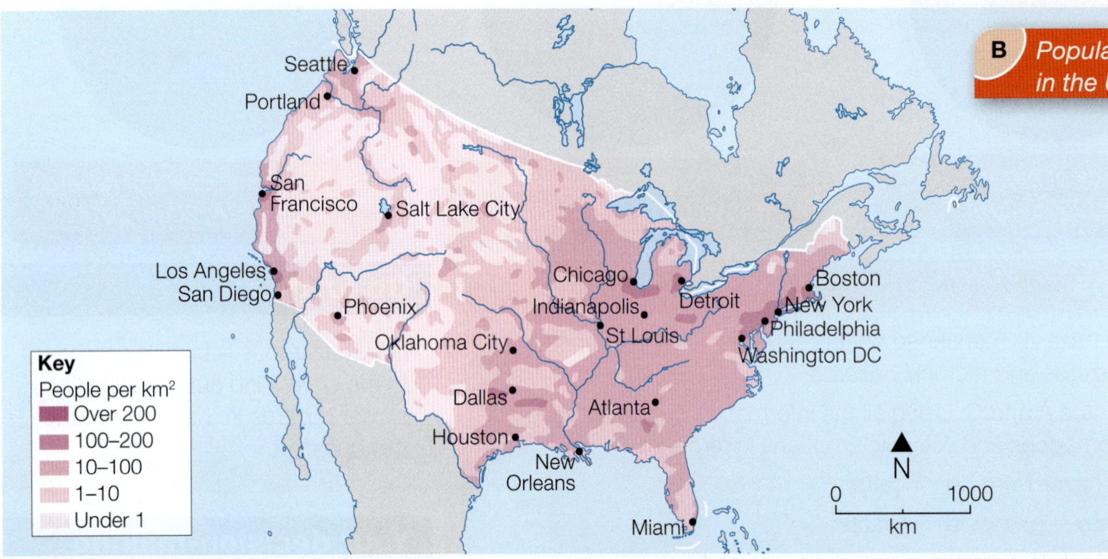

B Population distribution in the USA

Hazard risk in ACs and LIDCs

The USA, as an **advanced country (AC)**, has already done a lot to reduce the risks posed by natural hazards to human life. The death toll from disasters in the USA is lower than it would be if similar hazards struck a **low-income developing country (LIDC)**. However, the economic cost of disasters in the USA is higher. This is because the value of individual properties and the infrastructure is greater than it would be in an LIDC (factfile C).

C USA economic factfile

USA (2014 data)	
Population	324 million
Population density	34.6 people/km^2
GDP	$16 800 billion ($16.8 trillion)
GDP/capita	$53.042
Urban population	81%
Number of natural disasters (2005–2014)	22
Deaths from natural disasters (2005–2014)	427
Percentage of GDP spent on disaster relief	0.03%

Keeping track of Sandy

Hurricane Sandy was one of the most costly natural disasters ever. The hurricane developed over the warm tropical waters of the Caribbean Sea (see spread 1.6). It then tracked north, across several Caribbean islands, leaving a trail of destruction in its path (map D).

Sandy continued north across the Atlantic, building in strength, until it hit the US coast on 29 October. A storm surge was caused by the intense winds over the ocean. It led to severe flooding along the coast, including the city of New York. The hurricane decreased in strength once it moved inland towards Canada.

Sandy's impact varied across the countries that it hit (table E). As you might expect, by far the greatest economic cost was in the USA. The death toll was also high in the USA, with most of the deaths in New York. However, other countries with much smaller populations, notably Haiti, also experienced deaths. Haiti is the poorest country in the region. Like many LIDCs, it is less well prepared for natural hazards. In 2010, Haiti also experienced a major earthquake, when 230 000 people were killed.

D *Hurricane Sandy's path from the Caribbean to the USA*

Country	Population (millions)	GDP/ capita ($)	Deaths	Damage (million $)
Jamaica	2.8	8 893	1	$100 million
Haiti	10.5	1 703	54	$750 million
Dominican Republic	10.5	12 186	3	$30 million
Cuba	11.3	18 796	11	$200 million
The Bahamas	0.4	23 464	2	$700 million
USA	314.0	53 042	73	$65 000 million
Canada	35.5	42 753	1	$100 million

E *The impact of Hurricane Sandy by country*

Activities

1. Look at photo A. Why did Hurricane Sandy cost so many lives and do so much damage?
2. Look at map B and compare it with map B on spread 9.1.
   - **a** Compare the population distribution in the USA with the distribution of hazards. How safe are densely populated areas?
   - **b** Which is the most dangerous city in the USA? Give reasons for your answer.
   - **c** Which is the safest city? Again, give reasons.
3. Study the data in factfile C.
   - **a** How great is the impact of natural disasters in the USA?
   - **b** Why is the USA in a good position to respond to disasters?
4. Look at sources D and E.
   - **a** Describe Hurricane Sandy's track and its impacts in each country.
   - **b** Try to explain the varied impacts of Sandy in different countries.

Go further

Find out about any hurricanes, or other natural disasters, that have affected the USA since 2012.

- How many people died?
- What was the cost of the damage?
- Which other countries were affected?

Drought – a hidden disaster

On this spread you will find out more about one of the USA's worst natural disasters.

USA drought, 2012–2015

The greatest natural disaster in recent US history is probably one you have never heard of. It is the drought that has affected many of the southern and western states of the USA since 2012 (photo A). The worst affected state is California, where the drought continued into 2015.

The reason you probably have not heard of the drought is that there were no deaths, so there has been little media coverage outside the USA. Yet by 2015 the drought had already become the most costly disaster in US history.

A *A farmer surveys a field affected by drought*

The extent of the drought

A drought is a long period of abnormally low rainfall that is harmful to growing or living conditions. In 2012 drought covered 80% of the USA. Some areas have since returned to normal rainfall levels, but by 2015, 30% of the country (particularly in the south and west) was still in drought (map B).

2013 was the driest year on record for many places in the USA. There have been droughts in the past, notably in the 1930s and 1950s. Large areas in the mid-west experienced many years of drought then, in what came to be known as 'the Dust Bowl'. The most recent drought has lasted longer than either of the previous worst droughts (graph C).

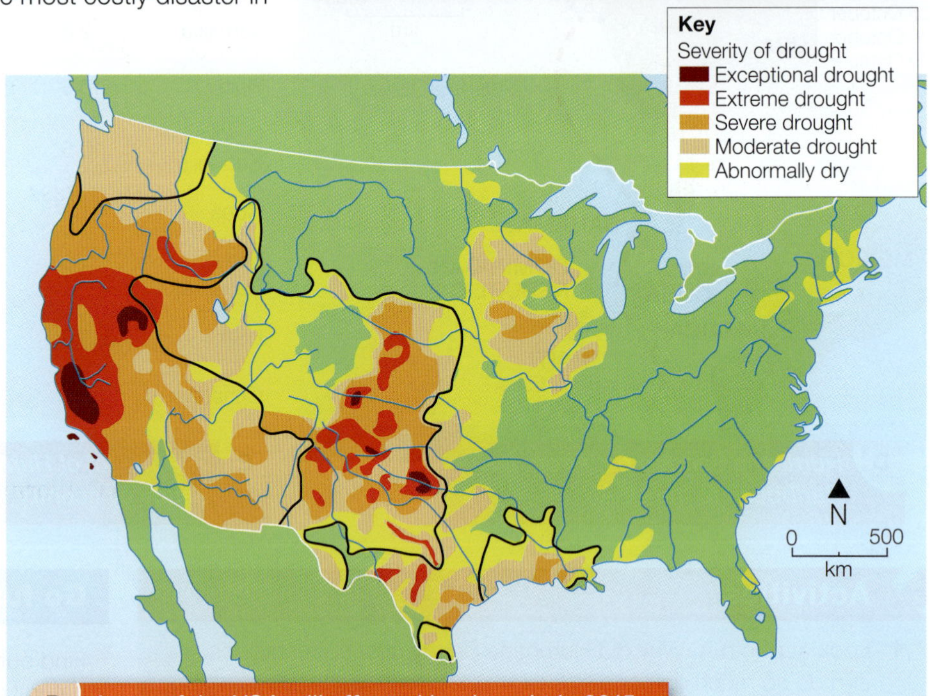

B *Areas of the USA still affected by drought in 2015*

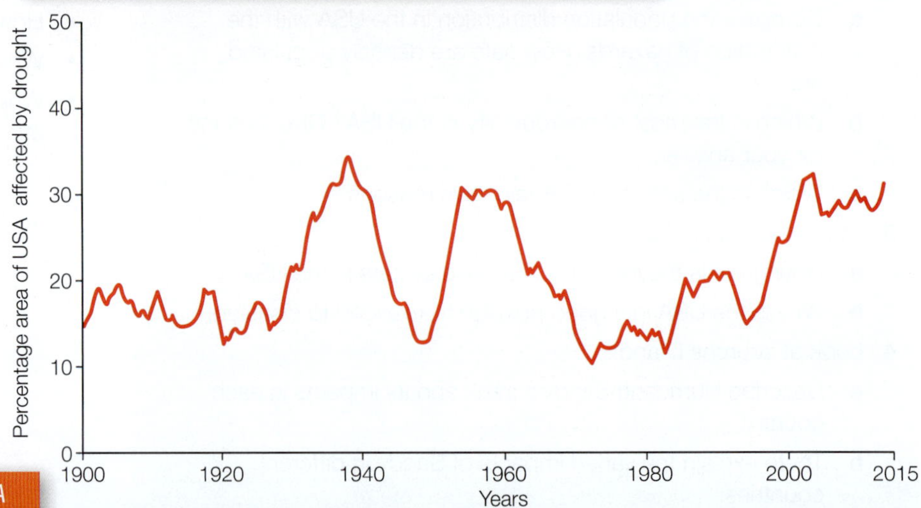

C *History of drought in the USA*

Causes of the drought

The reasons for the drought in the USA are not fully understood. Intense summer heat in the USA due to La Niña and warm, dry winters during El Niño are two possible causes (see spread 1.7). It is also possible that climate change has played its part. In the USA, the amount of rainfall is rising in the east but falling in the west. Increased temperatures due to global warming cause moisture to evaporate faster in places that are already dry.

D *Lawns in California may soon be a thing of the past*

Responses to the drought

In 2015 the US government announced financial help for those affected by the drought. It included farmers whose livelihood depends on water for growing crops and raising livestock. 80% of the water used in California is for farming. The government money was intended for:

- efforts to conserve water on farmland by reducing wind erosion and improving livestock access to water
- animal feed for livestock at risk of starvation
- drinking water for rural communities at risk of running out of water
- food banks for families in financial hardship because of the drought.

In addition, the governor of California imposed water restrictions to reduce consumption. For years, people in California, and other states in the USA, have defied the dry climate by continuing to water their lawns (photo D and article E).

California turns to fake grass in response to drought

Christopher Knight makes no apologies – he likes a green lawn. But he also wants to do his part to conserve water. The solution? Fake grass.

'It feels totally different', Knight marvels, stepping barefoot onto a deceptively life-like expanse of newly-installed plastic turf.

After four blistering years of drought in California, more people are doing it. The fake grass business is booming.

The benefits of fake grass are hard to deny. Live grass guzzles some 2200 litres per square metre annually, making the all-American lawn increasingly untenable in an era of skyrocketing water rates and excessive-use penalties.

E *Adapted from news reports, 30 May 21015*

Activities

1. Look at photo A. Is this your idea of a natural disaster? Explain your answer.
2. Look at map B. Describe the distribution of drought-affected areas in the USA. (Clue: find a map so you can refer to the names of the US states.)
3. Look at graph C.
   - **a** Describe the pattern of drought over time in the USA.
   - **b** How does the recent drought compare with the worst droughts in the past?
4. Look at photo D and article E. Does reducing water consumption by replacing grass lawns seem like:
   - **a** a sensible response to drought?
   - **b** an adequate response to drought?

   In each case, give reasons.

Go further

Find out if and when the USA drought that began in 2012 came to an end.

- How long did the drought last altogether?
- What was the final cost of the drought?
- What other impacts did it have?
- Apart from replacing lawns, has the USA made any other changes to prepare for droughts in future?

Waiting for the 'Big One'

On this spread you will examine the likelihood of a major earthquake in California and plan suitable responses to it.

The San Andreas Fault

Southern California lies on the San Andreas Fault, part of the boundary between the North American and Pacific plates (see spread 1.14). Earthquakes happen frequently along the fault. The last major quake near the fault was the 1994 Northridge earthquake, which killed 33 people and caused $40 billion worth of damage (photo A). The magnitude of the earthquake was 6.7 on the Richter scale.

A *Damage caused by the Northridge earthquake in 1994*

Predicting the next quake

The San Andreas Fault is due for another earthquake. It is predicted to be at least ten times more powerful than the Northridge quake, at about 7.8 on the Richer scale. Major earthquakes of this magnitude happen on the fault roughly every 150 years (map B). In 1906 an earthquake of magnitude 7.8 destroyed the city of San Francisco at the northern end of the San Andreas Fault.

The next 'Big One' is expected to happen at the southern end of the fault, where there hasn't been an earthquake for over 150 years. The city of Los Angeles is close to this section (diagram C). When the earthquake happens, it is likely to cause widespread damage and serious loss of life. In 1906, 3000 people died in San Francisco and 80% of the city was wiped out. The population of California has grown enormously since then and cities like Los Angeles are now many times bigger.

B *Historic earthquakes on the San Andreas Fault*

What will happen in the 'Big One'?

Geologists have estimated there is a 45% chance of a 7.5 magnitude earthquake, or larger, on the southern section of the San Andreas Fault within the next 30 years. The rocks on both sides of the fault here have become locked together. So tension between the rocks has increased. The force of the earthquake, when this tension is released, will be immense.

The consequences of the 'Big One', when it eventually happens, could be catastrophic (photo D).

- Water and sewage pipes will crack.
- Electric power will fail.
- The internet will be down.
- Major highways will be destroyed.
- Older concrete and brick buildings will collapse.
- Fires will burn as gas mains burst and cables break.
- Hospitals will be swamped with casualties (if ambulances can get to them).
- Little, if any, food or medicine will be able to reach the area.

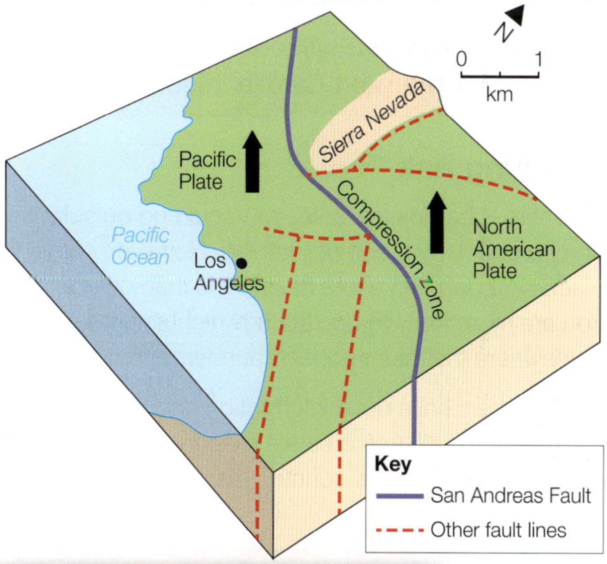

C *Cross-section of the San Andreas Fault close to Los Angeles*

D *The San Andreas Fault runs through residential areas*

Activities

1 Look at photo A.

  a Explain how the damage was caused by an earthquake.

  b Suggest how the damage could affect the response to the earthquake.

2 Look at photo D. Imagine the consequences of another earthquake on the San Andreas Fault. Draw a sketch to show this street after the earthquake.

3 Look at map B.

  a Describe the pattern of previous major earthquakes along the San Andreas Fault.

  b From the pattern, predict where the next major earthquake is likely to happen. Explain your prediction.

4 Look at diagram C. Explain why Los Angeles is threatened by a major earthquake.

5 Prepare an emergency kit that you could use in the event of an earthquake. Choose at least six items and explain how they would be useful after an earthquake.

6 How should a family in Los Angeles prepare for a major earthquake? Choose one of these options. Give reasons for choosing one option and a reason for not choosing the other two.

- Move away from Los Angeles now, before an earthquake happens.
- Wait until the earthquake happens and then evacuate the city.
- Stay in Los Angeles and use your emergency kit.

Be prepared!

On this spread you will consider the hazard risk across the USA before you prepare for the next natural disaster.

Mapping risk

Nowhere is completely safe. You could go outside and be struck by a bolt of lightning. However, some places are definitely more hazardous than others. You are more likely to be struck by lightning if you live somewhere that has frequent thunderstorms!

Geographers have used information about natural hazards in the USA to create a map that shows the level of risk around the country (map A).

The factors used to create the map include:

- the number of types of hazard in each area
- the frequency with which each type of hazard occurs
- the likely impact of hazards in lives lost and economic cost.

A *Hazard risk map for the USA*

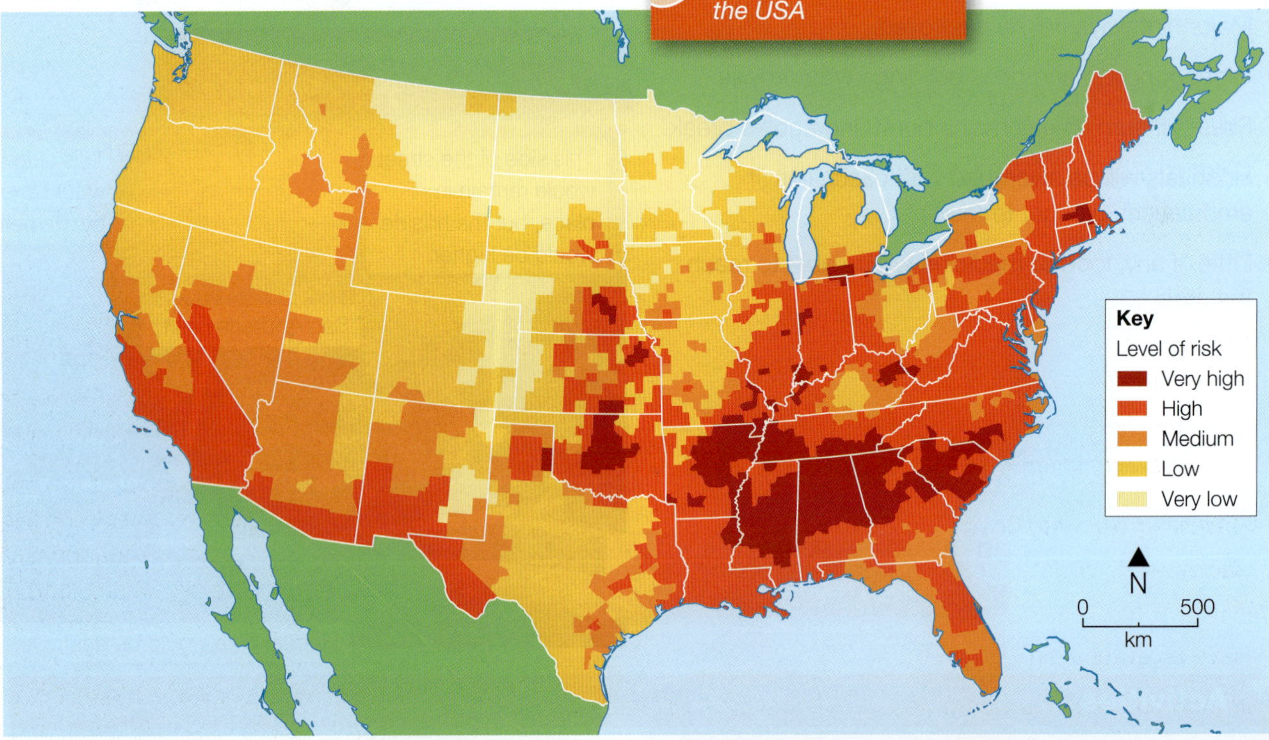

Preparing for natural hazards

It is impossible for us to prevent natural hazards from occurring, but it is possible to prepare, in order to reduce the risks. We can do this by:

- **Prediction:** using knowledge and experience of previous events to forecast the likelihood of future events.
- **Protection:** putting in place structures or infrastructure that will protect people and reduce the likelihood of injuries or deaths.
- **Planning:** putting in place plans for how to deal with the hazard when it happens and, if possible, keep people out of danger.

B *Planes extinguishing wildfires in the Rocky Mountains*

Preparing for wildfires

Taking wildfires as an example, we can see how this three-part strategy can be used to prepare for natural hazards.

1 Prediction of wildfires

There are three pre-conditions needed for wildfires.

- Ignition to light the fire. In the wild, this is usually lightning.
- A fuel supply in the form of trees and shrubs, often dry after a long drought.
- Hot, dry windy conditions to spread the flames.

Knowing this, it is possible to map the areas where wildfires are most likely to occur. These are going to be forested areas where hot, dry, windy conditions are found (see map B on spread 9.1).

2 Protection against wildfires

The most effective protection against wildfires is good ecosystem management.

- Clear trees from strips in the forest as natural firebreaks to stop the flames spreading.
- Use livestock to graze land to prevent shrubs growing to become a fuel supply.
- Clear low-growing vegetation in the forest that could act as a 'fire ladder' for the fire to climb trees.
- Create space around homes and don't build with flammable materials that would burn (drawing C).

3 Planning for wildfires

In places where wildfires could happen, the following precautions need planning.

- Have an evacuation plan in place, with escape routes for residents.
- Identify safety zones where people can stay until the fire is under control.
- Have fire-fighting equipment maintained and ready to use, including aircraft and water supply.

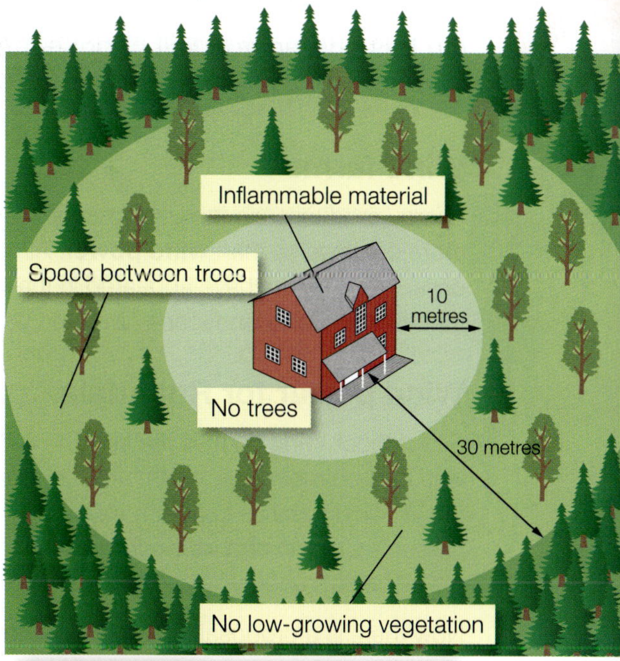

C *Advice for homeowners in areas where wildfires occur*

Activities

1. Look at map A. Describe the pattern of risk on the map. Which parts of the USA have the highest and lowest risks? (Clue: use the names of individual states to make the description easier, e.g. California.)
2. Try to explain the pattern on map A. Why do some areas of the USA have a higher risk than others? (Clue: look back at map B on spread 9.1.)
3. Look at photo B and drawing C. In each case, explain how it is reducing the risk from wildfires. Is it an example of predicting, protecting or planning?

Your decision-making enquiry

How should the USA prepare for the next natural disaster?

Read about how you can prepare for wildfires to reduce risk. Now choose at least three other natural hazards that occur in the USA. They could be hazards you have studied in this book, such as hurricanes (tropical storms), droughts or earthquakes. Or you could do some research on other hazards you would like to know more about, e.g. heatwaves, tornadoes and volcanoes.

Complete a large table to show how the USA could prepare for each hazard to prevent it turning into disaster. Use your own ideas as well as ideas you have read about.

Hazard	Prediction	Protection	Planning
Tropical storm			
Drought			
Earthquake			

10.1 DME A wind farm controversy

On this spread you will choose the best site for a new wind farm and consider the impact it would have.

Links

This decision-making exercise links to other topics:

- UK in the 21st Century (Chapter 7)
- Distinctive Landscapes (Chapter 3)
- Resource Reliance (Chapter 8)
- Changing Climate (Chapter 2)
- Sustaining Ecosystems (Chapter 4).

Wind farms

Energy production in the UK is a controversial topic and no form of energy causes more controversy than **wind power**. Almost wherever you travel in the UK these days, **wind farms** have appeared in the landscape. Many people object, saying that they spoil the landscape. What do you think?

Wind power for Rochdale

Scout Moor Wind Farm is the largest onshore wind farm in England (photo B). It was built on moorland, north of Rochdale in Greater Manchester in 2008. There are 26 wind turbines (diagram A), each 100 metres high – twice the height of Nelson's Column and visible from 30 km away. Together, they generate about 40% of Rochdale's energy, enough for 40 000 homes.

Now there is a proposal for a new wind farm near Rochdale and some people are objecting.

Where to build a wind farm

There are a few factors to consider when choosing the site for a wind farm.

- It needs to be windy! The best sites are on flat areas, hilltops or open moorland.
- It should be located away from buildings or natural barriers like trees, which block the wind.
- It should be well away (at least 1 km) from settlements, where the noise could disturb residents.
- It should not be located in an area of natural beauty where it could spoil the landscape.

A *How a wind turbine works*

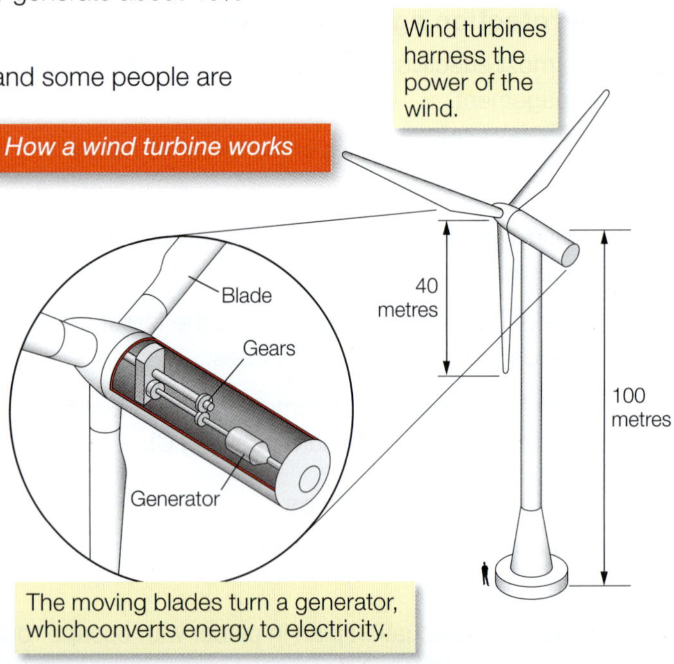

B *Scout Moor Wind Farm near Rochdale*

C Rochdale and Scout Moor; OS map, scale 1:50 000

Activities

1 Look at photo B. What do you think about wind farms? Do they spoil the landscape or improve it?

2 Look at map C. Find Scout Moor Wind Farm.

   a Give a four-figure grid reference for the squares covering the area of the wind farm.

   b Estimate the total area of the wind farm in km². (Clue: each square on the map is 1 km².)

   c Suggest why this site was chosen. Give at least two reasons.

3 A new wind farm was proposed near Rochdale. Choose the best site for the wind farm, centred at one of the six-figure grid references below. Remember you are looking for a flat site on a hilltop or open moorland.

- 855 188
- 847 177
- 854 170
- 868 153.

Give at least two reasons for choosing this site and one reason for not choosing each of the other three sites.

Go further

Find out whether the proposed wind farm at Rooley Moor was ever built or not. You could search the internet for 'Rooley Moor Wind Farm decision'.

Your decision-making enquiry

Where should the UK get its energy from in future?

As you work through this chapter, make notes about anything that will help you to answer the question. You will bring your ideas together to do an activity at the end of the chapter.

10.2 DME Wind power in the UK

On this spread you will examine the growth in the number of UK wind farms and think about future requirements.

Onshore and offshore wind farms

The number of wind farms in the UK grows by the week. In 2015 there were 923 wind farms, most of them **onshore**. However, a growing number of wind farms are located **offshore** and usually these are larger (map A).

People are concerned about the impact of wind farms on the landscape. So the government wants to switch wind power generation from onshore to offshore. In future it will be harder to get planning permission to build onshore wind farms (see spread 10.1).

Many of the issues that concern people about onshore wind farms become less of a problem when they are located offshore (panel B). However, it is more expensive to build offshore wind farms because it is difficult to erect wind turbines on the seabed.

A UK wind farms

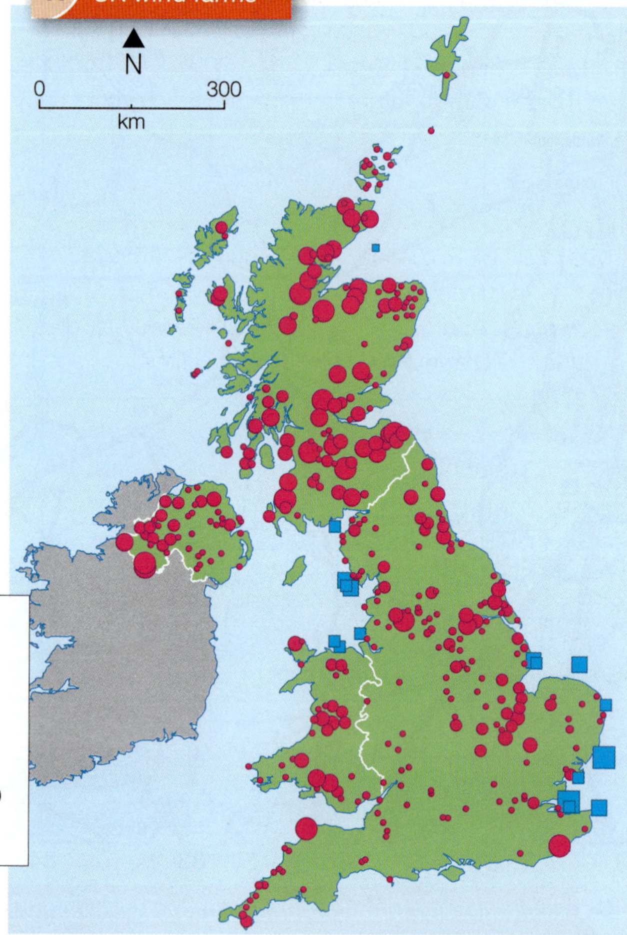

B Advantages and disadvantages of onshore wind farms

Advantages

- Wind power is relatively cheap when compared to other sources of energy (see spread 10.3) and is getting cheaper as technology improves.
- It is more efficient in winter when more energy is required.
- It provides a high output of energy for the amount of space the turbines use.
- The land around each turbine can still be used for agriculture even when the turbine is generating electricity.

Disadvantages

- Wind farms spoil the view and can be seen from far away.
- Wind turbines create noise, although technology is reducing the problem.
- There are a limited number of sites on land with high, steady wind speeds.
- Wind turbines have higher maintenance costs than other types of renewable energy.
- The moving turbine blades can cause injury to birds and bats.

Future wind farms

The UK is aiming to meet the target of generating 30% of its electricity from **renewable sources** by 2020. These sources can be used over and over again without being used up. By far the largest share of this will come from wind. The percentage of our electricity from wind power rose from 1.5% in 2008 to over 10% in 2015 (table C). In future, a higher proportion of wind power is likely to come from offshore wind.

Nine zones around the UK coast, up to 200 km from the coastline, have been identified for offshore wind farms (map E). Potentially the UK has the largest offshore wind resource in the world. It is an island nation surrounded by the relatively shallow water of the North Sea and the English Channel (photo D). Wind turbines can be constructed on the seabed up to 50 metres below the surface. Much of the North Sea is within this depth.

Year	Electricity generation (MW)	Percentage of total electricity generated
2008	2 974	1.5
2009	4 051	2.0
2010	5 204	2.3
2011	6 540	3.8
2012	8 871	5.5
2013	10 976	7.2
2014	12 440	9.3
2015	13 312	10.4

C *UK wind generation, 2008–2015*

D *An offshore wind farm*

E *Offshore wind farm zones around the UK*

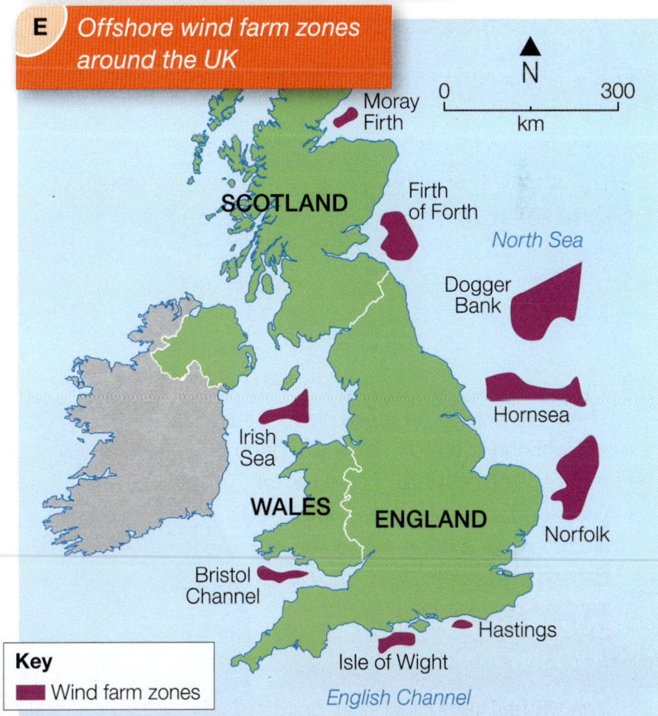

Activities

1. Look at map A. Describe the distribution of wind farms in the UK. What proportion are offshore wind farms?
2. Compare map A with a relief map of the UK (see spread 3.2).
   - **a** What do you notice about the distribution of wind farms compared with relief?
   - **b** Explain the distribution of wind farms.
3. Study the information in panel B.
   - **a** How would the advantages and disadvantages of offshore wind compare with onshore wind?
   - **b** Explain why the UK government now favours offshore wind.
4. Look at table C. Draw two graphs to show:
   - **a** the growth in the amount of electricity generated by wind power since 2008
   - **b** the increase in the percentage of electricity from wind power. Extend your graph to estimate the percentage by 2020.
5. Look at photo D and map E.
   - **a** Some people think wind farms will spoil the view from the coast. What do you think?
   - **b** On balance, do you think offshore wind is a better or worse alternative to onshore wind? Give reasons for your opinion.

UK energy sources

On this spread you will compare the impacts of the main sources of energy in the UK.

Renewable and non-renewable energy

The UK generates electricity in a number of ways. It is important to have different fuel sources and technologies to ensure a constant supply and so that we don't become over-reliant on a single type of power.

Most of the UK's electricity is still generated in power stations by burning **fossil fuels** – coal, oil and gas – although that proportion is declining. In addition, we also obtain **nuclear energy** from a nuclear fuel called uranium. Fossil fuels and uranium are **non-renewable sources** of energy. Once the fuel has been used, it is gone forever and cannot be re-used.

Almost a quarter of our energy is now generated from renewable sources of energy, including wind, waves, tides, hydro (water), biomass (organic matter) and solar. The supply of these sources of energy is endless and can be used over and over again.

A *Coal-fired power station at Drax, Yorkshire*

There are only ten of these power stations still working in the UK. Coal has been the main source of energy in the UK for 200 years and most of it is now imported. Burning coal is our main source of carbon emissions.

B *Gas terminal at Milford Haven, Pembrokeshire*

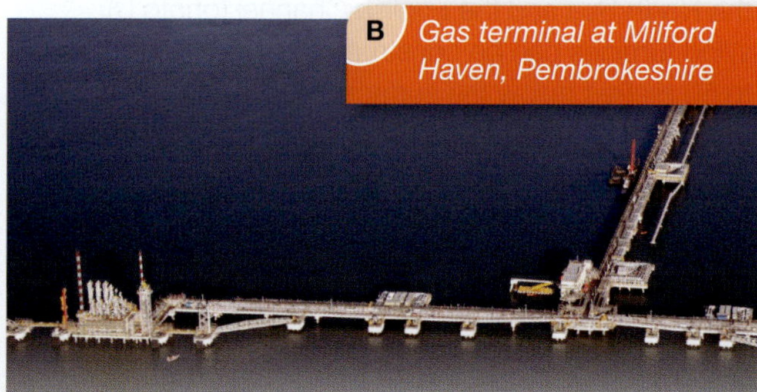

Gas is imported by pipeline from Europe and shipped from other parts of the world as liquefied natural gas (LNG). Gas arrives at nine terminals around the UK coast. In addition, there are plans to use shale gas found underground in the UK (see spread 10.4).

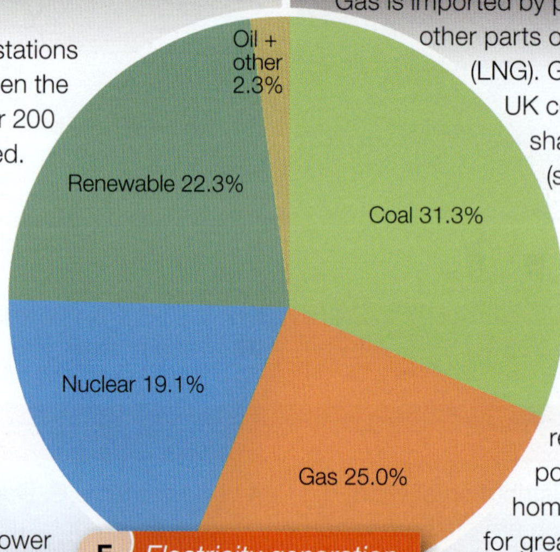

E *Electricity generation in the UK, 2015*

C *Nuclear power station at Hinckley Point, Somerset*

The UK has 19 nuclear reactors, located at ten coastal sites around the country. There are plans to replace the old reactors with new ones so that a quarter of the UK's electricity will come from nuclear power by 2025.

D *Offshore wind farm in the Thames Estuary*

There are about a thousand wind farms in the UK. Wind is not our only renewable source of energy. Solar power is becoming widespread on homes and in fields. There are also plans for greater use of tidal and wave power (see spread 10.5).

Impact on...	Coal	Gas	Nuclear	Renewable, e.g. wind
Climate change	It emits more CO_2 than other fossil fuels.	It emits less CO_2 than coal or oil.	There are low CO_2 emissions from power station construction.	There are very low CO_2 emissions from construction.
Natural ecosystems	Burning coal produces acid rain, which affects trees and plants.	Drilling for gas affects sea life.	It makes little impact, but the process uses a lot of water.	Turbines can injure birds and bats.
Landscape	Power stations are large and dirty.	Gas storage terminals are large.	Power stations are large but isolated.	Wind farms are visible from far away.
Resources	The UK imports coal but still has plenty of its own.	North Sea gas is running out but shale gas could replace it.	The UK imports uranium and it may only last 50 years.	It will never run out.
Health and safety	Coal-mining is dangerous and emissions affect health.	Drilling for gas is dangerous.	Accidents are rare but very serious. Waste is also dangerous.	There is no risk to people.
Reliability	It is a very reliable form of energy.	It is reliable for now but not if we have to import.	It is a very reliable form of energy.	Wind is unpredictable.

Key
Level of impact
Very high High Medium Low

F *Types of energy in the UK compared*

G *Costs of electricity generation*

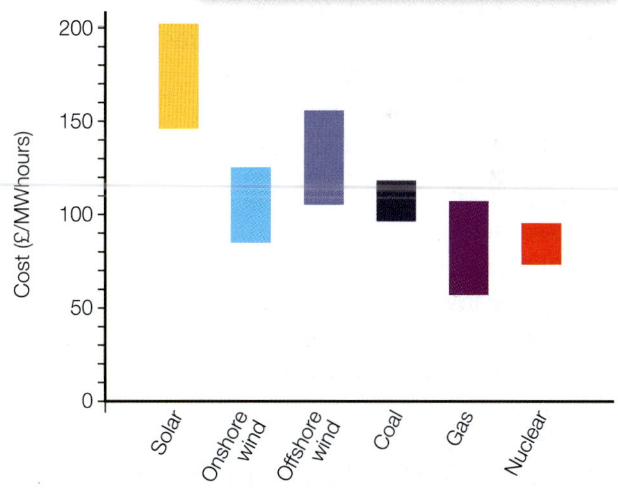

Activities

1. Look at chart E.
   a. Rank the sources of energy in order of importance.
   b. How do you think the order could change by 2030? Justify your answer.
2. Look at photos A–D. Carry out a simple environmental impact assessment for each source of energy, using a table like the one below.
   a. Give a score for the landscape impact where 0 is no impact and 5 is a major impact.
   b. Give a score for the number of sites in the UK, where 0 is none and 5 is for hundreds of sites.
   c. Multiply your scores together to obtain an overall environmental impact score for each source of energy.
   d. Which source of energy has most impact on the environment in the UK? Explain why.
3. Study table F and graph G.
   a. Using all the information on this spread, which would be the best source of energy for the UK to use? Give reasons for your answer.
   b. Why might it still be a good idea to use other energy sources rather than just one?

	Landscape impact (0–5)	Number of sites in the UK (0–5)	Overall environmental impact score (0–25)
Coal			
Gas			
Nuclear			
Renewable			

Shale gas – a non-renewable option

On this spread you will consider the benefits and problems of fracking for shale gas.

North Sea gas

For the past 50 years the UK has relied on gas from beneath the North Sea. About 40% of our total energy needs are met by burning gas. Most of us use gas to heat our homes and it is also used to generate electricity. But North Sea gas is beginning to run out, so we are now more dependent on foreign imported gas. This makes the UK's energy supply more vulnerable, if there was a war or terrorist attack, for example.

Shale gas

The UK has another source of gas called **shale gas** that it could exploit. Shale gas is found in shale, a sedimentary rock (see spread 3.3) that underlies large areas of the UK (map A). The gas is contained in small spaces within the rock, but is difficult to extract. It is estimated that over 3 trillion cubic metres of shale gas could be extracted in the north of England alone. That would be enough to supply the UK for ten years.

Shale gas is another controversial source of energy. The technique used to extract the gas is known as **fracking**, or hydraulic fracturing. Water, sand and chemicals are pumped into the rock under high pressure to obtain the gas (diagram C). So far, only one site in Lancashire has been fracked – in 2011. This led to two earth tremors and a temporary government ban on fracking.

But, with the discovery of shale gas at more sites in the UK (photo B), the pressure on the government to allow fracking is growing.

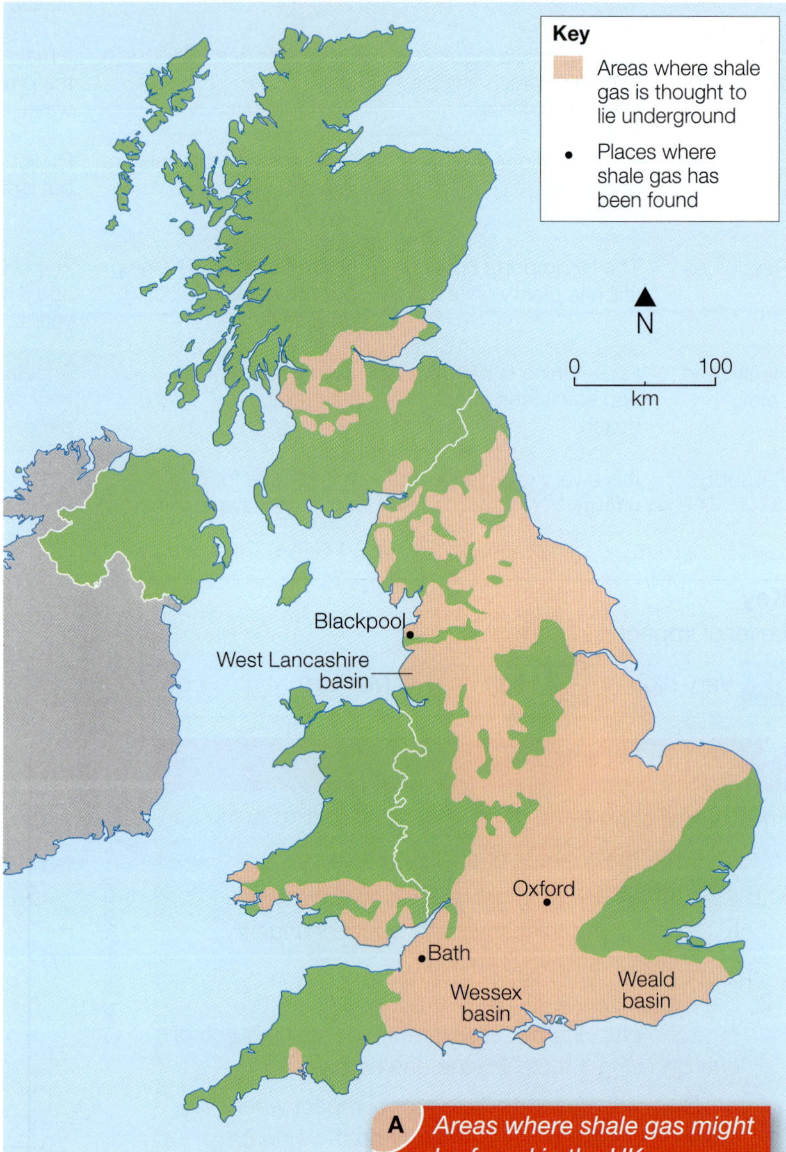

A *Areas where shale gas might be found in the UK*

B *A test-drilling site for shale gas in Lancashire*

C *How fracking works*

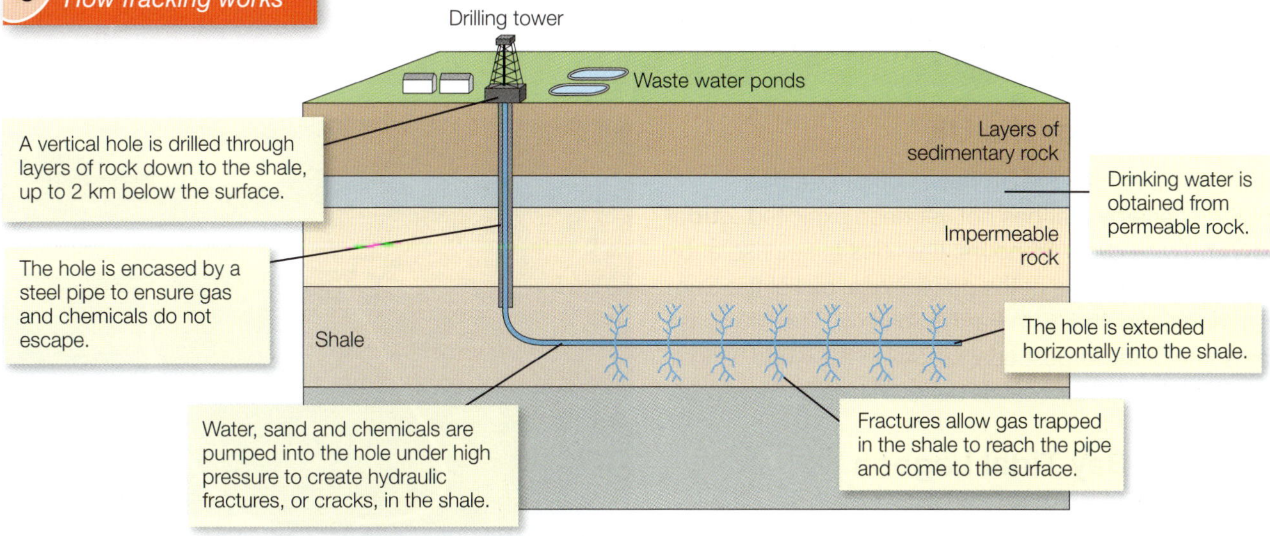

The problems with fracking

In recent years, shale gas in the USA has helped to boost the American economy by providing cheap energy and creating new jobs. There are hopes that it could do the same in the UK. However, many people here object to fracking, especially if it is going to happen near them (photo D). There are reasons for people to be concerned:

- Fracking can lead to unexpected consequences like earth tremors.
- In the USA, careless drilling led to water supplies being contaminated (poisoned) with chemicals.
- The USA is a much larger country than the UK, with lower population density. Any problems in the UK would be worse, because they would be closer to people.
- The world already has more oil and gas than can be burned if we are to avoid the worst impacts of climate change (see Chapter 2).

D *Protests against fracking in the UK*

Activities

1. Look at map A. Compare it with where people live in the UK (see spread 7.1).
   - **a** How does the distribution of shale compare with where people live?
   - **b** Which towns or cities are close to the sites where shale gas has been found?
2. Look at photo B.
   - **a** What is the purpose of test-drilling?
   - **b** If you lived near a test-drilling site, would you be excited or worried? Explain your answer.
3. Look at diagram C. Explain how fracking might be connected with:
   - **a** earth tremors
   - **b** contaminated drinking water.
4. Look at photo D. How might a fracking company respond to these protests? Write a leaflet from the company to inform local people:
   - **a** what precautions the company will take to ensure there are no problems
   - **b** what the benefits will be for the local area and for the rest of the UK economy.

Go further

Find out from online news reports if fracking has begun at any sites in the UK. If so, how many sites are there? Where are they? You could also check the activities of fracking companies like Cuadrilla, iGas and Third Energy by going to their websites.

Tidal power – a renewable option

On this spread you will learn about tidal power before you make a decision about the energy future of the UK.

Tidal power for the UK

As an island nation, the UK has another potential source of energy that has not yet been exploited – **tidal power**. Twice every 24 hours, the tides at sea rise and fall naturally because of the gravitational pull of the Sun and moon, together with Earth's rotation. The UK coast has some of the largest tidal ranges in the world, with a difference of up to ten metres between high and low tides.

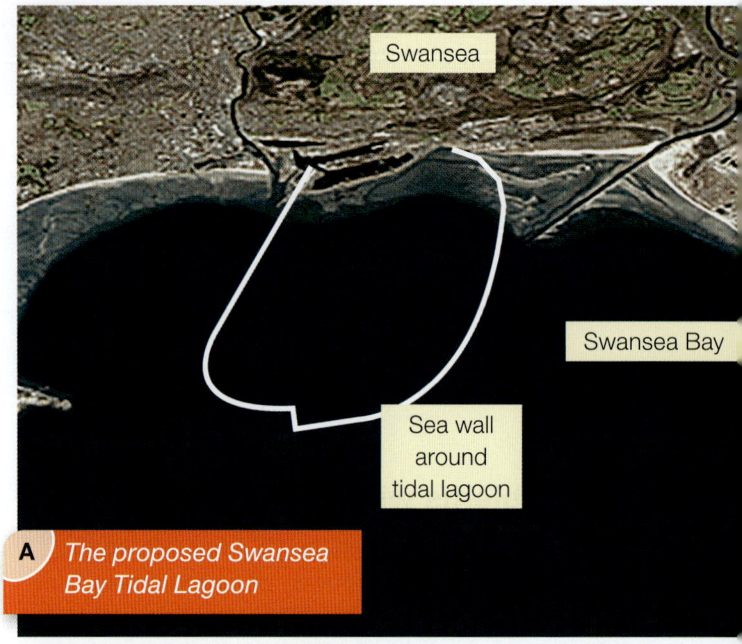

A *The proposed Swansea Bay Tidal Lagoon*

Swansea Bay Tidal Lagoon

The first tidal lagoon (saltwater lake) in the UK could be built in Swansea Bay if it is given the final go-ahead by the government (image A). It would generate renewable power for 150 000 homes (90% of the city's electricity) at a cost of £1 billion. The future of the project, however, lies in doubt.

The Swansea Bay scheme could be one of four tidal lagoons on the Bristol Channel, which has the largest tidal range in the UK (map B). The cost of further lagoons is likely to come down as the technology develops.

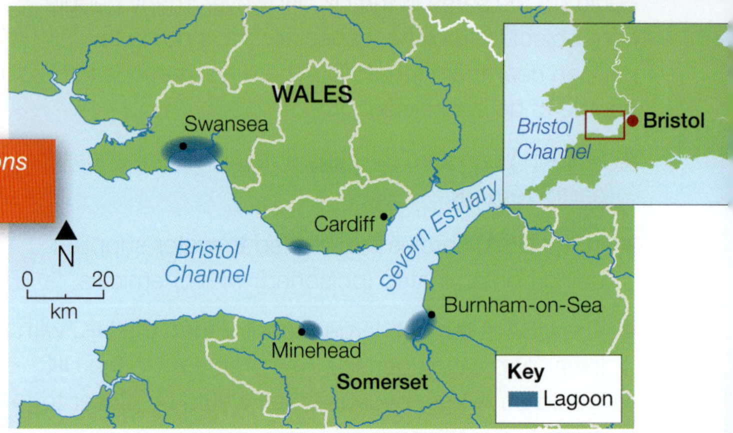

B *Four proposed tidal lagoons in the Bristol Channel*

How a tidal lagoon works

A sea wall is built to enclose an artificial lagoon, with gates controlling the water level inside. The wall houses underwater turbines to generate electricity on the rising and falling tides (diagram C).

High tide
Gates in the wall are closed as the tide rises. At high tide the gates open and water rushes in through the turbines to fill the lagoon.

Low tide
Gates in the wall are closed as the tide falls. At low tide the gates open and water rushes out through the turbines to empty the lagoon.

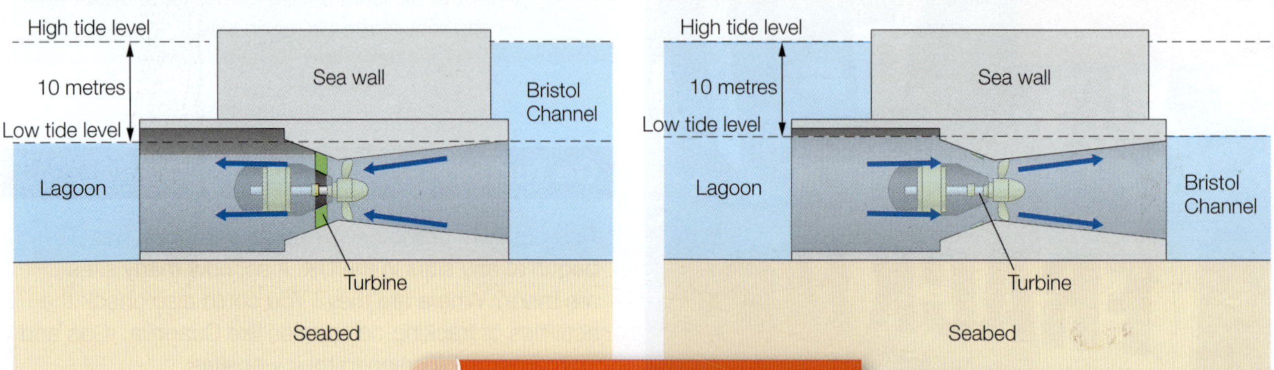

C *How a tidal power scheme works*

Advantages and disadvantages of tidal power

Tidal power, like other energy sources, has advantages and disadvantages (panel D).

D *Advantages and disadvantages of tidal power*

Advantages

- Tidal power, unlike wind or solar power, is predictable. High tide and low tide happen twice a day.
- The UK has a long coastline with many potential sites for tidal power.
- Tidal lagoons do not block river estuaries as a wider, tidal barrage would.
- The impact of a tidal lagoon on natural ecosystems is minimal.

Disadvantages

- Initially, tidal power would be expensive, though costs would fall as more tidal lagoons were built.
- It would require government subsidy (money) at the start to compete with other forms of electricity.
- Tidal lagoons could have a negative impact on local businesses such as tourism and fishing.

Activities

1 Look at image A and map B.
  a Describe the location of the proposed Swansea Bay Tidal Lagoon.
  b Explain why this would be a good location for the project.
2 Compare the advantages and disadvantages of tidal power in panel D. Overall, do you think a tidal lagoon is a good idea? Give reasons for your opinion.

Your decision-making enquiry

Where should the UK get its energy from in future?

1 Look at the drawing below. Identify seven potential sources of energy at sites A–G.
2 Which of the sources you have identified would be:
  a most reliable? **b** most sustainable?
  Give reasons.
3 Think of any energy sources not shown on the drawing.
  a Which sources are they?
  b How important will they be in future?
4 Plan an energy strategy for the UK, using one or more of the energy sources. Write a short report of about 500 words to justify your strategy.

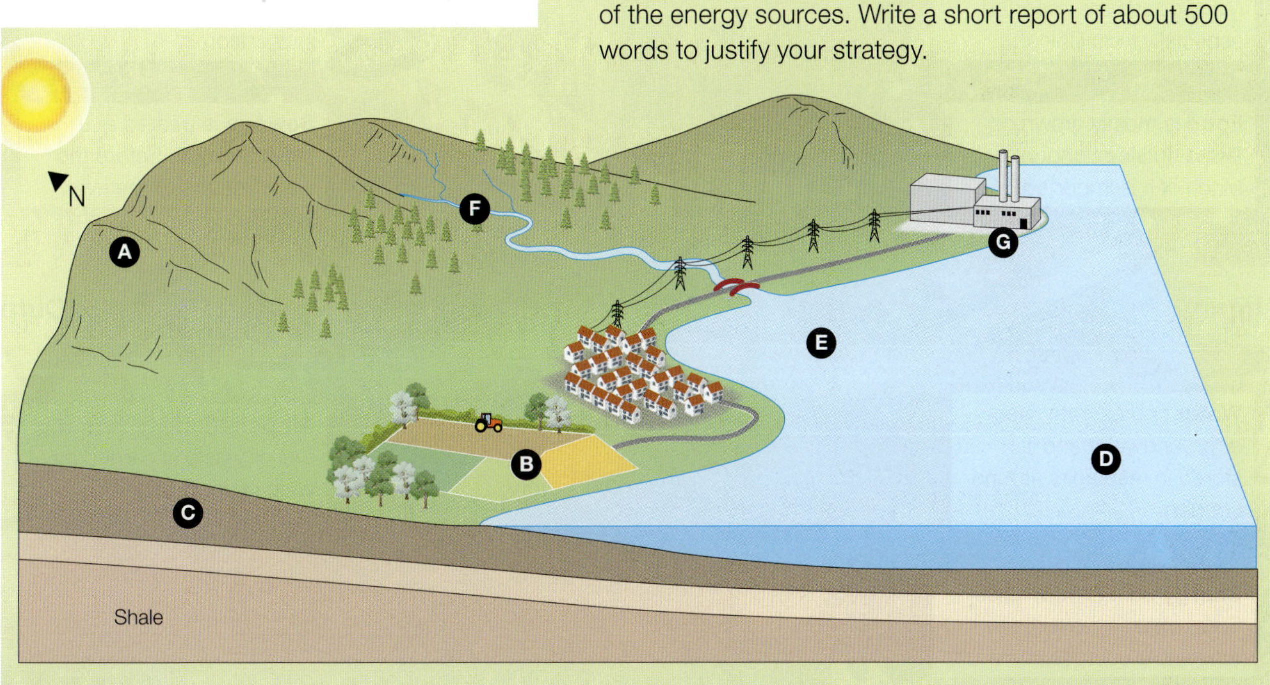

London's ecological footprint

On this spread you will think about the city as a system and examine London's footprint.

Links

This decision-making exercise links to other topics:

- Urban Futures (Chapter 5)
- Sustaining Ecosystems (Chapter 4)
- Resource Reliance (Chapter 8)
- UK in the 21st Century (Chapter 7)
- Changing Climate (Chapter 2).

Unsustainable cities

Over half the world's population lives in cities (see Chapter 5). They consume 75% of the world's energy and produce 80% of all greenhouse gas emissions. Clearly this is not sustainable and cities need to reduce their environmental impact. In the UK, over 80% of us live in cities and London is home to 12% of the UK's population.

Those of us who live in cities often forget how dependent we are on the natural environment. For example, when we go shopping we expect the supermarket shelves to be full of food, but we don't think about the farms where it came from or how it was produced. It's the same with energy. We flick a switch for light or heat, but don't think about the fuel that was burnt to give us power or about the CO_2 it emits.

The city as a system

A long time ago, when cities began to grow, they were surrounded by the resources they needed from farms, forests, grazing land and mines. As cities expand, they become increasingly separated from the places where their resources are produced and where their waste ends up.

London now has a population of 8.6 million. It outgrew the original city a long time ago. Today, we can think of London – or any city – as a **system** (photo A). The resources it uses are **inputs** and the waste it produces are **outputs**. Most cities are linear systems, with resources flowing through the system. For cities to become more sustainable, they need to turn from linear systems into circular systems, more like natural ecosystems (see Chapter 4). There, resources are recycled within the system, with fewer inputs or outputs.

A *London as a system*

Ecological footprints

We can measure the environmental impact of a city by its **ecological footprint**. This is the total area of land (and sea) required to supply its resources and to absorb its waste. London's ecological footprint is estimated to cover an area 125 times the size of the city – an area bigger than the whole of the UK (map B)!

Of course, not all of the UK's land is devoted to producing resources for London. But, if you added up all the land, globally, from which London gets it resources, the footprint represents the total area that land would cover.

London's ecological footprint is not sustainable. Each person in London requires 5.8 global hectares (gha) of land to support them (graph C). This is slightly above the UK average and well above the world average. In fact, there is only about 1.9 gha available for each person on the planet. This is what we should be aiming towards in order to be sustainable.

B *London's ecological footprint compared to the UK*

Activities

1 Look at photo A. Complete a large diagram like this to show the city as a linear system.

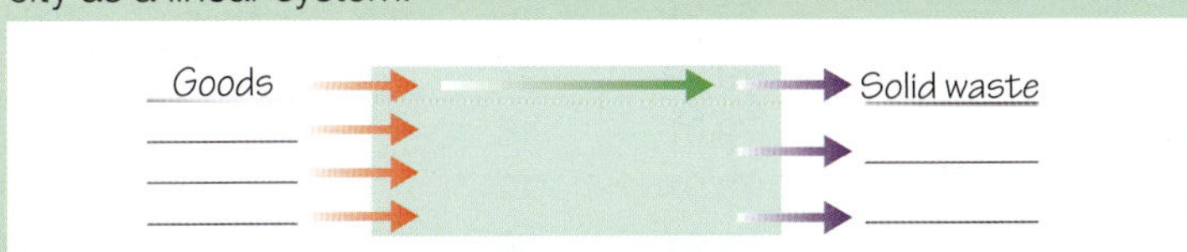

- **a** Name the inputs and outputs.
- **b** Draw arrows to show the links between inputs and outputs. One is done for you.
- **c** Describe the link on each arrow, e.g. *Goods are thrown away when they get old.*

2 Now design a circular system for a more sustainable city. Think about all the ways in which resources could be recycled, e.g. *Goods are repaired when they get old. Materials can be recycled.*

3 Look at map B.

- **a** Compare the size of London's ecological footprint with the map of the UK.
- **b** Try to explain the size of London's footprint. Why is it so large?

4 Look at graph C. Compare the average land required by a person in London to:

- **a** the UK average **b** the world average
- **c** the area of land available in the world.

5 **a** Explain why cities like London are unsustainable places to live.

- **b** Can you think of any ways in which living in a city could be more sustainable? Explain how.

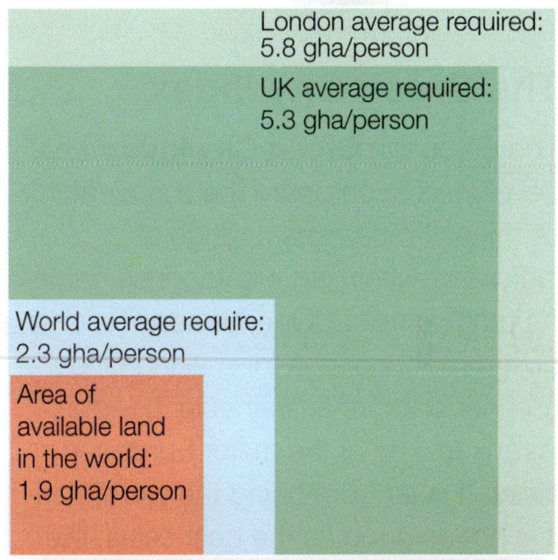

C *Ecological footprints per person*

Your decision-making enquiry

How could living in London (or another UK city) be more sustainable?

As you work through this chapter, make notes about anything that will help you to answer the question. You will bring your ideas together to do an activity at the end of the chapter.

11.2 DME More sustainable cities

On this spread you will compare sustainability in London and Copenhagen to find out what London could improve.

Towards a sustainable city

A sustainable city is one that can meet the needs of the present without sacrificing the ability of future generations to meet their needs. The key word here is 'needs', not 'wants'.

Cities like London can become more sustainable by reducing the size of their ecological footprint. They can do this by:

- reducing the amount of resources they consume
- recycling more resources, in order to consume less
- using more renewable sources of energy (see spread 10.3)
- producing less waste.

The Green City Index

The European **Green City Index** was developed to compare the sustainability of cities. It measures a range of indicators within eight categories (table A), ranging from CO_2 emissions to government policies for making cities greener or more sustainable. London is ranked 11th of the major European cities. It is far from being the worst, but it is not as good as the best cities, like Copenhagen (graph B).

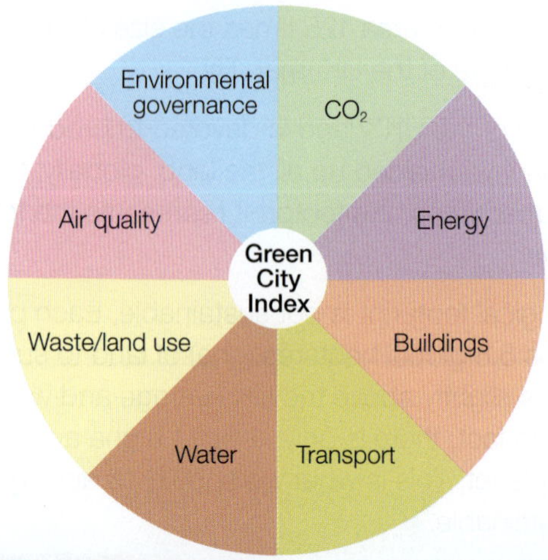

Rank	City	Score	Rank	City	Score
1	Copenhagen	87.31	11	London	71.56
2	Stockholm	86.65	12	Madrid	67.08
3	Oslo	83.98	13	Vilnius	62.77
4	Vienna	83.34	14	Rome	62.58
5	Amsterdam	83.03	15	Riga	59.57
6	Zurich	82.31	16	Warsaw	59.04
7	Helsinki	79.29	17	Budapest	57.55
8	Berlin	79.01	18	Lisbon	57.25
9	Brussels	78.01	19	Ljubljana	56.39
10	Paris	73.21	20	Bratislava	56.09

A The Green City Index used to compare European cities

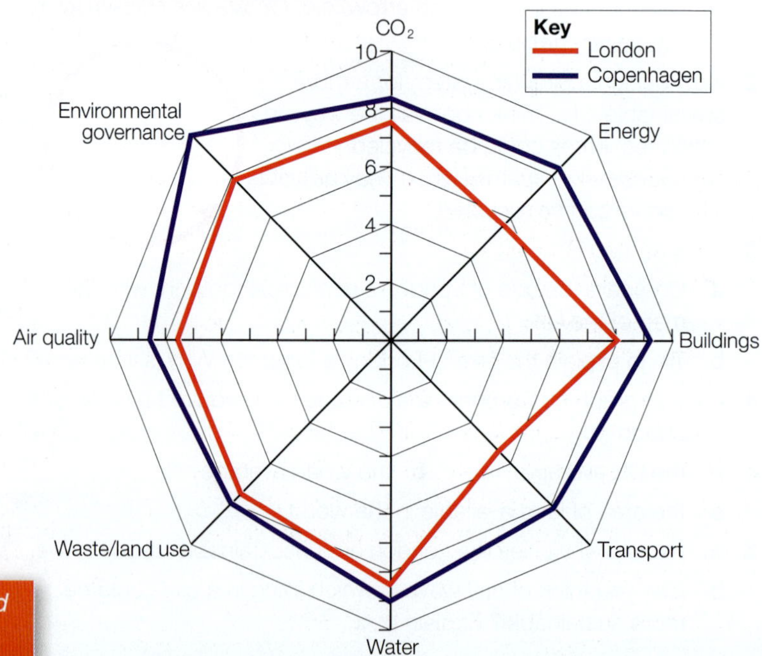

B London and Copenhagen compared on the Green City Index

Copenhagen's achievements

Top of the European Green City Index league table is Copenhagen, capital of Denmark. The city has set very high standards of sustainability for other cities like London to aim for.

Copenhagen expects to become the world's first **carbon-neutral city** by 2025. Carbon emissions will either be reduced to zero or balanced by removing the same amount of CO_2 from the atmosphere that the city produces.

One of the city's first flagship policies was to make the water in the city harbour clean enough for people to swim in (photo C). Imagine swimming in the River Thames in London!

C Copenhagen harbour

	Problems	Solutions
Water pollution	Water in the harbour was badly polluted by the city's old sewage system and local industry. The area was neglected and run down.	The sewage system was modernised. Water quality improved and three public baths plus a swimming beach have opened in the harbour.
Water supply	There was a lack of high-quality drinking water. Pesticide pollution around the city meant water had to be piped long distances.	Better management of water resources, including water meters, leak reduction and better storage, led to improved supply.
Cycling	City streets were congested with cars and cycling was dangerous. This deterred people from cycling and caused slower journeys.	New cycle lanes were introduced to separate cyclists from traffic to make cycling safer. There are now 400 km of cycle lanes in the city.
Public transport	The number of daily car journeys into the city almost doubled from 1970 to 2010. CO_2 emissions increased.	A better, integrated public transport system was developed, including bus, train and metro. People were encouraged to stop driving cars.
Waste	In 1980, 40% of the city's waste was sent to landfill sites. Burning waste in the city created dangerous air pollution.	By 2009 less than 2% of waste was sent to landfill. Most waste is now recycled and there are charges for landfill or for burning waste.
Heating	The city depended on burning fossil fuels for its heating. With rising prices, this was expensive and also caused air pollution.	The city now has one of the world's largest combined heat and power (CHP) systems, producing both heat and electricity for 98% of the city.

D Copenhagen's solutions to its environmental problems

Activities

1 Look at table A.
   - a Which countries are these cities in? Mark and label them on a map of Europe.
   - b What do you notice about the location of the top ten cities in Europe compared to the next ten?

2 Look at graph B. Compare sustainability in London and Copenhagen using the Green City Index. On which indicators does London do better or worse?

3 Look at photo C.
   - a What evidence is there in the photo of a sustainable city?
   - b How was this change achieved? (Clue: look at table D.)

4 Study graph B and table D.
   - a What are the main lessons that London (or your city) could learn from Copenhagen? Mention three.
   - b Suggest what changes London (or your city) would have to make to be more sustainable.

Improving London's transport

On this spread you will investigate transport improvements in London that are helping to reduce carbon emissions.

Public transport in London

London has a well-integrated public transport system, with buses and underground and overground trains. But it is struggling to cope with the increase in passenger numbers. As London's population grows, so does the number of commuters. In 2014, 75 million passengers used buses and underground trains in London each week. Driving a car is not a sensible option for most Londoners. There is limited space to park and traffic congestion means longer journey times.

A *A crowded London Underground station*

Crossrail (Elizabeth Line)

Crossrail, a new high-speed east–west rail route across London, is due to open in 2018 (map B). It will tunnel 26 km under central London, linking Reading and Heathrow Airport in the west with Shenfield and Abbey Wood in the east. It is expected to increase rail passenger numbers by 10% – or 200 million people a year.

Transport infrastructure in London requires long-term planning and investment. Crossrail 2 is already being planned for 2030, linking the north and south of London.

Crossrail is expected to bring a number of benefits to London. It will:

- increase the capacity (the maximum number of passengers) that the rail network is able to carry
- reduce journey times by up to a half, e.g. the time from Liverpool Street to Heathrow will fall from one hour to 35 minutes
- bring an extra 1.5 million people into central London within 45 minutes, increasing the number of people who can commute to work
- raise property values around stations along the route by up to 25%
- encourage regeneration across London, creating thousands more jobs
- improve the integration of the transport network in London by providing more interchange stations
- provide better disabled access with lifts from platforms to street level.

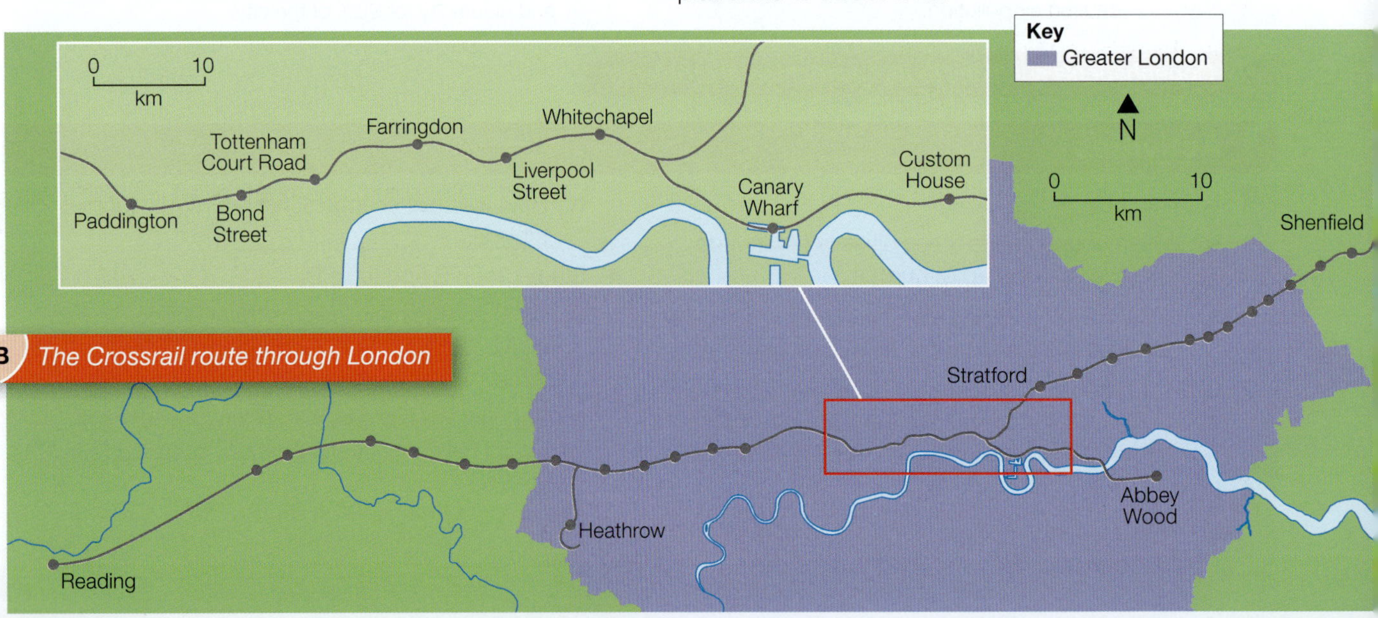

B *The Crossrail route through London*

Transport emissions

By encouraging more people to use public transport, Crossrail will also help to keep cars off the road. This, in turn, will help to:

- reduce traffic congestion
- speed up road journey times
- cut the number of road accidents
- reduce air pollution
- and, crucially, cut CO_2 emissions.

Graph C shows the CO_2 emissions for different modes of transport. It is clear that emissions per passenger from all types of cars are far greater than for public transport. Of course, the lowest emissions of all are for walking and cycling.

Cycling is quickly growing in popularity in London. 15% of all journeys to work are now by bicycle. To encourage more cycling, new cycle superhighways are being built to make cycling safer (photo D).

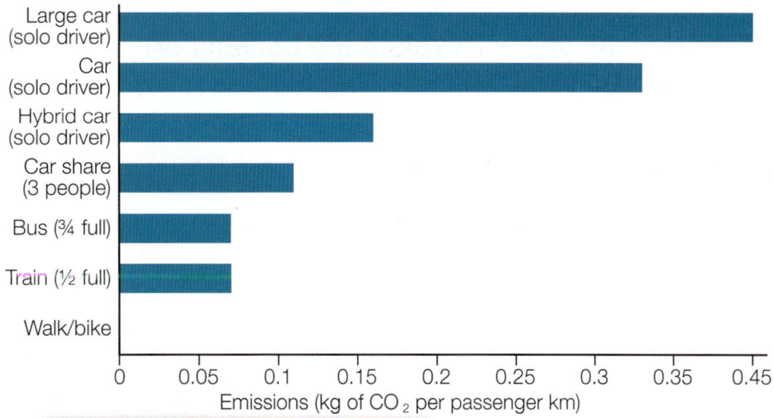

C *CO_2 emissions from different modes of transport*

D *Plans for a new cycling superhighway in London. Notice the separate lane for cyclists and reduced width for traffic*

Activities

1. Look at photo A.
   - **a** Explain why trains in London are so crowded.
   - **b** Why don't people drive a car instead?
2. Look at map B. Explain how Crossrail will:
   - **a** reduce journey times in London
   - **b** increase the capacity of the rail network
   - **c** improve integration of transport in London. (Clue: it may also help to look at a rail map of London online).
3. Look at graph C. In order to reduce carbon emissions, why would it be good to:
   - **a** share a car with other people?
   - **b** use public transport?
   - **c** go by bicycle?
4. Look at photo D.
   - **a** What often stops people from cycling in cities?
   - **b** How could cycle superhighways help?

Go further

- **a** Find out what transport improvements are planned for your city.
- **b** What other transport improvements would you recommend?

Using London's green spaces

On this spread you will explore the benefits of London's green spaces, including food security.

London's green spaces

London is one of the world's greenest cities. Almost half the city – 47% – is green space (graph A). The area would have been 62% if so many gardens had not been paved over. Green spaces include:

- parks
- unpaved domestic gardens
- woodland
- cemeteries
- allotments (plots of land rented by people to grow fruit and vegetables)
- farmland.

You might be surprised that London includes farmland. About 10% of London's area is farmland, though most of this is found near the edge of the city. Increasingly, farmland is put to other uses, e.g. for keeping horses, car boot sales, motorbike scrambling and caravan sites.

Green benefits

There are many benefits to having green spaces in cities.

- Trees produce oxygen to help us breathe and they absorb CO_2, reducing our emissions. London has 8.1 million trees – almost one per person!
- Trees and plants reduce the danger of flooding by absorbing water and slowing the rate at which it drains from the land (see spread 3.16).
- Woodland and other green spaces provide a habitat for wildlife, including birds, insects and mammals.
- People enjoy green spaces and they help to keep us fit and healthy, e.g. by running or playing football.
- People use green spaces for growing food (panel B and photo C).

A *Land use in London*

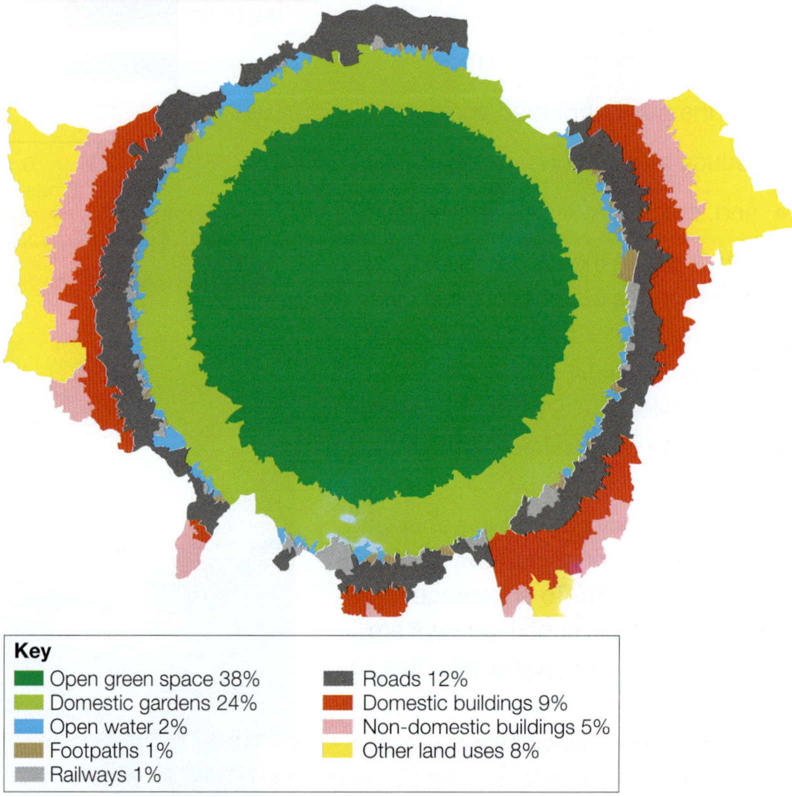

B *London food facts*

- Londoners eat over 2.5 million tonnes of food a year. Most of it is bought in supermarkets.
- Up to 80% of London's food is imported. 40% of food is imported in the rest of the UK, but London grows less of its own food and has a more diverse population.
- There are 472 farms in London, but the number is falling. There are only seven dairy farms producing milk.
- There are over 700 allotment sites with 20 000 individual plots.
- There are five wholesale food markets, which supply 20% of London's fresh meat, fish, fruit and vegetables.
- There are 70 street markets and 27 farmers' markets in London.
- Food is responsible for about 40% of London's ecological footprint (see spread 11.1).

Food security for London

Most people in London, like other cities, don't often worry about food security (see spread 8.5). They assume that the supermarket will always have the food they want on the shelves. However, even a small disruption to any part of the **food supply chain** can affect city dwellers.

In 2008, food prices rose by 25% because of global food shortages, putting some food prices beyond the reach of people on low incomes. The number of people relying on **food banks** trebled from 2012 to 2014, mainly due to changes and delays in the benefit system (money the government provides for people who need it). About 70 000 people in London – a third of them children – use food banks.

Many people in London want to grow their own food (photo C). There has been an increase in the number of people using allotments, so there are now long waiting lists to get one. Allotments began in the early 20th century, before financial assistance was given to the poor, as a means to help families feed themselves.

Of course, there will never be enough allotments to feed the 8.6 million people in London. However, other things can be done at each stage of the food supply chain to improve food security (diagram D).

C *Growing food on an allotment in London*

D *Ways to improve food security through the food supply chain*

Activities

1 Look at graph A. What do you find interesting or surprising about the data? Why?

2 London's green space has:

- **a** environmental benefits
- **b** social benefits
- **c** economic benefits.

Give at least one example of each, e.g. *Trees helping to reduce CO_2 emissions is an environmental benefit.*

3 Study panel B. Choose four important facts about London's food. Turn them into graphs or diagrams about food in London.

4 Look at photo C.

- **a** What makes this type of food production sustainable?
- **b** Is this the way to achieve food security in London? Explain your answer.

5 Study diagram D.

- **a** Explain why reliance on supermarkets to buy all our food could lead to food insecurity.
- **b** Describe what an alternative food supply chain that created more food security might look like. Mention all four stages of the chain – production, processing and distribution, consumption and waste.

Sustainable urban living

On this spread you will explore a new community in London, designed as a model of sustainable urban living.

East Village

One of the newest communities in London is East Village, formerly known as the Athletes Village. It is in Stratford, east London, next to Queen Elizabeth Olympic Park. It was designed as part of the 2012 Olympic legacy to become a sustainable new community (photo A).

2800 new homes have been created for local residents and newcomers on a 27-hectare site. There is a 50/50 mix of privately rented homes and cheaper, more affordable rented homes, all sharing the same buildings and facilities. Nobody in the village owns their own home.

A *East Village was the Athletes Village in the 2012 Olympics*

Water
- Buildings have green roofs (with plants on them) to absorb water.
- Water is recycled for irrigating plants and flushing toilets.

CO_2
- Homes use 30% less energy than normal homes.
- Trees have been planted to absorb CO_2.

Energy
- The energy comes from a combined heat and power station.
- LED lighting uses less energy.

Buildings
- The buildings are high-rise and use up less land.
- They are well insulated and more energy-efficient.

Waste and land use
- The village was built on a derelict brownfield site.
- There is 10 hectares of open green space in the village.

Transport
- A light railway station is linked to London Underground.
- There are bus connections to other local areas.

Air quality
- Hundreds of trees produce oxygen and absorb CO_2.
- Reduced traffic means less harmful gases.

Environmental governance
- East Village is managed by two organisations that own the homes.
- There is strict environmental management.

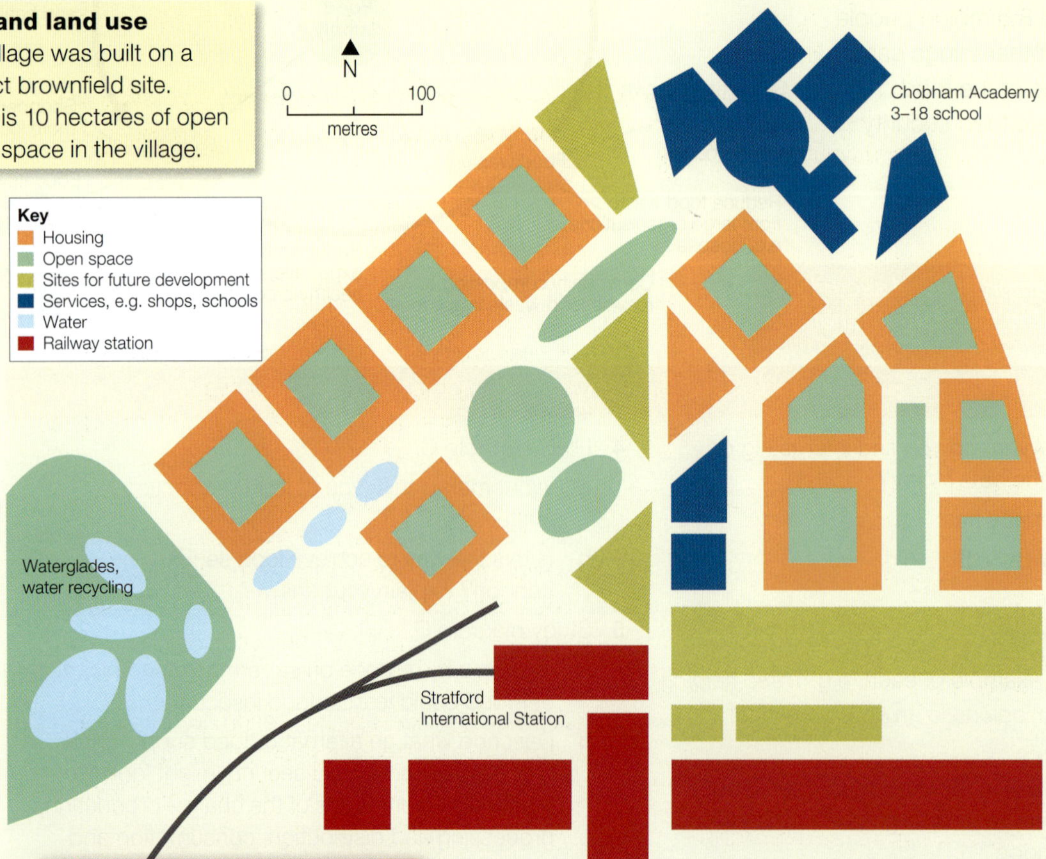

B *Sustainability in East Village*

The Energy Centre

East Village has the advantage of being built recently, using the latest urban design ideas. Among the features that make it more sustainable is a **combined heat and power (CHP)** energy system (photo C).

The Energy Centre in the Olympic Park, close to the village, generates electricity by burning biomass (wood chips from felling trees). Heat from burning the fuel provides a central supply of hot water that is piped to the village. As a result, houses in East Village are 30% more energy efficient than average and do not need boilers.

C *The Energy Centre in the Olympic Park*

Activities

1. Look at photo A. What evidence can you see that this is a sustainable community? Mention at least two things.
2. Study the aspects of sustainability around plan B. How well do you think East Village would score on the Green City Index (see spread 11.2)?
   - **a** Draw a simple kite diagram like this. Label it with the aspects of sustainability.

   - **b** Give a score for each aspect of sustainability for East Village. Estimate each score based on how East Village compares with London and Copenhagen (see graph B on spread 11.2). Then plot the scores on your diagram.
   - **c** Join your scores with a line to create a kite. Overall, how does East Village compare with London or Copenhagen?
3. Look at photo C.
   - **a** Describe how a combined heat and power system works.
   - **b** How does this reduce energy consumption in East Village?

Your decision-making enquiry

How could living in London (or another UK city) be more sustainable?

Either for London, or for another city you know well (like the one you live in), plan how it could be made more sustainable.

a Look at the graph you have drawn for East Village in activity 2.
- How do you think your city might compare with East Village?
- Which aspects of sustainability do you think your city might need to improve?
- Which aspects do you think it might already do well?

b Carry out some research to find out what efforts are being made to make your city more sustainable. For example, if you chose London, you could read about transport and land use on spreads 11.3 and 11.4. Then you could do more research online.

c Write an eight-part plan to recommend how your city could be more sustainable. Write a short paragraph for each aspect of sustainability in the Green City Index.

12.1 DME China's economic development

On this spread you will investigate how economic development is changing China.

Links

This decision-making exercise links to other topics:

- Dynamic Development (Chapter 6)
- Changing Climate (Chapter 2)
- Urban Futures (Chapter 5)
- Resource Reliance (Chapter 8)
- Sustaining Ecosystems (Chapter 4).

China's importance

With a population of 1.4 billion people, China is the world's most populated country. This fact alone makes it hugely important. But for the past 30 years or so China's economy has also been growing.

Measured by GDP, China is the second largest economy after the USA. However, measured by **purchasing power parity (PPP)** (see spread 6.3), it has already overtaken the USA and is now the world's largest economy (graph A). Comparing PPP, rather than GDP, between countries is a more accurate reflection of their wealth since it compares what money can actually buy in each country.

China's recent history

China has come a long way in a short time. From 1949 until 1978 it was a **centrally planned economy** under the total control of a communist government. In 1978 the government changed to a **market-based economy**, with more control by companies and individuals. This led to rapid economic and social development (timeline B). From being a low-income developing country (LIDC) in 1978, today China has become an **emerging and developing country (EDC)**. More than 500 million people have been lifted out of poverty and China has now reached all the Millennium Development Goals (MDGs) (see spread 6.11).

However, China's economic development is not complete and about 100 million people, mainly in rural areas, remain in poverty. The country is still controlled by the communist government, which is not elected by the people, unlike countries in the West.

A *Economic growth in China and the USA compared, using PPP*

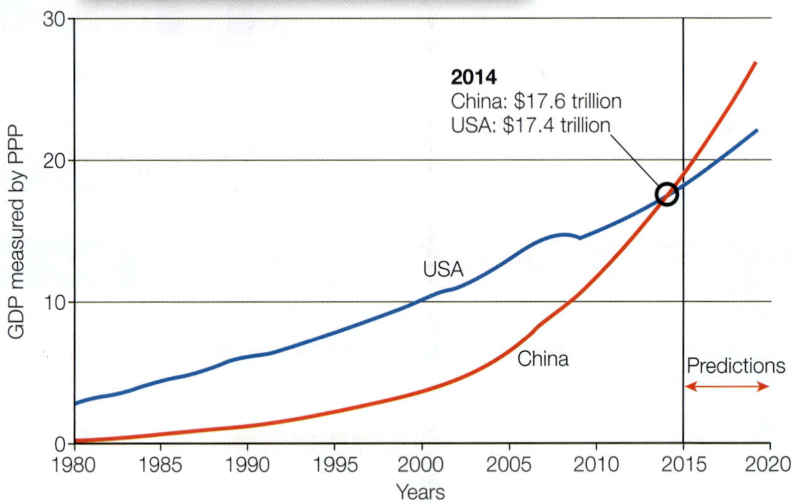

B *Timeline of economic development in China*

1978	First economic reforms begin.
1980	Five special economic zones are set up to bring in foreign investment.
1984	Coastal cities are opened up to foreign investment.
1989	Shanghai Stock Market opens again for the first time in 40 years.
1997	Asian financial crisis slows growth in China, but it soon recovers.
2001	China joins the World Trade Organisation.
2005	Private companies contribute more to the Chinese economy than government investment for the first time.
2011	China overtakes Japan as the world's second largest economy.
2014	China overtakes the USA as the world's largest economy (measured by PPP).

China urbanises

Urbanisation in China has happened at an incredible rate over the past 30 years. There are now six megacities with populations over 10 million – Shanghai (photo C), Beijing, Guangzhou, Tianjin, Shenzhen and Wuhan. The UK has none. There are also 40 cities with over a million people, compared to two in the UK.

Economic growth has happened at the same time as urbanisation. Most new jobs are created in cities, leading to rural–urban migration. About 500 million people have been added to China's urban population in 30 years. It is no coincidence that this is the same number of people who have been lifted out of poverty.

C *Shanghai, China's largest city, today and as it was in 1990*

Economic growth

Economic growth in China has averaged about 10% a year since 1978 (graph D). This high rate over a long period has never happened before anywhere in the world. By comparison, average economic growth in the UK over the same period was below 1%.

China's economic growth has not come without challenges:

- growing inequality between rich and poor, urban and rural
- rapid urbanisation
- environmental pollution (see spreads 12.2 and 12.3)
- imbalances in the world economy, with China growing faster than most other countries.

D *Economic growth rates in China, 1979–2014*

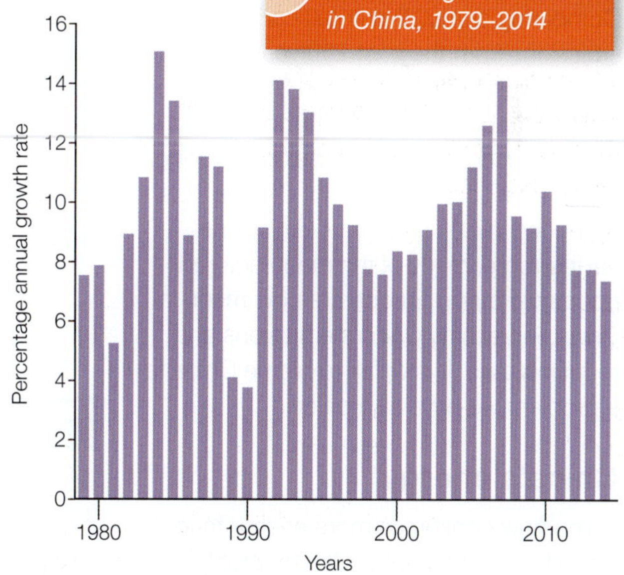

Activities

1 Look at graph A.

a Compare economic growth in China with the USA. Mention when China overtook the USA as the largest economy.

b What is the advantage of using PPP, rather than GDP, to measure economic growth?

2 Study timeline B. Choose three events that help to explain China's economic growth. In each case, explain how that event might have helped economic growth.

3 Look at the photos in C.

a Describe how Shanghai has changed since 1990.

b Explain the connection between these changes and China's economic growth.

4 Look at graph D.

a Describe the rate of economic growth in China over 35 years.

b What evidence is there that economic growth is slowing down?

Your decision-making enquiry

Which way should China develop now?

As you work through this chapter, make notes about anything that will help you to answer the question. You will bring your ideas together to do an activity at the end of the chapter.

China's carbon emissions

On this spread you will consider China's contribution to global carbon emissions.

Should we blame China?

China's economic growth has led to a rapid increase in the country's energy consumption and carbon emissions. Between 1990 and 2010, China's emissions trebled, so that it is now the world's leading emitter of greenhouse gases (graph A). As we come to realise the consequences of climate change (see Chapter 2), there are growing calls for countries to reduce their greenhouse gas emissions. Many fingers point to China, as the worst offender.

Cumulative emissions

Through economic growth, China is hoping to achieve levels of development that advanced countries (ACs) in the West have already achieved. In order to do so, it has developed manufacturing industries in the same way that these countries did in the past – notably the UK during the Industrial Revolution in the 19th century.

Comparing cumulative (total) carbon emissions from the past 250 years, China is not the worst offender (graph B). The USA, with only a quarter of China's population, has produced far more emissions. Europe, with a total population half of China's, has produced three times more emissions. Furthermore, many of the manufactured goods we buy in Europe are now made in China, increasing China's emissions and reducing ours. It is unfair to blame China for a problem we have all created.

China's energy

China now consumes more energy than any other country (table C). The problem is that most of the energy it uses comes from coal, the most polluting fossil fuel of all (chart D). China has vast reserves of its own coal and has 12 000 coal mines. As well as burning coal in power stations, coal is also used by the country's manufacturing industries, especially the steel industry.

A *Growth of China's greenhouse gas emissions, 1990–2011*

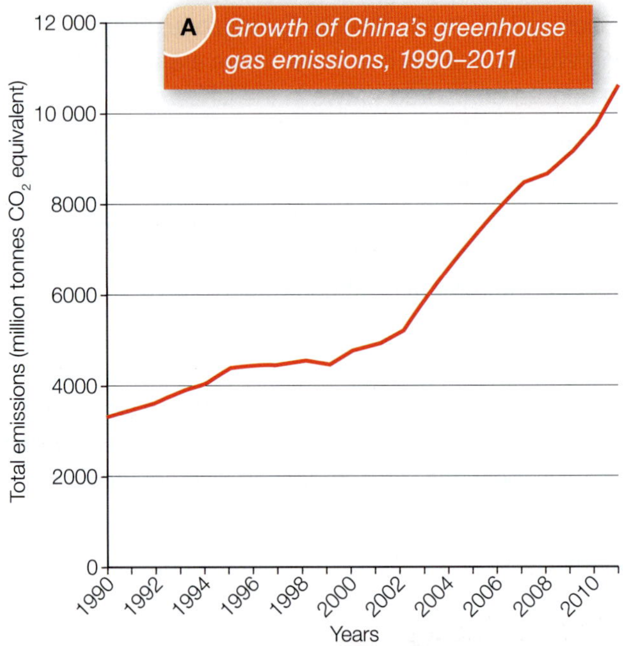

B *Cumulative (total) carbon emissions, 1750–2010*

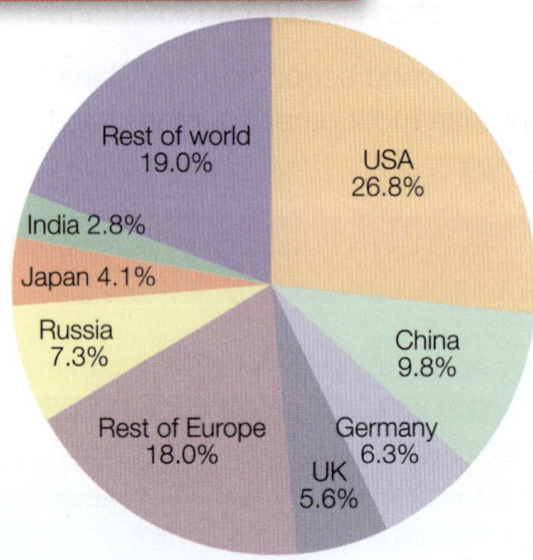

Country	Million tonnes of oil equivalent
China	3034
USA	2224
India	872
Russia	751
Japan	437
Germany	307
Brazil	306
South Korea	277
Canada	251
France	243

C *The world's top ten energy consumers, 2015*

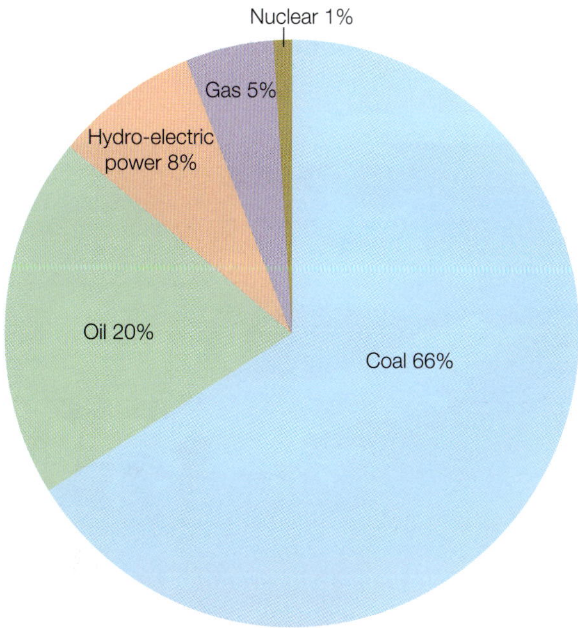

D China's sources of energy

E A steelworks in China

China's steel industry

No material is more important to modern civilisation than steel. Modern buildings, ships, cars, planes, railways and bridges cannot be made without it. And the leading producer of steel in the world is China.

In 2000, China produced 15% of the world's steel, but by 2015 this had risen to 50%. Steel has become one of China's most important exports. It earns money for the country and contributes to its economic growth. But China does not only produce steel, it also consumes it. Over half the world's steel for building construction is used in China. Just think of all those new skyscrapers in Shanghai.

The steel industry is dependent on burning coal (photo E). All the energy used in the production of steel comes from coal and, so far, no low-carbon alternative has been found. Each tonne of steel requires 0.7 tonnes of coal to make it. China's steel industry alone uses 7% of all the world's coal. If the Chinese steel industry were a country, it would rank sixth in the world for energy consumption, ahead of Germany and Canada.

Activities

1 Look at graph A.
- **a** Describe how China's greenhouse gas emissions have increased since 1990.
- **b** Explain why this is a problem.

2 Look at pie chart B.
- **a** What is China's contribution to global cumulative carbon emissions?
- **b** Explain why China's cumulative carbon emissions are less than those of the USA or Europe.

3 Study table C.
- **a** Draw a graph to illustrate the data in the table. Choose a line graph, bar graph or pie chart.
- **b** Justify your choice of graph. Why did you not choose the other two?

4 Look at pie chart D.
- **a** Compare the sources of energy used in China with the UK (see chart E on spread 10.3).
- **b** Suggest reasons for the differences.
- **c** How could China reduce its carbon emissions?

5 Look at photo E. What are the advantages and disadvantages of Chinese steel production for China and for the rest of the world? Complete a large table like the one below.

	Advantages	Disadvantages
China		
Rest of the world		

Environmental problems in China

On this spread you will examine the environmental impact of economic growth on China.

Climate change in China

China is not just a contributor to global climate change through its greenhouse gas emissions. It also experiences the impacts of climate change. Over the past century, average temperatures in China have risen 1.1°C, causing glaciers in the Himalayas to start melting (photo A). Temperatures are predicted to rise by up to 3.5°C in the 21st century.

China's economic growth over the past 30 years has created widespread environmental problems, including pollution and damage to natural ecosystems. Climate change adds to the environmental problems China already faces (map B).

A *Glaciers in the Himalayas melt with rising temperatures*

B *Impacts of climate change in China*

Grow first, clean up later!

For years, China's priority has been economic growth. The policy of the Chinese government was 'grow first, clean up later'. But now, environmental problems are making the government think again about this policy. These are some of the main problems China faces:

- **Habitat and biodiversity loss:** Ecosystems have been destroyed and species driven from their natural habitats by:
  - major engineering projects on rivers, like the Three Gorges Dam on the Yangtze (see spread 8.4)
  - the conversion of wetlands to farming
  - urban expansion.

  Climate change makes the problem worse. For example, lower rainfall threatens the habitat of the giant panda in south-west China.
- **Air pollution:** Lung and heart disease, both related to air pollution, are the leading causes of death in China. Dependence on burning coal and the growing number of cars in cities are the main causes. It is estimated that air pollution costs the Chinese economy up to 10% of GDP each year.
- **Water pollution:** Decades of waste poured from factories and cities into China's rivers have turned them into open sewers (photo C). About 40% of river water is unfit for humans to drink and leads to higher cancer rates (extract D).
- **Desertification and erosion:** Around 30% of China's land has been desertified and is no longer suitable for human habitation. As the desert expands, soil is eroded and land can no longer be used for farming. The problem is worst in northern China where villages have been abandoned.

C *A Chinese river, polluted by industry*

Living in China's 'cancer villages'

Xie Zhengqiang is no stranger to death but he trembles every time he thinks about his nephew. When doctors diagnosed three-month-old Xie Yuling with a rare disease in 2011, Xie went to the Shanghai hospital where his nephew was undergoing treatment. Three times a week he cycled from his village, past the miles of adjacent chemical, pharmaceutical and power plants, along the Yangtze River where his family collects drinking water.

Then one day, he got the call. 'The doctor tried everything to make him better, but he died anyway,' said Xie, gazing up at the paper mill which towers above his home. 'It was definitely the pollution. What else could it have been?'

Yanglingang residents count their home as one of China's 'cancer villages' – small communities near polluting factories where cancer rates have soared far above the national average.

D *Adapted from news reports, 5 June 2013*

Activities

1 Look at photo A.
   - **a** Why are glaciers in China threatened by climate change?
   - **b** What impact could this have on people in China?

2 Look at map B. Explain the connection between climate change and the following hazards. (Clue: look back at Chapter 1 for ideas.)
   - **a** Typhoons
   - **b** Rising sea level
   - **c** Drought

3 Look at photo C and extract D.
   - **a** What is the evidence that pollution is the cause of high cancer rates in Xie's village?
   - **b** Suggest how you could carry out a research project to prove the connection.

4 To what extent do you think the policy of 'grow first, clean up later' has been:
   - **a** successful? (Clue: look back at spread 12.1.)
   - **b** desirable?

   Write two short paragraphs.

China's water crisis

On this spread you will find out how a large-scale water transfer project could solve China's water crisis.

Water shortage in China

China's economic growth, together with the environmental problems it has caused, has led to water shortage. More than half of China's rivers have disappeared and half of the remaining rivers are polluted. The problems are most severe in the semi-arid North China Plain (see map B on spread 12.3), where there has been rapid economic growth around the cities of Beijing and Tianjin.

A *A canal moving water for the South–North Water Transfer Project*

China's South–North Water Transfer Project

The Chinese government aims to solve the water shortage problem in the North China Plain with the South–North Water Transfer Project (SNWTP). It is the world's largest **water transfer scheme**, built at a cost of $62 billion (photo A). Water is moved from the humid south, with high rainfall, to the more arid north (map B).

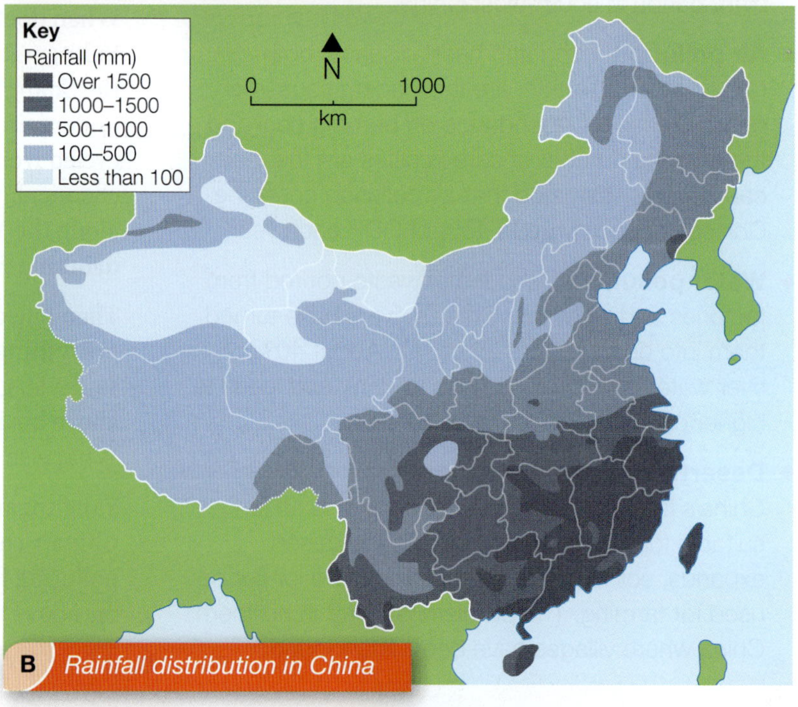

B *Rainfall distribution in China*

The plan is to increase the water supply to Beijing and other cities to maintain industrial and agricultural production and to allow continued urbanisation (chart C). With low rainfall, most of the water supply for the North China Plain has previously come from groundwater. The **water table** (the level of water in the underlying rock) has been sinking by about 5 metres every year. Wells have to be drilled to a depth of 1 km in order to obtain water.

The idea of a water transfer scheme arose in the 1950s when the former Chinese leader, Mao Zedong, said, 'The south has plenty of water but the north is dry – if we could borrow some that would be good.' But, as with many large-scale projects, it turned out to be more complicated than it seemed.

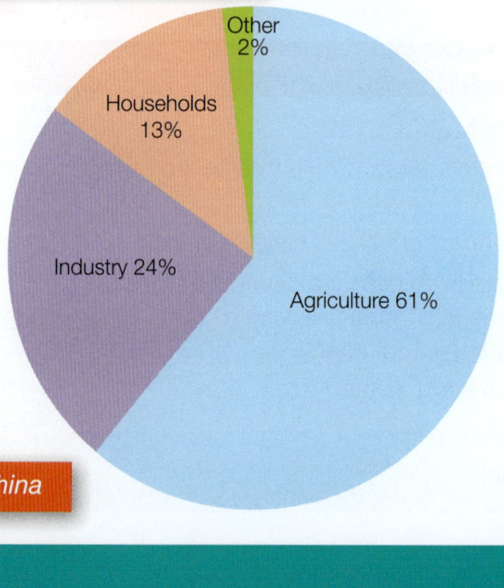

C *Water use in China*

How the project works

The idea of the SNWTP is to transfer water along three routes between China's two largest river basins – the Yangtze River in the south to the Yellow River in the north (map D).

- **Eastern route:** Completed in 2012, this route is 1150 km long from the lower Yangtze to Tianjin via canals, aqueducts and tunnels. The problem is that the lower Yangtze is already polluted and there is a risk of polluted water being transferred.
- **Central route:** Completed in 2013, a large reservoir has been created at Danjiangkou on the Han River, a tributary of the Yangtze. 330 000 people were forced to leave their villages to make way for the reservoir. The water is moved 1260 km along a canal to Beijing.
- **Western route:** This route is planned but may never be built. It would involve transferring water from the upper Yangtze to the upper Yellow River by building several dams and tunnels through the mountains. It could prove too costly, even for China.

D *China's South–North Water Transfer Project*

Uncertain future

In practice the SNWTP has several drawbacks. It may be that the Chinese government will abandon plans for the Western route.

- Changing climate means that the south of China will not be able to supply as much water as expected.
- Water evaporates from canals and this is wasteful.
- Focusing on the water supply for cities means that villages and farms miss out.
- The Yangtze's flow is reduced by transferring water from it, so there is less floodwater for natural ecosystems.
- The cost of the project has risen above $62 billion and there may be cheaper, more sustainable ways to solve China's water crisis.

Activities

1 Look at photo A.
- **a** Explain the purpose of China's South–North Water Transfer Project.
- **b** From the photo, suggest one problem the project may face.

2 Look at map B.
- **a** Describe the distribution of rainfall in China.
- **b** Compare this with the distribution of China's major cities (Clue: look at map B on spread 12.3 or use a more detailed atlas map.)

3 Look at chart C.
- **a** Explain how agriculture, industry and households each use water.
- **b** How do you explain the high use of water for agriculture?

4 There are mixed opinions in China about the SNWTP. Suggest what each of these people might think about the scheme. In each case, give reasons.
- **a** A factory worker in Beijing
- **b** A villager in Danjiangkou
- **c** A fisherwoman on the Yangtze River
- **d** A Chinese government minister

Sustainable development

On this spread you will consider whether it is possible to have economic growth and sustainable development.

China's priorities

China's priorities are changing. The Chinese government now emphasises the importance of the 'quality of growth', not simply the economic growth rate. It now expects the Chinese economy to grow at 7% each year rather than the average of 10% over the past 35 years. This is still faster than most other economies and certainly a lot faster than the UK's.

People in China are concerned about the environment. For many years, the government ignored, or even denied, China's environmental problems. Grow first, clean up later! Now the government openly talks about the importance of tackling climate change and the need for sustainable development.

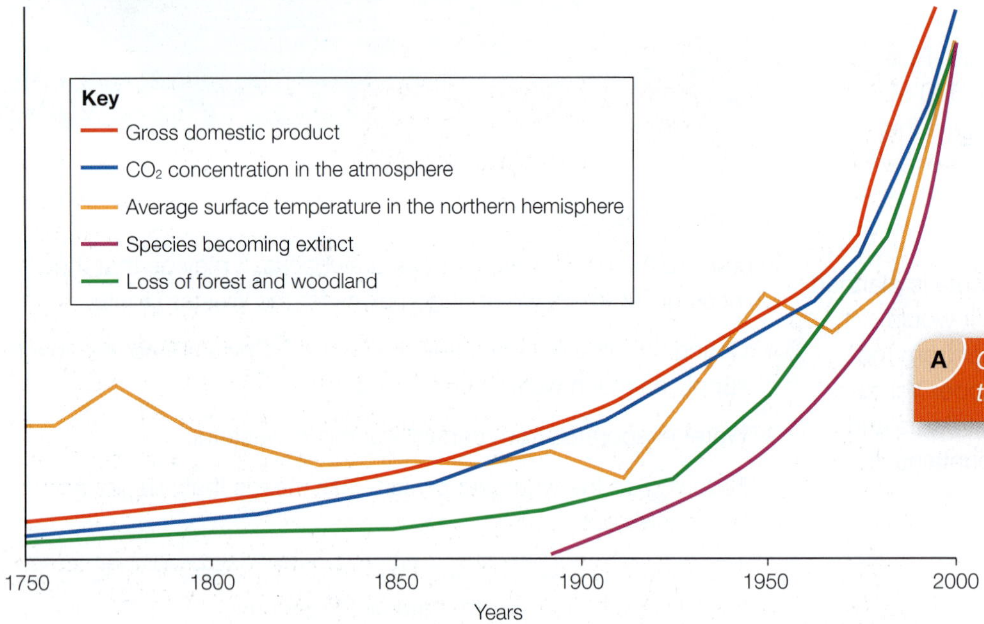

A *Global changes over the past 250 years*

Economic growth vs sustainable development

Is it possible to have both economic growth and sustainable development? Graph A might give us a clue. It shows connections between some of the global changes over the past 250 years. Remember, China's economic development has only taken off in the past 35 years, just as these global changes have begun to get quicker.

Sustainable development is development that meets people's present needs without compromising the ability of future generations to meet their needs. It means that we must look after the planet that we and our descendants depend on, including:

- the atmosphere
- rivers and lakes
- oceans
- natural ecosystems.

Sustainable water solutions

We can apply the idea of sustainable development to China's water supply (see spread 12.4). Rather than expensive engineering solutions to transfer water from one region of China to another, there may be other ways of improving water supply without damage to the environment. Some sustainable water solutions are:

- reducing water loss through evaporation and leaks
- using water more efficiently, e.g. showering rather than bathing
- reducing unnecessary use of water by increasing the price
- re-using water by cleaning it and removing sewage or pollution
- maintaining groundwater supplies and trying to restore them to previous levels.

Sustainability – back to the future

There was a time, only 40 years ago, when China was known for its bicycles (photo B). In cities, bicycles were more common than cars. Now it is the other way round (photo C). From a sustainability viewpoint, this is not a good thing. Cars are far more polluting than bicycles and they cause more traffic congestion. Even worse, cars also lead to more carbon emissions than bicycles (see graph C on spread 11.3).

In UK cities bicycles are growing more popular again as people realise they can get fitter and healthier riding a bicycle. We realise they are more sustainable too. It may take people in China longer to reach that conclusion, having only started to drive cars more recently.

B *Then… bicycles in China in the 1970s*

C *…and now. Traffic in Beijing today*

Activities

1 Look at graph A. Describe what the graph shows. Write a sentence about each line on the graph.

2 Explain the connections between each of the lines on graph A. One way to do this is to draw a large diagram, like the one below. Draw lines to link the changes that you think are connected. Explain the connection on each line, e.g. *CO_2 concentrations increase with GDP because as countries get richer they use more energy and burn more fuel.*

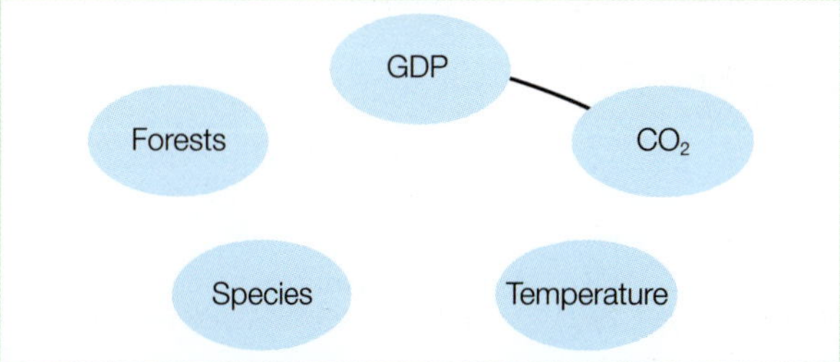

3 Is it possible to have economic growth and sustainable development? Explain your answer.

Your decision-making enquiry

Which way should China develop now?

Decide how China should develop. For each of the following pairs of options, decide which you think is the more sustainable. This will give you a four-point plan for the way China should develop. In each case, give reasons for your decision. Remember, there are 100 million people in China still living in poverty.

Economic growth

- Switch from a manufacturing to a service economy (like the UK), by closing factories and opening offices.
- Keep factories open but where possible switch to cleaner fuels and technology.

Energy

- Switch from burning coal to cleaner non-renewable energy sources like gas and nuclear power.
- Switch from burning coal to renewable forms of energy, like wind and solar power, which are less reliable.

Water

- Transfer water from wet regions of the country to dry regions by a network of canals and tunnels.
- Charge more for water. Use the extra money to repair leaks and clean up pollution.

Transport

- Ban cars in cities and make people go back to using bicycles.
- Improve public transport, like trains and buses, so people will choose to use them rather than cars.

Aiming for success in Geography

About the course

The GCSE course in this book is for the OCR Geography B specification. One thing makes it stand out from other GCSE courses. The clue is in the name – Geography for Enquiring Minds. This course is designed to make you think like a geographer, by asking interesting geographical questions – How can weather be hazardous? Why should tropical rainforest matter to us? Why are some countries richer than others?

Here are some other important features of the course:

- Eight topics cover aspects of physical and human geography under the headings, Our Natural World, and People and Society.
- Links and interconnections are made between the topics, particularly through decision-making exercises in the section on Geographical Exploration.
- Fieldwork is built into the course. You do at least two fieldwork investigations, one into physical geography and one into human geography.
- Other skills are also built into the course, including the use of a wide range of maps, graphs and statistics.

Plan your revision

- Don't leave all your revision to the last minute. Start a few weeks before the assessments.
- Split revision into chunks. Don't try to do too much at one time. You could split it into at least eight chunks – one for each topic in the course.

The structure of the assessment

At the end of the course you will be assessed through three question papers.

Paper	Content	Assessment
1 Our Natural World (physical geography)	There will be questions on all themes: • Global Hazards • Changing Climate • Distinctive Landscapes • Sustaining Ecosystems • Fieldwork and Geographical Skills	35% of GCSE total (70 marks) 1 hour 15 minutes
2 People and Society (human geography)	There will be questions on all themes: • Urban Futures • Dynamic Development • UK in the 21st Century • Resource Reliance • Fieldwork and Geographical Skills	35% of GCSE total (70 marks) 1 hour 15 minutes
3 Geographical Exploration (physical and/or human geography)	• Geographical Skills • Decision-making Exercise	30% of GCSE total (60 marks) 1 hour 30 minutes

Note: Paper 3 carries fewer marks, but you get more time. This is because there is a longer Resource Booklet for you to read.

How to answer an assessment question

The assessments are quite straightforward. You have to answer all the questions. What could possibly go wrong?

> **1** Study **Fig. 4** in the separate Resource Booklet, showing a climate graph.
>
> **(a)** The climate graph in **Fig. 4** shows the climate of a:
>
> **A** hot desert ☐
>
> **B** polar region ☐
>
> **C** temperate forest ☐
>
> **D** tropical rainforest ☐
>
> Write the correct letter in the box. **[1]**
>
> **(b)** Describe the impact of human activities in tropical rainforest.
> Mention **four** activities. **[4]**
>
> **(c)** **CASE STUDY – an attempt to sustainably manage an area of tropical rainforest.**
> Examine how successful **one** example of sustainable management of tropical rainforest has been. **[8]**

You will be given a separate Resource Booklet with data, graphs, diagrams, photos and/or maps. You need to refer to the booklet to answer the questions.

Questions are broken down into parts, (a), (b), (c), etc. The questions start easier and get more difficult, so there are more marks for the later parts.

'Describe' and 'Examine' are command words. They tell you what to do (see the table below). Only do what you are asked. You won't get any extra marks by doing more than you are asked.

'Case study' means you have to refer to an example you have studied in the classroom. There is a table of case studies in this book on page 324.

The marks are allocated for each part of the question. The more marks, the more time you should spend on that part and the more you need to write. Some of the tasks that need an extended response will be allocated 3 marks for spelling, punctuation and grammar, and the use of geographical terminology. These will be clearly indicated on the paper.

Command word	What it means	Example
Compare	Describe similarities and differences between two or more things.	Compare the characteristics of cities in an LIDC and an AC.
Define	Give the meaning of a word, phrase or idea.	Define 'sustainable development'.
Describe	Give the main characteristics of something. No explanation is needed.	Describe how a river changes from its source to its mouth.
Discuss	Give the points on both sides of an argument and come to a conclusion.	Discuss the effectiveness of coastal management.
Evaluate	Weigh up the good and bad points to make a judgement.	Evaluate a development strategy in an LIDC.
Examine	Investigate in detail.	Examine the evidence for climate change in the UK.
Explain	Give reasons why something happens.	Explain why earthquakes occur at plate boundaries.
Identify	Name an example, sometimes from a map, photo or graph.	Identify three countries with food shortages from the map.
Justify	Give evidence to support your ideas.	Does the UK still have any global influence? Justify your opinion.
Outline	Summarise the main points.	Outline how the population of the UK is changing.
Suggest	Give an explanation for something when you can't be sure.	Suggest why the number of polar bears in the world has declined.
To what extent...?	Judge the importance of something.	To what extent are people the cause of flooding?

Case studies

In this book you have learned about a number of case studies. You will need to revise these for your assessments. When you are asked about a case study, you should be able to refer to one you know about.

Case study in the specification	Case study used in this book	Spreads
1 Global Hazards		
Weather hazard outside the UK	Typhoon Haiyan, Philippines, 2013	1.8–1.9
Weather hazard in the UK	UK drought, 2012	1.10
Tectonic hazard	Earthquake, Nepal, 2015	1.17–1.18
3 Distinctive Landscapes		
Coastal landscape in the UK	Jurassic Coast, Dorset and Devon	3.9–3.11
River basin in the UK	River Thames	3.14–3.17
4 Sustaining Ecosystems		
Sustainable management of tropical rainforest	Ecotourism, Costa Rica	4.7–4.8
Small-scale sustainable management in a polar region	Sustainable whaling, northern Canada	4.12
Global sustainable management in a polar region	Proposed sanctuary/treaty, Arctic	4.13–4.14
5 Urban Futures		
City in an LIDC or EDC, including one initiative to make it more sustainable	Istanbul, Turkey, including the Marmaray Rail Project	5.15–5.19
City in an AC, including one initiative to make it more sustainable	Birmingham, UK, plan to become a leading green city	5.8–5.12
6 Dynamic Development		
An LIDC, including one top-down development strategy and one bottom-up strategy	Zambia, including the Kariba Dam and Room to Read literacy project	6.9–6.17
7 UK in the 21st Century		
The UK in the 21st century, including one economic hub, the UK's role in one global conflict and the contribution of ethnic groups to one aspect of cultural life of the UK	The UK, including Cambridge as an economic hub, the UK's role in the Middle East and the contribution of ethnic groups to food in the UK	7.11, 7.15, 7.17
8 Resource Reliance		
Attempts to achieve food security in one country, including one attempt at a local scale, and one past and one present attempt at a national scale	Tanzania, including Goat Aid in Babati, the Tanzania–Canada Wheat Programme (1967–93) and the Southern Agricultural Growth Corridor of Tanzania	8.8–8.11

Tips for extended writing

To aim for top grades, you need to be able to do extended writing. Here are a few writing tips:

1. Read the question carefully. The command words tell you what to do.
2. Plan your answer before you start writing. Jot down the key points on a piece of paper to remind you.
3. Structure your writing into clear sentences or paragraphs. Don't forget – there are extra marks for spelling, punctuation and grammar.
4. Don't panic! If you are running out of time, just put down the key points. You will still get some marks.

Skills matching grids

This student book presents all the geography you need to study your course. As far as possible, it also gives you the opportunity to practise and development the skills you need to 'think geographically', as detailed in these grids. For more practise in all the required skills, see the worksheets on **GCSE Geography for OCR B Kerboodle**.

Cartographic skills

Skills	Spreads
Select and construct maps, using appropriate scales and annotations, to present information	1.16, 1.19; 2.2, 2.9, 2.11; 3.7, 3.11; 4.7; 5.5, 5.6; 6.3, 6.9, 6.16; 7.6, 7.9, 7.11; 11.2
Interpret cross-sections and transects	3.3
Use and understand co-ordinates, scale and distance	3.13, 3.15; 5.5; 7.13; 10.1
Extract, interpret, analyse and evaluate information	1.1–1.5, 1.7, 1.8, 1.10, 1.12–1.14, 1.16, 1.17; 2.2, 2.3, 2.6–2.12; 3.2–3.4, 3.7, 3.9, 3.10, 3.11, 3.13, 3.14, 3.16, 3.17; 4.2, 4.6–4.8, 4.10–4.12, 4.13; 5.1, 5.2, 5.4, 5.6–5.9, 5.11, 5.15, 5.17–5.19; 6.2–6.9, 6.12–6.14, 6.16; 7.1, 7.2, 7.5–7.7, 7.9–7.15; 8.6, 8.8–8.11; 9.1–9.5; 10.1, 10.2, 10.4, 10.5; 11.1–11.3, 11.5; 12.3, 12.4
Use and understand gradient, contour and spot height (on OS and other isoline maps)	3.13, 3.15; 10.1
Describe, interpret and analyse geo-spatial data presented in a GIS framework	1.8, 1.17; 3.9, 3.17; 5.4, 5.7; 6.6, 6.7; 9.1, 9.4; 10.2, 10.5
Study atlas maps	1.4, 1.16; 2.10; 3.2, 3.3, 3.9–3.11, 3.14, 3.17; 4.6–4.8, 4.10, 4.11; 5.1, 5.2, 5.8, 5.15, 5.17; 6.9, 6.16; 7.1, 7.2, 7.7, 7.11, 7.14; 8.8–8.10; 10.1, 10.5; 11.2; 12.3, 12.4
Study OS maps (1:50 000 and 1:25 000 scales)	3.13; 7.13; 10.1
Study base maps	1.8, 1.17; 6.7; 9.1; 10.2; 11.1
Study chloropleth maps	1.2, 1.4, 1.7, 1.10; 2.3, 2.6, 2.8–2.9, 2.11; 3.4, 3.10, 3.16; 5.1, 5.9, 5.11; 6.3; 7.2, 7.5; 9.2, 9.3, 9.5; 12.4
Study isoline maps	1.4; 3.13; 10.1
Study flow-line maps	5.2, 5.8; 6.5, 6.14
Study desire-line maps	6.4
Study sphere of influence maps	1.17; 7.9
Study thematic maps	1.1–1.3, 1.5, 1.7, 1.10, 1.12–1.14; 2.7, 2.10; 3.2, 3.10, 3.14, 3.16, 3.17; 4.2, 4.7, 4.11–4.13; 5.2, 5.4, 5.6, 5.7, 5.18; 6.2, 6.6–6.8, 6.12; 7.2, 7.9–7.11, 7.14, 7.15; 8.6, 8.11; 9.1, 9.4; 10.2, 10.4, 10.5; 11.5
Study route maps	1.8; 3.10; 4.11, 4.12; 5.8, 5.19; 6.13; 7.11; 9.2; 11.3; 12.4
Study sketch maps	5.14; 7.12; 11.5

Graphical skills

Skills	Spreads
Select and construct appropriate graphs and charts, using appropriate scales and annotations to present information	1.11,1.12; 2.5, 2.12; 3.13; 5.11, 5.14, 5.16; 6.3, 6.4; 7.1, 7.5, 7.6, 7.17; 8.4, 8.6; 10.2; 11.4, 11.5; 12.2
Effectively present and communicate data through graphs and charts	1.11; 2.5, 2.12; 3.13; 5.11, 5.14, 5.16; 6.3, 6.4; 7.1, 7.6, 7.17; 8.4, 8.6; 10.2; 11.4, 11.5; 12.2
Extract, interpret, analyse and evaluate information	1.3–1.5, 1.10, 1.11; 2.1, 2.2, 2.5, 2.6; 3.4, 3.5; 4.2, 4.4, 4.6, 4.7, 4.9, 4.12; 5.1, 5.2, 5.4, 5.9, 5.11, 5.14, 5.16, 5.17, 5.19; 6.2–6.4, 6.7, 6.9–6.11, 6.13, 6.17; 7.1, 7.3, 7.4, 7.6–7.8, 7.10, 7.14, 7.15, 7.17; 8.1, 8.5–8.8, 8.10, 8.12, 8.13; 9.1, 9.3; 10.2, 10.3; 11.1–11.5; 12.1, 12.2, 12.4, 12.5
Study bar graphs (horizontal, vertical and divided)	1.4, 1.5, 1.11; 3.4; 4.2, 4.4, 4.9; 5.11, 5.14, 5.17, 5.19; 6.2, 6.7, 6.10, 6.11, 6.13, 6.17; 7.3, 7.4, 7.7, 7.8, 7.10, 7.14, 7.15; 8.1, 8.10, 8.12, 8.13; 10.3; 11.1, 11.3; 12.1
Study histograms (with equal class interval)	*You will find worksheets to practise this skill on GCSE Geography for OCR B Kerboodle.*
Study line graphs	1.5; 2.1, 2.2, 2.6; 3.4, 3.5; 4.4, 4.6, 4.7, 4.9, 4.12; 5.1, 5.9, 5.16; 6.10, 6.13; 7.3, 7.6–7.8; 8.1, 8.7, 8.8, 8.10; 9.3; 12.1, 12.2, 12.5
Study scatter graphs (including best fit line)	6.4; 8.6
Study dispersion graphs	*You will find worksheets to practise this skill on GCSE Geography for OCR B Kerboodle.*
Study pie charts	1.10; 2.5, 2.6; 3.5; 6.9, 6.13; 7.1, 7.6, 7.17; 8.5; 9.1; 10.3; 11.4; 12.2, 12.4
Study climate graphs	1.4; 3.4; 4.4, 4.9
Study proportional symbols	1.17; 5.2, 5.4, 5.5, 5.7; 6.6, 6.7; 10.2; 11.2
Study pictograms	7.14, 7.17
Study cross-sections	1.12–1.16, 1.19; 3.2, 3.4, 3.6, 3.8, 3.9, 3.13–3.15; 4.10; 6.16; 7.2; 10.4, 10.5
Study population pyramids	5.9; 7.3, 7.4, 7.6
Study radial graphs	11.2, 11.5
Study rose charts	1.3

Numerical and statistical skills

Skills	Spreads
Demonstrate an understanding of number, area and scale	1.2–1.12, 1.17, 1.19; 2.1–2.7, 2.10–2.12; 3.2–3.5, 3.12, 3.13, 3.15, 3.16; 4.2–4.4, 4.6, 4.7, 4.9, 4.12–4.13; 5.1, 5.2, 5.4–5.7, 5.9–5.14, 5.16, 5.17, 5.19; 6.2–6.11, 6.13, 6.17; 7.2–7.10, 7.14–7.16; 8.1, 8.4, 8.6–8.10, 8.12, 8.13, 8.15; 9.1–9.3; 10.1–10.3; 11.1–11.3, 11.5; 12.1, 12.2, 12.4
Demonstrate an understanding of quantitative relationships between units	2.6; 5.1, 5.9, 5.16, 5.17; 6.2, 6.4, 6.6, 6.13; 7.3, 7.4, 7.6; 8.6–8.8, 8.10; 12.1, 12.5
Understand and correctly use proportion, ratio, magnitude and frequency	1.8, 1.10, 1.15, 1.17; 5.9, 5.14; 6.8, 6.9, 6.17; 7.4, 7.6; 9.4
Understand and correctly use appropriate measures of central tendency, spread and cumulative frequency, including median, mean, range, quartiles and inter-quartile range, mode and modal class	1.4; 5.16; 8.8
Calculate and understand percentages (increase and decrease) and percentiles	1.5, 1.8, 1.10, 1.17; 2.5, 2.6, 2.8, 2.11; 3.5; 4.5, 4.7, 4.13, 4.14; 5.1, 5.4, 5.6, 5.9–5.12, 5.14, 5.16, 5.19; 6.6, 6.7, 6.9, 6.10, 6.11, 6.13, ,6.15, 6.17; 7.1, 7.4–7.8, 7.10, 7.17; 8.5, 8.8–8.10; 9.1–9.3; 10.2, 10.3; 11.1, 11.4, 11.5; 12.1–12.4
Design fieldwork data collection sheets and collect data with an understanding of accuracy, sample size and procedures, control groups and reliability	1.11; 2.12; 5.13, 5.14; 7.12
Interpret tables of data	1.6, 1.8, 1.9, 1.15; 2.3, 2.12; 3.12, 3.13; 5.5, 5.6, 5.11, 5.13, 5.14, 5.16; 6.2–6.5, 6.7, 6.9; 7.5, 7.6, 7.9, 7.10, 7.16; 8.4, 8.6, 8.8, 8.9; 9.2; 10.2; 11.2; 12.2
Describe relationships of bivariate data	2.6; 6.4, 6.13; 8.8, 8.10
Sketch trend lines through scatter plots	6.10
Draw estimated lines of best fit	6.3, 6.4; 8.6
Make predictions; interpolate and extrapolate trends from data	2.2, 2.3; 3.13; 7.3; 9.4; 10.2, 10.3
Be able to identify weaknesses in statistical presentations of data	6.10; 7.3
Draw and justify conclusions from numerical and statistical data	1.4, 1.6, 1.10, 1.15, 1.17; 2.3; 3.13; 4.3; 5.5–5.7, 5.10, 5.11, 5.14; 7.4–7.6

Skills for formulating enquiry and argument

Skills	Spreads
Deconstruct, interpret, analyse and evaluate visual images including photos, cartoons, pictures and diagrams	1.1–1.19; 2.1–2.10, 2.12; 3.1–3.17; 4.1–4.14; 5.1–5.6, 5.8, 5.10–5.13, 5.16–5.19; 6.1, 6.3–6.17; 7.1, 7.2, 7.4, 7.5, 7.7–7.10, 7.13, 7.15, 7.17; 8.1– 8.4, 8.8, 8.12, 8.15; 9.1–9.5; 10.1–10.5; 11.1–11.5; 12.1–12.5
Analyse written articles from a variety of sources for understanding, interpretation and recognition of bias	2.3; 3.10; 4.8, 4.12; 7.5, 7.16; 8.12; 9.3; 12.3
Suggest improvements to, issues with or reasons for using maps, graphs, statistical techniques and visual sources, such as photos and diagrams	1.2, 1,10–1.12; 2.3, 2.5, 2.6, 2.12; 3.16; 4.9; 5.11, 5.13, 5.14; 6.3, 6.10; 7.1, 7.3, 7.5–7.7, 7.12, 7.17; 8.4, 8.6; 11.4; 12.1, 12.2

Fieldwork skills

Skills	**Spreads**
Understand the kinds of question capable of being investigated through fieldwork	1.11; 2.12; 3.12, 3.13; 5.13, 5.14; 7.12, 7.13
Understand the geographical enquiry processes appropriate for investigating fieldwork questions	1.11; 2.12; 3.12, 3.13; 5.13, 5.14; 7.12, 7.13
Understand the range of techniques and methods used in fieldwork, including observation and different kinds of measurement	1.11; 2.12; 3.12, 3.13; 5.13, 5.14; 7.12, 7.13
Process and present fieldwork data in various ways, including maps, graphs and diagrams	1.11; 2.12; 3.13; 5.14; 7.12
Analyse and explain data collected in the field, using knowledge of relevant geographical case studies and theories	1.11; 2.12; 3.12; 5.14
Draw evidenced conclusions and summaries from fieldwork transcripts and data	1.11; 3.12, 3.13; 5.13, 5.14; 7.12
Reflect critically on fieldwork data, methods used, conclusions drawn and knowledge gained	1.11; 2.12; 5.13; 7.12

Symbols on Ordnance Survey maps

ROADS AND PATHS

- M I or A 6(M) — Motorway
- A 35 — Dual carriageway
- A 31(T) or A 35 — Trunk or main road
- B 3074 — Secondary road
- Narrow road with passing places
- Road under construction
- Road generally more than 4 m wide
- Road generally less than 4 m wide
- Other road, drive or track, fenced and unfenced
- Gradient: steeper than 1 in 5; 1 in 7 to 1 in 5
- Ferry — Ferry; Ferry P – passenger only
- Path

PUBLIC RIGHTS OF WAY

(Not applicable to Scotland)

1:25 000 1:50 000

- Footpath
- Road used as a public footpath
- Bridleway
- Byway open to all traffic

RAILWAYS

- Multiple track
- Single track
- Narrow gauge/Light rapid transit system
- Road over; road under; level crossing
- Cutting; tunnel; embankment
- Station, open to passengers; siding

BOUNDARIES

- National
- Dictrict
- County, Unitary Authority, Metropolitan District or London Borough
- National Park

HEIGHTS/ROCK FEATURES

- 50 — Contour lines
- ·144 — Spot height to the nearest metre above sea level

ABBREVIATIONS

P	Post office	PC	Public convenience (rural areas)
PH	Public house	TH	Town Hall, Guildhall or equivalent
MS	Milestone	Sch	School
MP	Milepost	Coll	College
CH	Clubhouse	Mus	Museum
CG	Coastguard	Cemy	Cemetery
Fm	Farm		

ANTIQUITIES

- VILLA — Roman
- Castle — Non-Roman
- *Battlefield* (with date)
- *Tumulus/Tumuli* (mound over burial place)

LAND FEATURES

- ruin — Buildings
- Public building
- Bus or coach station
- Place of Worship: with tower; with spire, minaret or dome; without such additions
- Chimney or tower
- Glass structure
- Heliport
- Triangulation pillar
- Mast
- Wind pump / wind generator
- Windmill
- Graticule intersection
- Cutting, embankment
- Quarry
- Spoil heap, refuse tip or dump
- Coniferous wood
- Non-coniferous wood
- Mixed wood
- Orchard
- Park or ornamental ground
- Forestry Commission access land
- National Trust – always open
- National Trust, limited access, observe local signs
- National Trust for Scotland

WATER FEATURES

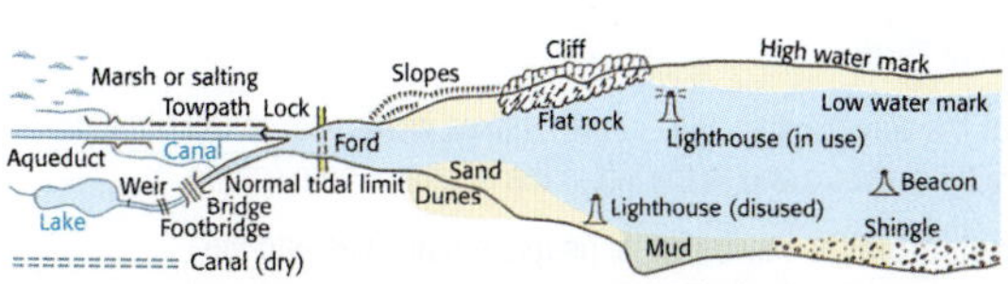

TOURIST INFORMATION

- Parking
- P&R — Park & Ride
- Visitor centre
- Information centre
- Telephone
- Camp site/ Caravan site
- Golf course or links
- Viewpoint
- PC — Public convenience
- Picnic site
- Pub/s
- Museum
- Castle/fort
- Building of historic interest
- Steam railway
- English Heritage
- Garden
- Nature reserve
- Water activities
- Fishing
- Other tourist feature
- Moorings (free)
- Electric boat charging point
- Recreation/leisure/ sports centre

Glossary

A

abrasion the process of erosion in which rocks and other hard materials in water wear away the water channel or coast

adult literacy rate the percentage of adults able to read

advanced country (AC) country that is wealthy, has a wide range of jobs and many services

afforestation planting trees to create new forest

agribusiness farming by large companies that grow food for profit on huge areas of land

agroforestry growing trees and crops together

albedo effect the amount a surface reflects the Sun's rays back into space

attrition the process of rocks in water bashing into and wearing each other down into smaller, rounder particles

B

backwash the movement of waves down a beach

bi-lateral aid international aid given by one country to another

biodiversity variety of plant and animal species

biofuel fuel made from renewable plant material

biological weathering breaking up of rock caused by plants or animals

biomass the mass, or weight, of living material

biome large ecosystem characteristic of a specific part of the world

bottom-up aid/development aid/development funded and carried out by NGOs in co-operation with local communities

brownfield site a site in an urban area, either derelict or with old buildings, that could be used for future development

by-catch the unwanted fish and animals caught in huge fishing nets

C

carbon footprint the amount of carbon emissions produced by individuals, organisations or communities

carbon sink an area, such as rainforest, that uses up carbon dioxide in the atmosphere

chemical weathering the breakdown of material caused by the action of rainwater in rock

climate change the long-term change in global or regional climate due largely to increased levels of CO_2 in the atmosphere

collision boundary where tectonic plates move towards each other, creating mountains

colonialism the ruling of one country by another country

composite volcano a steep volcano formed by alternate layers of lava and ash on destructive plate boundaries

conservation protection

conservative boundary where tectonic plates move alongside each other

constructive boundary where tectonic plates move apart and new land is created

continental crust the part of Earth's crust that makes continents

convection current pockets of warm air or liquid that rise, then cool and fall, in a circular motion

convectional rainfall rainfall, often heavy, that occurs when thick clouds form because of warm air rising quickly

core the centre of Earth

coriolis effect the way the spinning of Earth makes winds veer to the right in the northern hemisphere and to the left in the southern hemisphere

counter-urbanisation the movement of people out of cities into the countryside and smaller towns

D

debt relief the total or partial cancellation of debts owed by LIDCs

deep focus earthquake an earthquake that starts 70–700 km below Earth's surface

deforestation the clearing of trees to transform forest into cleared land

Demographic Transition Model a model that shows the changes a population is likely to go through over time

dependency ratio the proportion of people under 16 or over 65 who depend on the working population

destructive boundary where tectonic plates are forced together

development an improvement in living standards through better use of resources

drought a period when there is much less precipitation over a specific time than is usual for an area, leading to water shortages

dry valley valley left after the river that formed the valley has disappeared

E

economic development an increase in the total goods and services a country produces

economic hub the focal point for the economy of a country or region

ecosystem a community of plants and animals, and the environment in which they live

ecotourism travel that aims to conserve the natural environment and local communities

El Niño an event that occurs when weak trade winds blowing east to west allow surface temperatures to increase

emerging and developing country (EDC) country that is in transition from being a low-income developing country to an advanced country

emissions greenhouse gases put into the atmosphere

employment sectors different types of industry that employ people

enhanced greenhouse effect the trapping of more of the Sun's energy because of the cumulative effect of greenhouse gases, causing Earth's temperature to increase

epicentre the point on Earth's surface directly above an earthquake's focus

ethical consumerism buying products that do not have negative social, economic or environmental impacts

evapotranspiration evaporation of water from plants and soil

eye of the storm the area of calm, clear conditions where cool air sinks at the centre of a tropical storm

F

fairtrade trade that gives farmers a fairer price for their products

fault a crack in the Earth's crust

fauna animal life

flash flooding flooding that appears very quickly as a result of heavy rain

floodplain flat land, along the sides of a river, formed by silt and clay when the river floods

flora plant life

focus the source of an earthquake beneath the Earth's surface

fold mountains mountains formed when continental plate buckles

food chain producers and consumers linked within an ecosystem

food web interconnected food chains that make up a large ecosystem

food security when all people at all times have access to sufficient, safe, nutritious food to maintain a healthy and active life

fracking the method of obtaining gas and oil by blasting huge amounts of water mixed with sand and chemicals into the ground

freeze-thaw weathering the breakdown of rock caused by large variations between day and night temperatures

front the boundary at which warm air and cool air meet

frontal rainfall rainfall that occurs when warm air rises over cool air

G

genetically modified (GM) crop a crop developed by taking DNA from one species and inserting it into another

gentrification the improvement of areas, leading to the influx of wealthier people to replace poorer residents

geographical information system (GIS) a computer system that stores, analyses and presents different types of geographical data in map form

geomorphic processes the ways in which the landscape is shaped

glacial period a colder period in Earth's history

global circulation system the movement of air around the world

global warming an increase in the average temperature of Earth's atmosphere and oceans

GNI/capita gross national income divided by the population

gorge a valley with high, steep sides

green belt open space around cities on which no further development is allowed

Green Revolution a period in the 1960s when scientists produced rice and wheat seeds by cross-breeding, which gave very high yields

greenhouse gas a gas in the atmosphere, such as CO_2, that acts like the glass roof on a greenhouse, increasing temperature

gross national income (GNI) the sum of all the money earned in a country

growth corridor a region of economic growth

H

hard engineering human-made structures that help to deal with natural hazards, e.g. sea walls to stop coastal flooding

herbicide chemical used to kill weeds

high pressure (belt) a cold air mass pressing down on Earth

hotspot a place where lava rises up through very thin oceanic crust

Human Development Index (HDI) a measurement of development based on the indicators of wealth, health and education

hydraulic action the process of rock being broken up by air being forced into cracks

hydroponics growing plants indoors in water

I

ice age an period of time when Earth has permanent ice sheets

ice core a tube of ice drilled out from the depth of an ice sheet

indigenous people the people who originated in a particular place

infrastructure a country's transport and communication networks and essential services, such as train routes, water supply and hospitals

insolation the radiation from the Sun that reaches Earth

intensive farming farming that uses large amounts of machinery, chemicals or labour to maximise the amount of food produced

interglacial period a warmer period in Earth's history

internal growth growth of a population as a result of a higher live birth rate than death rate

international aid help that one country gives to another in times of need, taking the form of food, technology, money or advice

international migration the movement of people from one country to another

J

jet stream strong wind high in the atmosphere

L

La Niña an event that occurs when strong trade winds blowing east to west reduce surface temperatures

lahar a mudflow of melted snow and volcanic ash released from a glacier melted by a volcano

landslide a large body of soil and rock that slides down a steep slope

levee a river embankment caused by material deposited by river floods

life expectancy the average age to which people can expect to live

lithosphere the outer part of Earth, consisting of the crust and upper mantle

litter rotting leaves on the ground below growing vegetation

longshore drift the process by which pebbles and sand are transported along a beach

long-term development assistance international aid to help people to improve their lives and develop their country

low-income developing country (LIDC) a country that is poor, has a narrow range of jobs and few services

low pressure (belt) a warm air mass rising up from Earth

M

mantle the layer of very hot rock below Earth's crust

meander a bend along the course of a river

mechanical weathering the breakdown of rock caused by temperature changes, especially on mountains

megacity a city with a population over 10 million people

mid-ocean ridge a ridge of new rock under the ocean formed by eruptions of lava

migrant a person who moves from one place to live in another

Millennium Development Goals (MDGs) minimum acceptable standards of living agreed by world leaders from 189 countries in 2000 to cut world poverty in half by 2015

mitigation action taken to make something less severe

multi-lateral aid international aid given by countries through international organisations like the International Monetary Fund

N

natural greenhouse effect the natural process by which Earth is kept warm by gases that trap the Sun's energy in the atmosphere

net migration the different between immigration and emigration

nutrient cycle the stores and flows of nutrients in an ecosystem

O

oceanic crust Earth's crust beneath the ocean

organic farming the method of farming that bans the use of chemical fertilisers and pesticides

overseas development assistance (ODA) international aid given by governments and paid for by taxes

oxbow lake a crescent-shaped lake formed when a river meander is cut off from the river and isolated

P

peak oil when oil production reaches its maximum level before declining

periglacial around the edge of areas covered with ice

permaculture a way of growing food that tries to achieve food security by copying nature

permafrost ground that is frozen throughout the year

pesticide chemical used to kill insects

plate boundaries where the edges of tectonic plates meet

political unrest protest or violence arising from dissatisfaction with the government

population density a measure of the number of people who live within a given area

population pyramid a graph showing population structure, including the breakdown of age and sex within a population

prevailing wind wind from the most common direction in any given place

pull factor a factor that encourages people to move to a place

push factor a factor that encourages people to move from a place

pyroclastic flow torrent of hot ash, rock, gas and steam from a volcano

Q

quaternary industry industry based on high-level information and research

Quaternary period the time period in Earth's history that began about 2.6 million years ago and continues today; includes the Pleistocene epoch and the Holocene epoch

R

rain shadow the leeward side of mountains, where there is little rain

rejuvenated river a river that has eroded down into its old floodplain due to a fall in sea level

relief rainfall rainfall that occurs when clouds rise up over mountains

re-urbanisation the process of people moving back into city centres

rift valley a block of land that falls between fault lines when tectonic plates pull apart

river terrace an old floodplain of a rejuvenated river

Rostow's model a model of the stages of economic development

rural–urban migration movement of people from the countryside to cities

S

saltation the process in which pebbles and small stones are bounced along in flowing water

sanitation infrastructure for drainage and sewerage

sea ice maximum/minimum the maximum/minimum area of the Arctic Ocean covered in ice in any year

secondary industry manufacturing industries

seismic gap an area along known faults that has not experienced earthquake activity for a long time

seismic (shock) wave fast waves of energy generated from the focus of an earthquake

shallow focus earthquake an earthquake that starts within 70 km of Earth's surface

shield volcano a gently sloping volcano formed by runny lava, usually at a constructive plate boundary

shifting cultivation a sustainable way for farming in a forest by moving from one area to another

short-term emergency relief international aid to cope with the immediate problems caused by disasters like earthquakes and wars

social development an improvement in human welfare to meet people's needs

soft engineering adaptations to work with nature to limit damage from natural hazards

soil profile different layers within soil

solution the process in which rocks slowly dissolve in water

spit a long ridge of sand, shingle and/or pebbles attached to the land at one end but not at the other

stack column of rock surrounded by sea, once but no longer attached to the mainland

storm surge abnormal rises in sea level that occur when low air pressure raises sea levels, especially when a tropical storm nears land

subduction when a dense oceanic plate is forced under a continental plate at a destructive boundary

subsistence farming when farmers grow crops and animals to feed their own families

suburb residential areas outside the centre of a city

suburbanisation the process by which suburbs grow as a city expands outwards

sun spot a temporary dark spot on the Sun's surface caused by magnetic storms

suspension the way tiny particles of sand and silt are carried along, suspended in water

sustainable development economic, social and environmental development to meet people's needs now without compromising the ability of future generations to meet their needs

sustainable management using resources carefully so that future generations can also use them to meet their needs

swash movement of waves up a beach

T

tectonic plates the huge segments of Earth's crust

tertiary industry service industries

top-down development strategy development funded and carried out by government or a large international organisation like the World Bank

tornado a very strong rotating wind

traction the process in which boulders and large stones are dragged along a river bed or sea floor

trade the buying and selling of goods and services

trade winds winds that blow from high- to low-pressure belts

trans-national company (TNC) a large company that operates in more than one country

tropical rainforest large expanses of lush forest growing in the tropics in high temperatures and very high rainfall

tropical storm a powerful, rotating storm in the tropics (also known as a cyclone, hurricane or typhoon)

tundra areas in polar regions with cold winters, cool summers and low rainfall

U

urbanisation growth in the proportion of people living in towns and cities

W

water cycle the stores and flows of water in an ecosystem

water stress when the demand for water uses up a high proportion of the rainfall

water transfer scheme an infrastructure project for moving water from areas with lots of water to areas of shortage

world city one of the most important cities in the global economy

Index